Torsion and Shear Stresses in Ships

Mohamed Shama

Torsion and Shear Stresses in Ships

Springer

Mohamed Shama
Faculty of Engineering
Alexandria University
Alexandria 21544
Egypt
e-mail: mafshama@yahoo.com

ISBN 978-3-642-14632-9 e-ISBN 978-3-642-14633-6

DOI 10.1007/978-3-642-14633-6

Springer Heidelberg Dordrecht London New York

Library of Congress Control Number: 2010933503

Cover design: WMXDesign GmbH

Printed on acid-free paper

Springer is part of Springer Science+Business Media (www.springer.com)

To my wife
For her love, patience, encouragement and support

To my late parents
For their continuous care and encouragement

To my students
Whose enthusiasm and hard work have encouraged me to prepare the course material of this book.

Preface

In the last few decades, much research work was conducted to improve ship structure analysis and design. Most of the efforts were directed to improve the strength of hull girder and to use the method of finite element analysis more efficiently and effectively. Because of the high degree of complexity of ship structures the interaction between hull girder strength and local strength require special attention. Any structural element of the ship hull girder is subjected to several types of stresses including the fabrication and residual stresses. The stresses induced by hull girder and local loadings include the primary stresses, secondary stresses and tertiary stresses. Local loading comprises tensile, compressive, lateral, shear and torsion loadings. This complex system of stresses could produce unacceptable deformations and high values of equivalent stresses. Most of the methods commonly used for ship structure analysis and design focus on the stresses induced by hull girder bending and shear as well as the stresses induced by local lateral loadings.

This book is intended to cover an area of ship structure analysis and design that has not been exhaustively covered by most published text books on ship structures.

Also, it addresses a very complex subject in the design of ship structure and presents it in a simple and suitable form for research students and practicing engineers.

In addition it presents the basic concepts of the methods and procedures required to calculate torsion and shear stresses in ship structures.

Moreover, it presents valuable analysis and design material on torsion and shear loading and stresses. The book therefore should be very useful for practicing naval architects and students of marine engineering and naval architecture. The book is enhanced with a set of some solved and unsolved problems.

Outline of the Book

The book is composed of three parts: *Part I* is devoted to Torsion stresses in ships; *Part II* is concerned with Shear stresses in ships, whereas *Part III* is specialized to modeling the aforementioned methodology as separate subjective modules managed by a main executive program.

Part I of this book introduces the basic elements of pure torsion of uniform thin-walled open sections are presented. The various cases of local torsion loading on beam elements are presented. The basic equations of torsion of thin-walled closed sections, multi-cell box-girders and the general case of combined thin-walled open and closed sections are given.

Torsion of a thin-walled variable section beam subjected to non-uniform torque for both cases of free warping and constrained warping are considered. Warping deformations and flexural warping stresses of thin-walled sections are also presented for different types of loading and end constraints. Warping deformations and stresses in deck structure of container ships are highlighted.

Solution of the torsion equation for an assumed two distributions of torsional loading of container ships traveling obliquely in a sea-way is presented. The method of calculation is based on using an idealized ship section for calculating the sectorial properties of the ship section (principal sectorial area diagram, sectorial static moment). The position of the shear centre, torsion constant J_t and the warping constant $J(\omega)$, the shear and flexural warping stresses are then calculated. The total stress in the deck plating of a container ship due to hull girder bending and torsional loading is discussed. A numerical example is given to clarify the calculation procedure.

Chapter 5 gives the basic concepts and calculation procedure of the sectorial properties of open sections. Chapter 6 gives a general solution of the torsion equation.

Part II of this book presents the basic principles and concepts of shear flow, shear stress, shear deformation and the application of these principles to ship structure. The shear lag effect in thin-walled structures resulting from the effect of shear stress on bending stress is presented.

Methods of calculating the distribution of shear flow and stresses over symmetrical and asymmetrical thin-walled open sections are given.

Shear centre for symmetrical and asymmetrical thin-walled sections is explained. The distribution of shear stresses over thin-walled single and multi-box girders is given. The methods of calculation are explained and supported by numerical examples.

Methods of calculation of the distribution of shear flow and stress over ship sections are addressed. The methods of calculation are based on the introduction of a simplified idealization of ship section using an effective thickness for the shell plating and the attached stiffeners. For ship sections having closed boxes, a correcting shear flow is introduced to eliminate any torsional distortions induced by the assumed shear flow distribution.

The method is used to calculate the shear flow distribution over ship sections of single and double deck cargo ships and oil tankers with one and twin longitudinal bulkheads. A method for calculating shear load carried by the side shell plating and longitudinal bulkheads is given. The importance of calculating the distribution of shear stresses over ship sections of the hull girder is emphasized so as to determine the maximum allowable shearing force for a given ship section.

A damage occurring in any part of the ship structure will cause redistribution of the shear and bending stresses over the remaining intact structural members. Some structural members will be over stressed and others may be lightly stressed. The shear stress distribution over ship sections experiencing local damages is examined so as to ensure adequate safety of the overloaded structural members.

Shear loading on ship hull girder is given together with shear force distribution for alternate hold loading in bulk carriers.

Bulk carriers experience unique problems which result from the particular structural configuration and loading of these ships (alternate hold loading system).

In bulk carriers, the longitudinal vertical shearing force is carried by the side shell plating, top wing tanks, and hopper tanks. The side shell, therefore, may carry a high proportion of the longitudinal vertical shearing force. Consequently, shear stresses in the side shell plating may reach unfavorable values. Shear buckling, or high values of combined stresses, may occur in some panels in the side shell plating. Adequate measures should be taken, therefore, to prevent the initiation of instability and high stresses. The effect of using alternate hold loading system on the magnitude and distribution of shear loading along ship length is presented.

In order to calculate the shear stress distribution over a typical ship section of a bulk carrier, the ship section is idealized by a simplified configuration so as to reduce the laborious calculations associated with shear flow distribution. The idealized structure should affect neither the magnitude nor the distribution of shear flow distribution around the top wing tanks, hopper tanks, and side shell. Procedures for calculating shear flow and stresses in bulk carriers are given in derail.

Part III of this book is specialized to modeling the aforementioned methodology as separate subjective modules managed by a main executive program namely ***PROP***®. ***FORTRAN***® is the programming language in which the modules are written. The program has been written from scratch by *Dr. K. A. Hafez,*

Department of Naval Architecture and Marine Engineering, Faculty of Engineering, Alexandria University, Egypt.

For any arrangement of rectangular cross sections, ***PROP***® calculates their physical, sectional, sectorial properties, and the shear center of closed, opened, and combined cross sections. Also, the interested researcher may use the attached standard mathematical subroutines to manipulate the distribution of the sectional and/or sectorial properties together with their first few derivatives.

The information gathered in this program is expected to be sufficient for the first glance without going into more detailed discussion. However, for further details of the formulations and their associated computer programs, the interested researcher may consult the appropriate subjective chapters of this book, or he may refer to the listed references at the end of the book. All of the surveyed formulations and their associated computer program are computationally fast using the standard *IBM*® compatible computers, without any special requirements of the hardware configuration.

One of the main goals of the ***PROP***® program is the easy possibility of addition to and deletion of functional modules as required. The program is not an optimization routine but still considered to belong to the preliminary structural design stage without any economical or optimization consideration.

Finally, this part includes three solved examples that surely help in tracing the algorithm of the ***PROP***® program and understanding the way of input and output. Also, for the interested students and/or researchers a collection of non-solved problems is introduced.

Acknowledgments

I would like to thank *Dr. K. A. Hafez*, for receiving and correcting the presentation of this book and also for enhancing the value of the material presented herein by adding the source list and output of his computer program in *Chapter 12*. Also, I wish to thank all my graduate and undergraduate students who inspired me to write this book.

List of Symbols

A	Sectional area
B	Ship breadth
b	Flange width
C	Torsion rigidity
C_1	Warping constant
C_b	Block coefficient at summer load waterline
C_w	Warping constant
d	Web depth
dL	Elementary length
du	Linear deformation
dV	Elementary volume
E	Modulus of elasticity
e	Distance of shear center
e_y	Vertical distance of the shear center
f	A factor representing the degree of constrained against warping
G	Shearing modulus of elasticity
GJ	Torsional rigidity
I_h	Second moment of area of ship section about the y-axis
I_p	Polar moment of inertia
J	Torsion constant
L	Ship length
M	Bending moment
M_h	Horizontal bending moment
m	Intensity of torque load
M_{sw}	Still water bending moment
$M(\omega)$	Bimoment
p	Pressure load
q	Shear flow
R	Radius
r	Radial distance

S	Length parameter
$S(\omega)$	Sectorial static moment
T	Torque
T_e	Torsional moment at the end of the member
T_ω	Warping torque
T_s	Saint Venant torque
t	Thickness
t_f	Flange thickness
t_w	Web thickness
u	Linear displacement
β	Angle of deformation
γ	Shear angle
δ	Flexural warping coefficient
φ	Angle of twist
θ	Rate of twist
ν	Poisson's ratio
σ_1	Stress at the inner point of the flange
σ_2	Stress at the outer edge of the flange
σ_p	Stress at the attached plating
σ_ω	Flexural warping stress
τ	Shear stress
ω	Sectorial area

SI Units

International System of Units

This system can be divided into basic units and derived units as given in Tables 1 and 2.

Table 1 Basic units

Quantity	Unit	Symbol
Length	Meter	m
Mass	Kilogram	kg
Time	Second	s
Electric current	Ampere	A
Thermodynamic temperature	Degree Kelvin	°K
Luminous intensity	Candela	cd

Table 2 Derived units

Quantity	Unit	Symbol
Force	Newton	$N = kg\ m/s^2$
Work, energy	Joule	$J = N\ m$
Power	Watt	$W = J/s$
Stress, pressure	Pascal	$Pa = N/m^2$
Frequency	Hertz	$Hz = s^{-1}$
Acceleration	Meter per second squared	$g = m/s^2$
Area	Square meter	m^2
Volume	Cubic Meter	m^3
Density	Kilogram per cubic meter	$\rho = kg/m^3$
Velocity	Meter per second	$v = m/s$
Angular velocity	Radian per second	rad/s
Dynamic viscosity	Newton second per meter squared	$N\ s/m^2$
Kinematic viscosity	Meter squared per second	m^2/s
Thermal conductivity	Watt per (meter degree Kelvin)	W/(m.deg.k)

Table 3 Summary of the quantities commonly used in naval architecture

Quantity	SI unit
Length	0.3048 m
	1,842 m
	1,609 m
Area	0.0929 m^2
Volume	0.02832 m^3
Velocity	0.3048 m/s
	0.5144 m/s
Standard acceleration	9.8066 m/s^2
Mass	0.4536 kg
	1,016 kg
	1.016 tonne
Force	4.4482 N
Pressure, stress	6.8947 kN/m^2
	15.444 MN/m^2
Energy	1.3558 J
Power	745.7 W
Density (SW)	0.975 m^3/tonne
	0.01 MN/m^3
Density (FW)	1.0 m^3/tonne
	0.0098 MN/m^3
Modulus of elasticity (E)	20.9 GN
TPI (SW)	$1.025A_w$ tonne/m
TPM	10^4 A_W (N/m)
MCT″ (sw)	$\Delta\overline{GM}_L/L(MN\,m/m)$
Displacement mass force (Δ)	9964 N
Displacement mass (Δ)	1.016 tonne
Wetted surface	Δ = tonne, L = m

Table 4 Power conversion

Quantity	Common unit	SI unit
BHP	P_B	W
SHP	P_S	W
DHP	P_D	W
EHP	P_E	W

Table 5 Multiples and sub-multiples

Prefix	Factor	Symbol
Tera	10^{12}	T
Giga	10^9	G
Mega	10^6	M
Kilo	10^3	k
Milli	10^{-3}	m
Micro	10^{-6}	μ
Nano	10^{-9}	n
Pico	10^{-12}	p
Femto	10^{-15}	f
Atto	10^{-18}	a

Table 6 General units

Gravity acceleration: g	9.807 m/s^2
Water density (salt water): ρ_{sw}	1.025 $tonne/m^3$
Modulus of elasticity: E	20.9 MN/cm^2
Atmospheric pressure: p_{at}	10.14 kN/m^2
1.0 ton displacement	9964 N

Table 7 Shipbuilding units

(a) General	
Dimensions/distances	m
Primary spacing	m
Secondary spacing	mm
Area	m^2
Volume	m^3
Mass	kg
Velocity	m/s
Acceleration	m/s^2
(b) Hull girder properties	
Dimensions	m
Area	m^2
Section modulus	m^3
Inertia	m^4
Moment of area	m^3
Dimensions	mm
Area	cm^2
Thickness	mm
(c) Loads	
Pressure	kN/m^2
Loads	kN
Shear force	kN

Contents

Part I Torsion Stresses in Ships

1 Torsion Stresses in Ships 3
1.1 Introduction 3
1.2 Torsion Loading of Beam Elements 3
1.2.1 Direct Torsion Loads 3
1.2.2 Induced Torsion Load 3
1.3 Variation of Torque and Angle of Twist along Beam Length 4
1.3.1 Beams Subjected to Concentrated Torques 4
1.3.2 Beams Subjected to Uniformly Distributed Torsion Loading 5
1.4 Torsion of Uniform Thin Walled Sections 6
1.4.1 Pure Torsion of Uniform Open Thin-Walled Girders 6
1.5 Torsion of Uniform Thin-Walled Closed Sections 9
1.6 Basic Equations of Torsion of Thin-Walled Closed Sections 10
1.6.1 Shear Flow and Stress 10
1.6.2 Rate of Twist 11
1.7 Torsion of a Uniform Thin-Walled Tube 12
1.7.1 Angle of Twist 12
1.7.2 Torsion Shear Stress 13
1.8 Comparison between Open and Closed Thin-Walled Sections 16
1.8.1 Circular Section 16
1.8.2 Square Section 18
1.9 Torsion Constant of Uniform Thin-Walled Closed Sections with Attached Open Sections 19

2 Torsion Stresses in Thin-Walled Multi-Cell Box-Girders 21

2.1 Torsion of Uniform Thin-Walled Two-Cell Box-Girders 21

2.2 The General Case of a Uniform Two-Cell Box Girder. 26

2.3 Torsion Stresses in a Two Identical Cells Box-Girder 29

2.3.1 Shear Flow q 29

2.3.2 Shear Stress τ 29

2.3.3 Rate of Twist θ 30

2.4 Torsion of Three-Cell Box-Girder. 30

2.5 Torsion of Uniform Thin-Walled Multi-Cell Box-Girder 33

2.6 Combined Open and Closed Thin-Walled Sections 34

2.6.1 Combined Open Section with One Closed Cell 34

2.6.2 Combined Open Section with Two Closed Cells . . . 35

3 Torsion Warping Deformations and Stresses 41

3.1 Torsion of Thin-Walled Variable Section Beams 41

3.1.1 Free Warping 41

3.1.2 Constrained Warping 41

3.1.3 Warping of Thin-Walled Sections 43

3.1.4 Flexural Warping Stresses 46

3.1.5 Development of the General Equation of Torsion . . . 47

3.1.6 Solution of the Torsion Equation. 54

4 Torsion of Container Ships 65

4.1 Torsion Loading on Ships 65

4.2 Torsion Loading of Open-Decked Ships. 66

4.3 Torsion Loading on Catamaran Vessels 70

4.4 Warping Deformations and Stresses in the Deck Structure of Container Ships. 72

4.5 Torsional Deformation of Ship Hull Girder 72

4.6 An Approximate Method for Torsion Analysis of Open Deck Vessels 74

4.7 Calculation of the Shear and Flexural Warping Stresses 75

4.8 Solution of the Torsion Equation 75

4.8.1 Boundary Conditions 75

4.8.2 Distribution of Torsional Loading 76

4.8.3 Solution of the Torsion Equation for Constrained Warping 77

4.8.4 Calculation of the Sectorial Properties of Ship Section 78

4.9 Total Stress in the Deck Plating of Container Ships due to Hull Girder Bending and Torsional Loading . . . 85
4.9.1 Hull Girder Stresses due to Vertical Bending . . . 86
4.9.2 Horizontal Hull Girder Bending Stresses . . . 87
4.9.3 Local Stresses . . . 87
4.9.4 Flexural Warping Stresses . . . 88
4.9.5 Total Stress Over the Deck Plating . . . 89

5 Sectorial Properties of Thin-Walled Open Sections . . . 91
5.1 Introduction . . . 91
5.2 Sectorial Properties of Thin-Walled Sections . . . 91
5.2.1 Principal Sectorial Properties of Thin-Walled Sections . . . 93
5.2.2 Position of the Shear Center . . . 94
5.2.3 Sectorial Area Diagram . . . 95
5.2.4 Procedure of Calculation . . . 96
5.3 Applications to Some Typical Sections . . . 96
5.3.1 Sectorial Properties for Thin-Walled Sections Free to Warp . . . 96
5.4 Sectorial Properties for a Thin-Walled Section with an Enforced Axis of Rotation . . . 101
5.4.1 A thin-Walled T-Section with an Enforced Axis of Rotation . . . 101
5.4.2 Enforced Center of Rotation for a Thin-Walled Angle Section . . . 102
5.4.3 Enforced Center of Rotation at a Point C on the Opposite Side of a Thin-Walled Asymmetrical Fabricated Section . . . 102

6 General solution of the torsion equation . . . 105

Part II Shear Loading and Stresses in Ships

7 Shear Stresses in Thin-Walled Structures . . . 111
7.1 Basic Principles . . . 111
7.2 Shear Stresses in Beams due to Bending . . . 111
7.2.1 Solid Beams . . . 111
7.2.2 Average Shear Stress . . . 114
7.2.3 Shear Flow and Stress in Thin-Walled Sections . . . 115
7.3 Shear Centre . . . 124

7.4 Shear Deflection 127
7.4.1 Shear Deformation. 129
7.5 Shear Lag. 130

8 Shear Flow and Stresses in Thin-Walled Box-Girders 133
8.1 Single Cell Box-Girder 133
8.2 Shear Flow in Asymmetrical Closed Box-Girders Subjected to a Vertical Shear Force F 135
8.3 Shear Stresses in Thin-Walled Two-Cell Box-Girders 142
8.4 Calculation of the Correcting Shear Flow for 3-Cell Box-Girders Subjected to Shear Load 146

9 Shear Flow and Stresses in Ships 149
9.1 Introduction 149
9.2 Procedure of Calculation of Shear Flow Distribution. 149
9.2.1 Ship Section Idealization 149
9.3 Determination of the Effective Thickness. 157
9.4 Shear Flow Calculation 157
9.4.1 Procedure of Calculation of Shear Flow Distribution 158
9.4.2 Shear Flow Distribution over a Ship Section of a Two-deck Cargo Ship 160
9.5 Calculation of Shear Stress Distribution. 161
9.5.1 Equivalent Stress. 161
9.6 Calculation of Shear Stress Distribution over a Ship Section . . . 162
9.6.1 Calculation of Shear Flow Distribution over a Twin DeckCargo Ship 163
9.7 Shear Flow Distribution over a Catamaran Section 164

10 Calculation of Shear Stresses in Tankers Subjected to Longitudinal Vertical Shear Forces. 167
10.1 Coastal Tankers Having One Longitudinal Bulkhead. 167
10.2 Calculation of Shear Flow Distribution for Twin Longitudinal Bulkhead Tankers 169
10.3 Shear Load Carried by Longitudinal Bulkheads and Side Shell Plating 173
10.3.1 Sea-Going Tankers with Two Longitudinal Bulkheads. 173
10.3.2 Coastal Tankers with One Longitudinal Bulkhead. . . 175
10.4 Shear Flow Distribution Over a Ship Section of an Oil Tanker Experiencing a Local Damage in the Shell Plating or Longitudinal Bulkhead. 176
10.4.1 Introduction 176

10.4.2 Shear Stress Distribution Over a Tanker Section Experiencing a Local Damage. 177
10.4.3 Scenarios of Assumed Damage Locations on the Tanker Section 177

11 Shear Loading and Stresses in Bulk Carriers 187
11.1 Introduction 187
11.2 Structural Configuration. 187
11.2.1 Upper and Lower Stools of Transverse Bulkheads . . 188
11.2.2 Double Bottom Structure 189
11.3 Hull Girder Loading 189
11.4 Longitudinal Vertical Shearing Force 190
11.4.1 Stillwater Component (FS) 191
11.5 Wave-Induced Component (*FW*). 195
11.5.1 The Distribution of the Largest Expected Vertical Wave-Induced Shearing Force. 197
11.6 Dynamic Component (*FD*). 198
11.7 Total Vertical Shearing Force F 200
11.8 Approximate Value to the Maximum Vertical Shear Force . . 201
11.9 Variation of Various Shear Stress Components with Time . . . 202
11.10 Shear Flow Distribution in Bulk Carriers 202
11.10.1 Structure Idealization. 202
11.10.2 Effective Thickness 203
11.10.3 Shear Flow Distribution 204
11.10.4 Shear Stress Distribution 205
11.10.5 Shear Flow Distribution Over the Hopper Tank . . . 205
11.10.6 Shear Flow Distribution Over the Top Wing Tanks. . . 207

Part III Programming Implementation

12 Programming Implementation 213
12.1 Introduction 213
12.2 Program List. 214
12.3 Solved Problems 258

13 Problems 265

References 273

Index 275

CV of the Author 277

Part I
Torsion Stresses in Ships

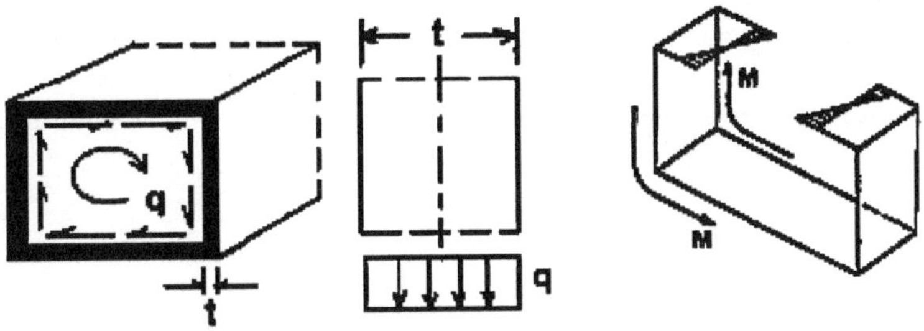

Chapter 1 introduces the basic elements of pure torsion of uniform thin-walled open sections. The basic equations of torsion of thin-walled closed sections are given together with the distribution of torsion shear flow and stress over the section as well as the angle and rate of twist. It also gives a comparison between open and closed thin-walled circular and box sections with regard to torsion stresses and rate of twist. Torsion of uniform thin-walled closed sections with attached open sections is also given. A list of the torsion and warping constants for some thin-walled open sections is given.

Chapter 2 introduces torsion of multi-cell box-girders. It covers torsion of uniform thin-walled two-cell box-girders, torsion stresses in a two identical cells box-girder and torsion of three-cell box-girder. It introduces the general case of torsion of uniform thin-walled multi-cell box-girders. The general case of combined thin-walled open and closed sections is presented. The special cases of combined open sections with one and two closed cell are presented.

Chapter 3 introduces torsion of a thin-walled variable section beam subjected to non-uniform torque. It is shown that plane sections no longer remain plane and torsion is associated with warping. Both free warping and constrained warping are considered. It covers also warping deformations and stresses in deck structure of container ships. Warping and flexural warping stresses of thin-walled sections are also considered for different types of loading and end constraints.

Chapter 4 introduces the definitions and differences between *St. Venant* torsion, warping torsion and the bi-moment. The main elements of torsion of open-decked ships are presented. Hull girder torsion loading and the various cases of local torsion loading on beam elements are considered.

Torsion of container ships traveling obliquely in a sea-may is highlighted. Solution of the torsion equation for assumed two distributions of torsional loading are presented. The method of calculation is based on a simplified idealization of the ship section, calculation of the sectorial properties of the ship section (principal sectorial area diagram, sectorial static moment), position of shear centre, torsion constant $\boldsymbol{J}_t$ and the warping constant $\boldsymbol{J}(x)$. The calculation of the shear and flexural warping stresses are carried out for two cases of boundary conditions, free end, i.e. free warping and constrained end, i.e. constrained warping. The total stress in the deck plating of a container ship due to hull girder bending and torsional loading is presented. A numerical example is given for a container ship to illustrate and clarify the calculation procedure.

Chapter 5 presents the calculation procedure of the sectorial properties of open sections.

Chapter 6 presents a general procedure for the solution of the torsion equation.

Chapter 1
Torsion Stresses in Ships

1.1 Introduction

The torsion analysis is devoted to determining the stress distribution in twisted single-span or continuous numbers with solid, thin-walled open or closed cross sections.

A prismatic member resists a twisting moment in two ways.

1. By producing a circulatory shear flow in the cross section.
2. By inducing sheer stresses resulting from the change in axial stresses.

The first is called *St. Venant* torsion and the second is called *warping* torsion or flexural twist. A flexural twist causes always some bending moments in a structure i.e., a pair or more of bending moments. These are called "*Bi-moment*".

1.2 Torsion Loading of Beam Elements

1.2.1 Direct Torsion Loads

- Concentrated torsion load, see Fig. 1.1.
- Uniformly distributed torsion loading, see Fig. 1.2.
- Linearly distributed torsion loading, see Fig. 1.3.

1.2.2 Induced Torsion Load

Torsion loads may be induced by lateral forces acting at an offset distance from the shear center of the beam section. For the thin-walled channel section shown in

M. Shama, *Torsion and Shear Stresses in Ships*,
DOI: 10.1007/978-3-642-14633-6_1, © Springer-Verlag Berlin Heidelberg 2010

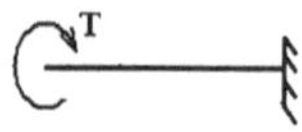

Fig. 1.1 Concentrated torque at the free end of the beam

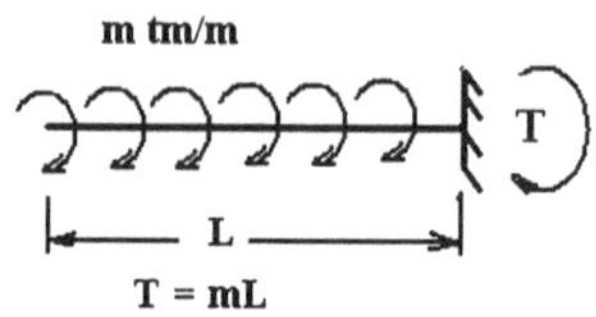

Fig. 1.2 Uniformly distributed torsion loading

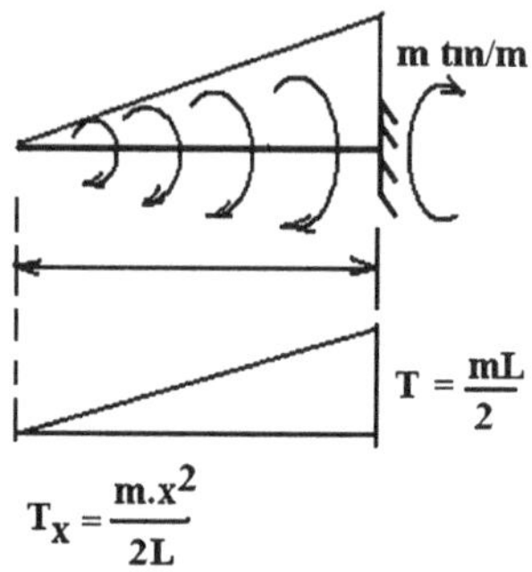

Fig. 1.3 Linearly distributed torsion loading

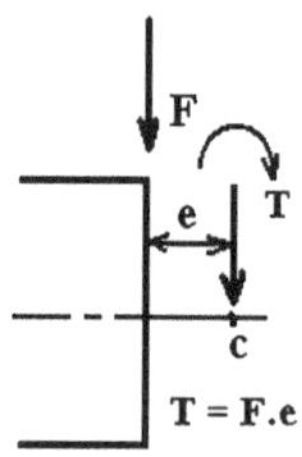

Fig. 1.4 Concentrated torque induced by the lateral force F

Fig. 1.4, the shear center of the section is located at a distance e from the web of the section. The induced concentrated torque is given by

$$T = F \cdot e$$

where F = lateral concentrated force; e = distance of shear center from the web of the section; T = induced concentrated torque.

1.3 Variation of Torque and Angle of Twist along Beam Length

1.3.1 Beams Subjected to Concentrated Torques

1.3.1.1 Concentrated Torque at End of Beam

The distribution of torque loading and the variation of angle of twist along a cantilever beam subjected to a concentrated torque at its free end are shown in Fig. 1.5.

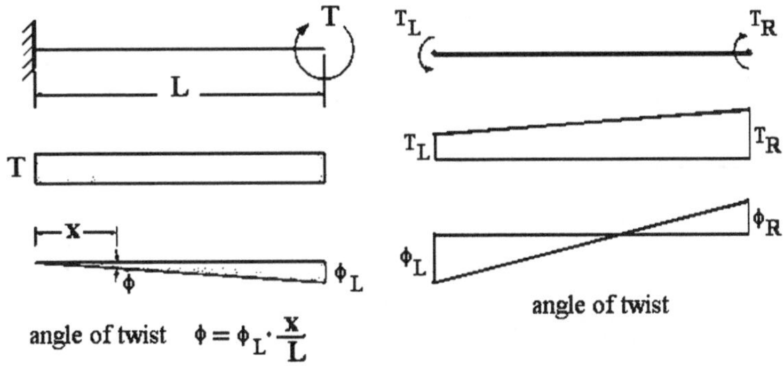

Fig. 1.5 Distribution of torsion loading and variation of angle of twist along a beam length subjected to end concentrated torque

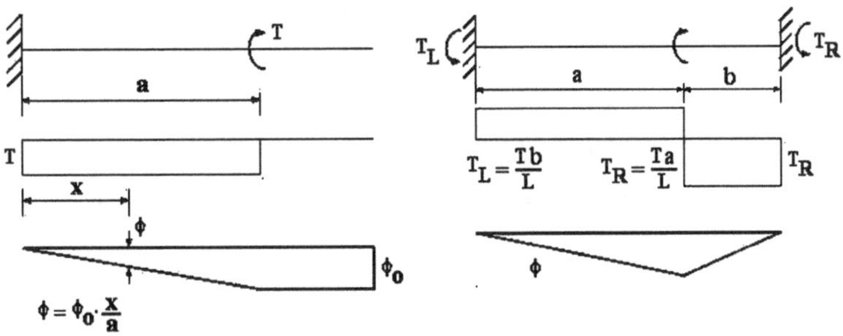

Fig. 1.6 Distribution of torsion loading and variation of angle of twist along a beam length subjected to a concentrated torque within beam length

1.3.1.2 Concentrated Torque within the Beam Length

The distribution of torsion loading and variation of angle of twist along a beam length subjected to a concentrated torque within the beam length is shown in Fig. 1.6.

1.3.2 Beams Subjected to Uniformly Distributed Torsion Loading

Figure 1.7 shows the distribution of torsional moment and the variation of angle of twist along a beam subjected to uniform torque along the beam length for different end conditions.

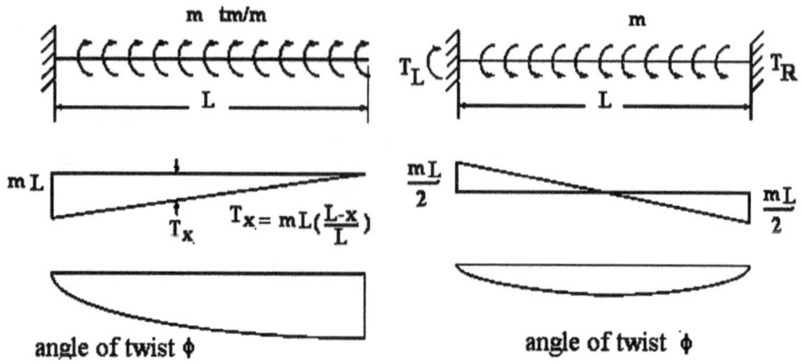

Fig. 1.7 Distribution of torsion loading and variation of angle of twist along a beam length subjected to uniform torque for different end conditions

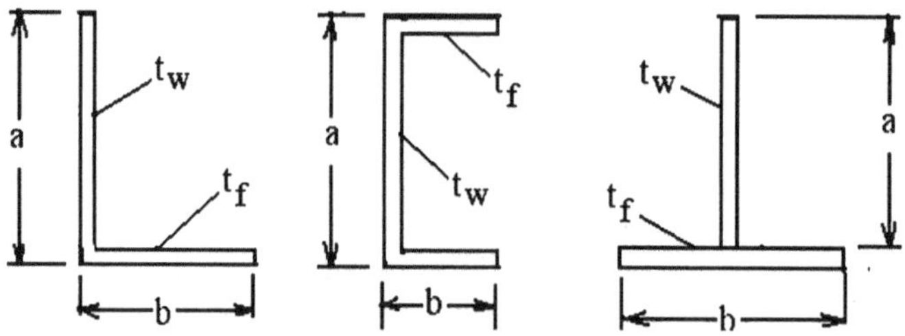

Fig. 1.8 Geometry and scantlings for three types thin-walled open sections

1.4 Torsion of Uniform Thin Walled Sections

1.4.1 Pure Torsion of Uniform Open Thin-Walled Girders

Pure Torsion is the case when a bar is twisted by couples applied at the ends and these ends are free to warp. Pure torsion of thin-walled sections with free ends produce only sheer stresses which are the same for all sections.

Plane sections do not remain plane and warping takes place. However, the effect of warping on the calculation of shear stresses and angle of twist could be neglected for simplicity. The distribution of the sheer stresses depends on the shape of the section.

For thin-walled open-sections, see Fig. 1.8, the angle of twist/unit length is given by

$$\theta = \mathrm{T/C} \tag{1.1}$$

where T = Torque; G = Sheering modulus of elasticity; J = Torsion constant; C = Torsional rigidity of the section; C = G/J.

The torque is related to the angle of twist by the following equation.

$$\mathrm{T} = \mathrm{G} \cdot \mathrm{J} \cdot \theta \tag{1.2}$$

1.4.1.1 Torsion Constant

The torsion constant J of an open thin-walled section is given by

$$J = \sum \left(s_i t_i^3/3\right)$$

where $s = a$ and b, see Fig. 1.8.

For the angle section shown in Fig. 1.8.

$$J = \sum \left(s_i t_i^3/3\right) = a t_w^3/3 + b t_f^3/3$$

For the channel section

$$J = \sum \left(s_i t_i^3/3\right) = a t_w^3/3 + 2b t_f^3/3$$

For the T section

$$J = \sum \left(s_i t_i^3/3\right) = a t_w^3/3 + b t_f^3/3$$

The torsion and warping constants for three thin-walled open sections are given in Table 1.1.

1.4.1.2 Distribution of Torsion Shear Flow and Stress Over a Thin-Walled Open Section

For open sections, the shear flow and stress are linearly distributed over the wall thickness of the open section and is given by, see Fig. 1.9.

The shear flow is given by

$$q = T \cdot t^2/J$$

The shear stress is given by

$$\tau = T \cdot t/J$$

The torsion constant

$$J = \Sigma\left(s_i t_i^3/3\right)$$

For the thin-walled rectangular section, see Fig. 1.9.

$$J = dt^3/3$$

Hence

$$\tau = 3T/dt^2$$

Table 1.1 Torsion and warping constants for some thin-walled open sections

O = shear center	J = torsion constant	C_w = warping constant
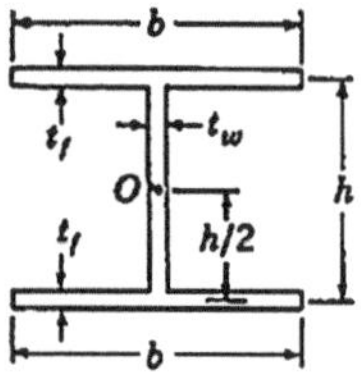	$J = \frac{2bt_f^3 + ht_w^3}{3}$ $C_w = \frac{t_f h^2 b^3}{24}$	If $t_f = t_w = t$: $J = \frac{t^3}{3}(2b + h)$
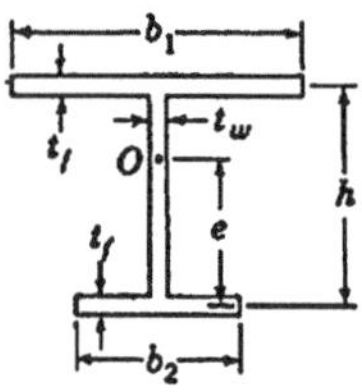	$e = h\frac{b_1^3}{b_1^2 + b_2^3}$ $J = \frac{(b_1 + b_2)t_f^3 + ht_w^3}{3}$ $C_w = \frac{t_f h^2}{12} \frac{b_1^3 b_2^3}{b_1^3 + b_2^3}$	If $t_f = t_w = t$; $J = \frac{t^3}{3}(b_1 + b_2 + h)$
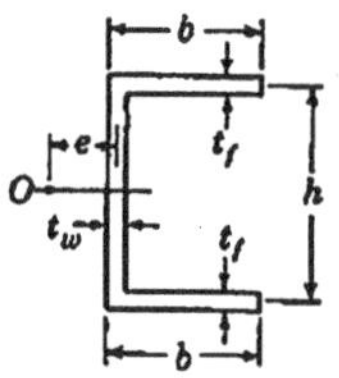	$e = \frac{3b^2 t_f}{6bt_f + ht_w}$ $J = \frac{2bt_f^3 + ht_w^3}{3}$ $C_w = \frac{t_f b^3 h^2}{12} \frac{3bt_f + 2ht_w}{6bt_f + ht_w}$	If $t_f = t_w = t$: $e = \frac{3b^2}{6b + h}$ $J = \frac{t^3}{3}(2b + h)$ $C_w = \frac{tb^3 h^2}{12} \frac{3b + 2h}{6b + h}$

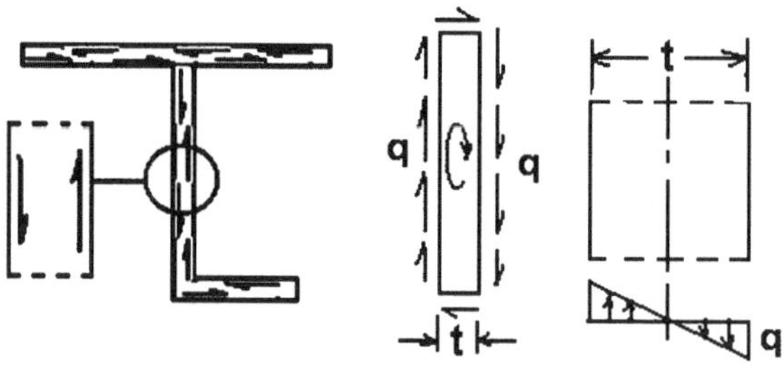

Fig. 1.9 Shear flow distribution over thin-walled open sections

For the thin-walled angle section, the torsion stress in the flange is given by

$$\tau_f = T \cdot t_f / J$$

The torsion stress in the web of the section is given by

$$\tau_w = T \cdot t_w / J$$

The angle of twist is given by

$$\varphi = TL/GJ$$

where G = shear modulus and is given by

$$G = E/2(1 + \nu)$$

1.5 Torsion of Uniform Thin-Walled Closed Sections

Let

q = shear flow in ***t/m*** over the periphery of the thin-walled closed section.

Then

$$\begin{aligned} T &= \oint q \cdot ds \cdot r \\ &= q \oint rds \\ &= 2q \oint dA \\ &= 2qA \end{aligned}$$

where r·ds = 2 sectional area, see Fig. 1.10; A = enclosed area of the section.

Hence

$$T = 2 \cdot q \cdot A$$

and

$$q = T/2A$$

Thus, when a torque *T* is applied to a closed thin-walled uniform member, the shear stress is given by

$$\tau = q/t = T/2At$$

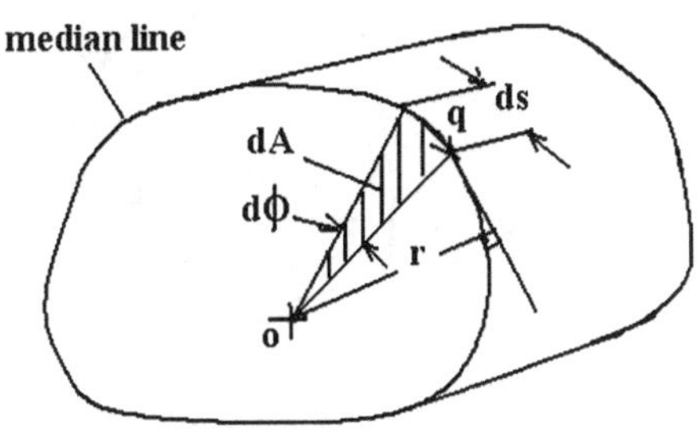

Fig. 1.10 Shear flow over a closed thin-walled section

The rate of twist, θ, is calculated by equating the internal energy to the external work, i.e.,

$$1/2 \cdot \mathrm{T} \cdot \theta = \oint \tau^2/2\mathrm{G} \cdot \mathrm{dV}$$

But $\mathrm{dV} = \mathrm{t} \cdot \mathrm{ds}$ for a unit length of the member and

$$\tau = \mathrm{T}/2\mathrm{At}$$

Hence

$$\theta = \mathrm{d}\varphi/\mathrm{dL} = 1/4\mathrm{A}^2\mathrm{G} \cdot \oint \mathrm{Tds}/\mathrm{t} = 1/2\mathrm{AG} \cdot \oint \mathrm{qds}/\mathrm{t}$$

In general

$$\theta = \mathrm{T}/\mathrm{GJ}_\mathrm{t}$$

where $\mathrm{J_t}$ = torsion constant of the thin-walled closed section and is given by

$$\mathrm{J_t} = 4\mathrm{A}^2 \Big/ \oint \mathrm{ds}/\mathrm{t}$$

1.6 Basic Equations of Torsion of Thin-Walled Closed Sections

1.6.1 Shear Flow and Stress

Let

q = shear flow in t/m around the periphery of the section, see Fig. 1.11. Hence the shear flow is given by

$$\mathrm{q} = \mathrm{T}/2\mathrm{At}$$

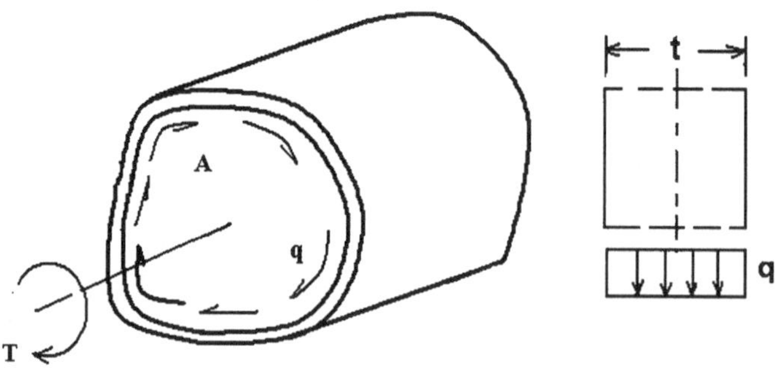

Fig. 1.11 Shear stress over the thickness of a closed thin-walled section

The shear stress is given by

$$\tau = q/t = T/2At$$

1.6.2 Rate of Twist

The rate of twist of the section is given by

$$\theta = d\varphi/dL = T/GJ_t$$

The angle of twist over the length *L* of the member is given by

$$\varphi = TL/GJ$$

where J_t = torsion constant of the section

$$J_t = 4A^2 \Big/ \oint ds/t$$

For constant *t*, the torsion constant is given by

$$J = 4A^2t/S$$

Hence the rate of twist is given by

$$\begin{aligned}\theta &= d\varphi/dx \\ &= TS/4GA^2t\end{aligned}$$

where S = perimeter of the section

Example 1.1 Assuming a thin-walled closed box-girder is subjected to a constant torque *T*

Solution Let

q = shear flow around the periphery of the closed section, see Fig. 1.12.
Hence the torque is given by

$$T = 2 \cdot q \cdot A$$

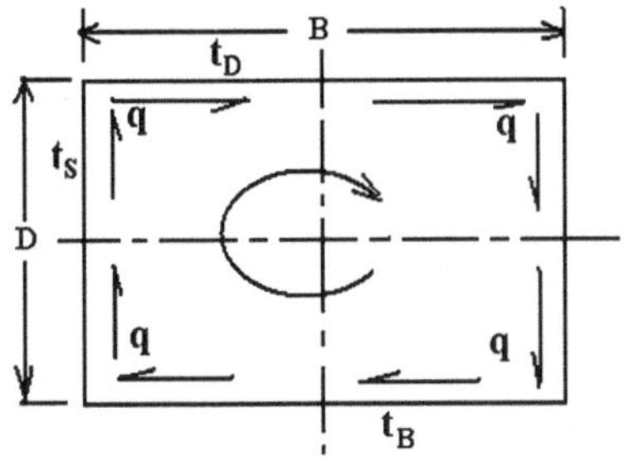

Fig. 1.12 Shear flow due to torsion of a closed rectangular thin-walled section

where

$$A = B \cdot D$$

The shear flow q could be calculated as follows

$$q = T/2A$$

The shear stress is given by

$$\tau = q/t = T/2At$$

Hence the shear stress in the sides, top and bottom plating are given by

$$\tau_s = T/2A\,t_s$$
$$\tau_D = T/2A\,t_D$$
$$\tau_B = T/2A\,t_B$$

Similarly the torque T is given by

$$T = GJ\theta$$

Hence

$$q = GJ\theta/2A$$

The angle of twist φ is given by

$$\varphi = T\lambda/4GA^2 \cdot \oint ds/t = T\lambda/GJ$$

where

$$J = 4A^2 \Big/ \oint ds/t$$

$$\oint ds/t = B\ (1/t_B + 1/t_D) + 2D/t_S$$

A = enclosed area of section; T = wall thickness; D_s = elementary length of contour; J = torsion constant; G.J = torsional rigidity.

1.7 Torsion of a Uniform Thin-Walled Tube

1.7.1 Angle of Twist

Assuming plane sections of the tube remain plane i.e., the section is not subjected to any warping.

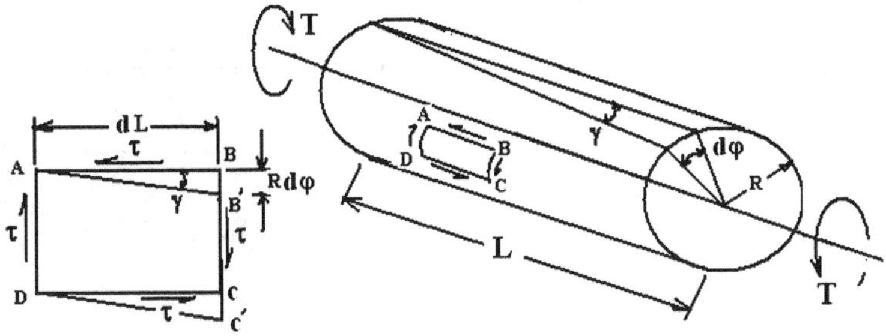

Fig. 1.13 Angle of twist due to torsion of a thin-walled tube

The angular deformation γ is given by see Fig. 1.13.

$$\gamma = \mathrm{R} \cdot \theta$$

Hence

$$\gamma = \mathrm{R} \cdot \mathrm{d}\varphi / \mathrm{dL}$$

where θ = rate of twist

$$\theta = \mathrm{d}\varphi / \mathrm{dL}$$

The angle of twist φ is given by

$$\varphi = \int_0^{\mathrm{L}} \theta \mathrm{dx}$$

1.7.2 Torsion Shear Stress

The torsion shear stress is given by

$$\tau = \mathrm{G}\gamma = \mathrm{GR}\theta$$

The applied torque is given by, see Fig. 1.13.

$$\delta \mathrm{T} = \tau \cdot \mathrm{t} \cdot \mathrm{R} \cdot \mathrm{d}\varphi \cdot \mathrm{R}$$

$$\mathrm{T} = \int_0^{2\pi} \tau \cdot \mathrm{tR}^2 \mathrm{d}\varphi = \int_0^{2\pi} \mathrm{GR}\theta \cdot \mathrm{tR}^2 \mathrm{d}\varphi = \mathrm{G}\theta \int_0^{2\pi} \mathrm{R}^3 \mathrm{d}\varphi = \mathrm{G} \cdot \theta \cdot \mathrm{I_P}$$

Hence

$$\theta = \mathrm{T} / \mathrm{GI_P}$$

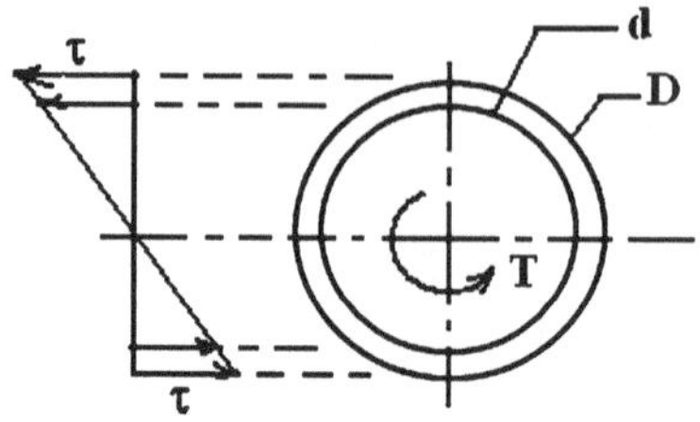

Fig. 1.14 Shear stress distribution over the thickness of a tube due to torsion

and

$$\varphi = \mathrm{T}/\mathrm{GI_P} \cdot \int_0^L \mathrm{dL}$$

where T and I_P are constants, then

$$\varphi = \mathrm{TL}/\mathrm{GI_P}$$

Example 1.2 Determine the torsion shear stress and angle of twist for the thin-walled tube of length L shown in Fig. 1.14 when the tube is subjected to a constant torque T at its end.

Solution The torsion shear stress is given by

$$\tau = \mathrm{TR}/\mathrm{GI_P}$$

The angle of twist is given by

$$\varphi = \mathrm{TL}/\mathrm{GI_P}$$

where L = length of member; G = Shear modulus;

$$\mathrm{G} = \frac{\mathrm{E}}{2(1+\upsilon)};$$

υ = Poisson's ratio; L = length of tube;

$$\mathrm{I_p} = \pi \mathrm{D}^4/32 \cdot \left[1 - (\mathrm{d}/\mathrm{D})^4\right].$$

Example 1.3 A solid circular shaft of 25 cm diameter is to be replaced by a hollow shaft, the ratio of the external to internal diameters being 2:1. Determine the size of the hollow shaft if the maximum shearing stress is to be the name as for the solid shaft. Calculate the reduction in the mass of the shaft?

Solution Let r = inside radius of the hollow shaft

Hence

$$\text{outside radius} = 2.r$$

Thus

$$J_{new} = \frac{\pi}{2}(16r^4 - r^4) = 7.5\pi r^4$$
$$J_{old} = \frac{\pi}{2}(0.125)^4 = 0.384 \times 10^{-3}\,m^4$$

Let T = applied torque

Hence τ_{max} for solid shaft is given by

$$\tau_{max} = 0.125T/0.384 \times 10^{-3}$$

and τ_{max} for the hollow shaft is given by

$$\tau_{max} = 2rT/7.5\pi r^4$$

Since τ_{max} is the same, we get

$$0.125T/0.384 \times 10^{-3} = 2rT/7.5\pi r^4$$

i.e.,

$$r^3 = 0.261 \times 10^{-3}\ m^3$$

Then

$$r = 0.046\ m$$
$$\text{Thus, internal diameter} = 0.128\ m$$
$$\text{And external diameter} = 0.256\ m$$

The relative weights of the two shafts are given by

$$\frac{\text{Area of new cross} - \text{section}}{\text{Area of solid section}}\frac{(0.128)^2 - (064)^2}{(0.125)^2} = 0.785$$

Thus, saving in weight $\approx 0.21 = 21\%$

Example 1.4 A hollow square thin-walled uniform member of length *L*, see Fig. 1.15, is acted upon by uniform torque *T*. Determine the torsion shear stress at points *A* and *B*, and the angle of the member twist.

Solution The torsion shear stress at point *A* is given by

$$\tau_A = T/2At = T/2a^2t_A$$

The torsion shear stress at point *B* is given by

$$\tau_B = T/2a^2t_B$$

Fig. 1.15 A thin-walled square section beam

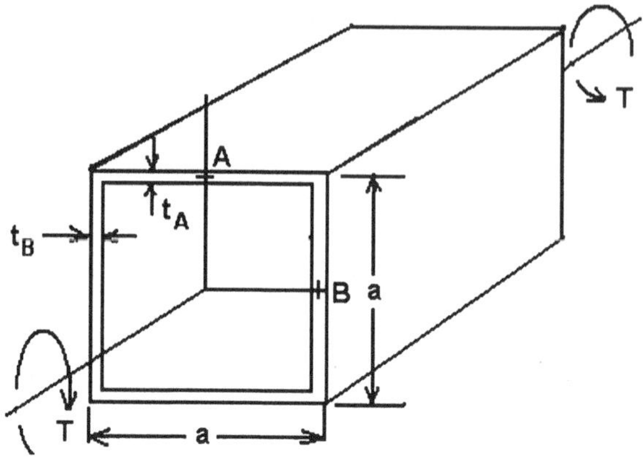

The torsion constant of the closed thin-walled section is given by

$$J_t = 4A^2 \Big/ \oint ds/t = 4a^4/(2a/t_A + 2a/t_B)$$
$$= 2a^3/[(t_A + t_B)/t_A t_B]$$

The angle of twist over the length L of the member is given by

$$\varphi = \frac{T \cdot L \cdot (t_A + t_B)}{2a^3 G t_A \cdot t_B}$$

where

$$t_A = t_B = t$$

The torsion constant is given by

$$J_t = a^3 t$$

and the angle of twist is given by

$$\varphi = T\lambda/GJ = TL/a^3 tG$$

1.8 Comparison between Open and Closed Thin-Walled Sections

1.8.1 Circular Section

For the open thin-walled section, see Fig. 1.16, the torsion constant is given by

$$J = \Sigma\left(s_i t_i^3/3\right)$$

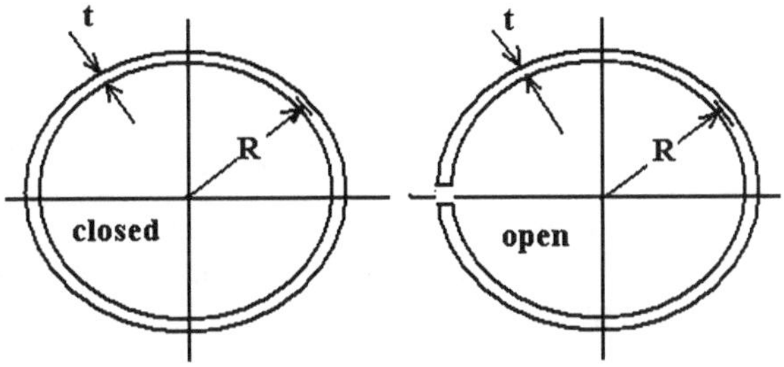

Fig. 1.16 Open and closed thin-walled tube

The torsion shear stress is given by

$$\tau = T \cdot t/J$$

Hence the torsion shear stress for the open thin-walled circular section is given by

$$\tau_O = \frac{3Tt}{2\pi R \cdot t^3}$$

The rate of twist is given by

$$\theta = T/GJ$$

Hence the angle of twist for the open thin-walled circular section is given by

$$\varphi_0 = 3T\lambda/2\pi RG_{t^3}$$

For a thin walled closed section, the torsion shear stress is given by

$$\tau = T/2At$$

Hence the torsion shear stress for the closed thin-walled circular section is given by is given by

$$\tau_C = T/2\pi R^2 t$$

The rate of twist is given by

$$\theta = T/GJ$$

Hence the angle of twist for the closed thin-walled circular section is given by

$$\varphi_C = T\lambda/GJ_c$$

The torsion constant J_c is given by

$$J_C = 4 \times \pi^2 R^4 \Big/ \frac{2\pi R}{t} = 2\pi R^3 t$$

Hence the angle of twist for the closed thin-walled circular section is given by

$$\varphi_C = T\lambda/2\pi R^3 tG$$

The ratio of the torsion shear stress for the closed circular section to the open section is given by

$$\tau_C/\tau_O = \frac{T}{2\pi R^2 t} \times \frac{2\pi R t^3}{3Tt} = t/3R$$

It is clear from this ratio that the torsion shear stress of closed sections is much smaller than the torsion shear stress for the open section having the same configuration and dimensions.

The ratio of the tangle of twist for the closed circular section to the open section is given by

$$\frac{\varphi_C}{\varphi_O} = \frac{T\lambda}{2\pi R^3 tG} \times \frac{2\pi RGt^3}{3T\lambda} = 1/3 \cdot (t/R)^2$$

Similarly, it is clear from this ratio that the angle of twist of a closed section is much smaller than the angle of twist for the open section having the same configuration and dimensions.

1.8.2 Square Section

For the closed section, see Fig. 1.17, the torsion shear stress and angle of twist for a closed thin-walled box-girder are given by

$$\tau_C = T/2a^2 t$$

and

$$\varphi_C = T\lambda \Big/ \left(\frac{4a^4 G}{4a/t}\right)$$

The torsion shear stress and angle of twist for an open thin-walled box-girder having the same dimensions are given by

For the open section

$$\tau_O = \frac{3Tt}{2a \cdot t^3}$$

$$\varphi_O = \frac{3T\lambda}{G \cdot 4at^3}$$

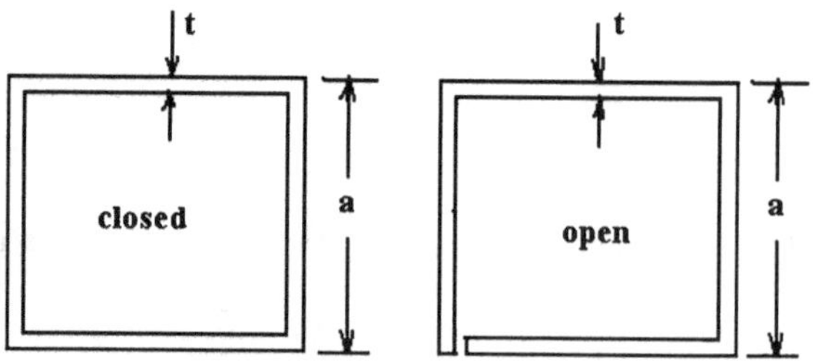

Fig. 1.17 Open and closed thin-walled square section beam

Hence, the ratio of the torsion shear stress for the closed box-girder to the open section is given by

$$\begin{aligned}\tau_C/\tau_O &= \frac{T}{2a^2t} \times \frac{4at^3}{3Tt} \\ &= 2/3 \cdot t/a\end{aligned}$$

The ratio of the tangle of twist for the closed box-girder to the open section is given by

$$\begin{aligned}\varphi_C/\varphi_O &= \frac{T\lambda \times 4a}{4a^4t \cdot G} \times \frac{4at^3G}{3T\lambda} \\ &= 4/3 \cdot (t/a)^2\end{aligned}$$

These results indicate clearly that the strength and stiffness of closed thin-walled sections are much superior the corresponding open sections.

1.9 Torsion Constant of Uniform Thin-Walled Closed Sections with Attached Open Sections

For a thin-walled closed section with attached open sections, as shown in Fig. 1.18, the torsion constant is given by

$$J_t = 4A^2 \bigg/ \oint ds/t + \sum \frac{b_i t_i^3}{3}$$

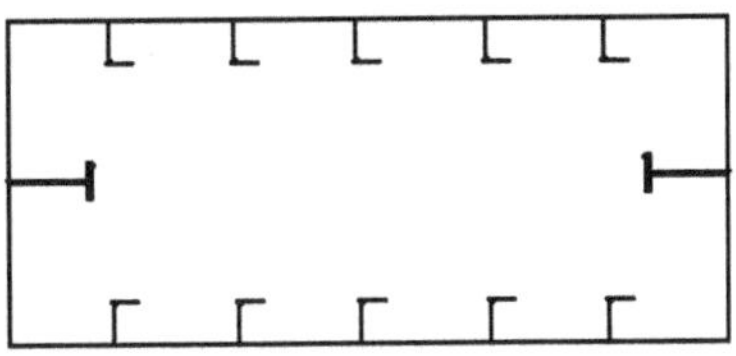

Fig. 1.18 A thin-walled closed section with attached open sections

Chapter 2
Torsion Stresses in Thin-Walled Multi-Cell Box-Girders

2.1 Torsion of Uniform Thin-Walled Two-Cell Box-Girders

The thin-walled box section with uniform thickness *t* as shown in Fig. 2.1, is subjected to a torsion moment T.

The shear flow and angle of twist for the thin-walled two cell structure shown in Fig. 2.1 could be determined as follows.

The flexural warping coefficients are given by

$$\delta_{11} = 1/G \cdot \oint_{s_1} ds/t$$

$$= 1/Gt \cdot (AC + CD + DB + BA)$$

$$\delta_{22} = 1/Gt \cdot (DC + CF + FH + HD)$$

$$\delta_{12} = -1/G \cdot CD/t$$

Since the angle of twist is the same for the two cells, then the basic equations are given by

$$\delta_{11} \cdot q_1 + \delta_{12} \cdot q_2 - 2A_1 \cdot \theta = 0 \tag{2.1}$$

$$\delta_{12} \cdot q_1 + \delta_{22} \cdot q_2 - 2A_2 \cdot \theta = 0 \tag{2.2}$$

From equations (2.2.1) and (2.2.2) we get

$$\delta_{11} \cdot \delta_{22} \cdot q_1 + \delta_{12} \cdot \delta_{22} \cdot q_2 = 2\delta_{22} \cdot A_1 \cdot \theta$$

$$\delta_{12}^2 \cdot q_1 + \delta_{12} \cdot \delta_{22} \cdot q_2 = 2\delta_{12} \cdot A_2 \cdot \theta$$

Hence

$$q_1(\delta_{11} \cdot \delta_{22} - \delta_{12}^2) = 2\delta_{22} \cdot A_1 \cdot \theta - 2\delta_{12} \cdot A_2 \cdot \theta$$

M. Shama, *Torsion and Shear Stresses in Ships*,
DOI: 10.1007/978-3-642-14633-6_2, © Springer-Verlag Berlin Heidelberg 2010

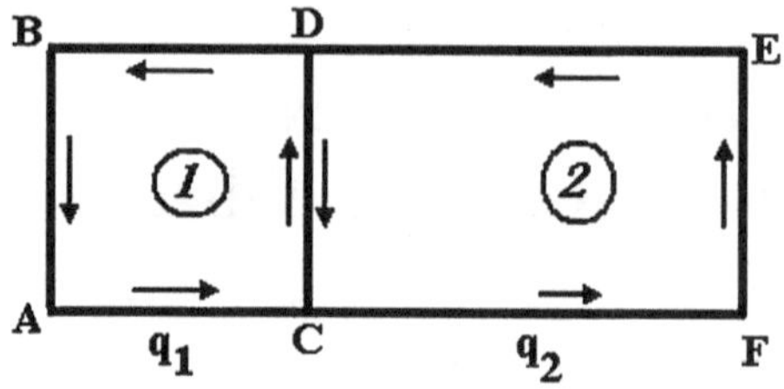

Fig. 2.1 Shear flow due to torsion of a thin-walled box girder with two unequal cells

The solution of equations (2.1) and (2.2) gives

$$q_1 = \theta \cdot (2\delta_{22} \cdot A_1 - 2\delta_{12} \cdot A_2)/(\delta_{11} \cdot \delta_{22} - \delta_{12}^2)$$
$$= \Delta_1 \cdot \theta$$

$$q_2 = \theta \cdot (2A_1 - \Delta_1\delta_{11})/\delta_{12}$$
$$= \Delta_2 \cdot \theta$$

where

$$\Delta_1 = (2\delta_{22} \cdot A_1 - 2\delta_{12} \cdot A_2)/(\delta_{11} \cdot \delta_{22} - \delta_{12}^2)$$

$$\Delta_2 = (2A_1 - \Delta_1\delta_{11})/\delta_{12}$$

The equilibrium condition gives

$$T = 2A_1 \cdot q_1 + 2A_2 \cdot q_2$$
$$= \Delta_3 \cdot \theta$$

where

$$\Delta_3 = (2A_1 \cdot q_1 + 2A_2 \cdot q_2)$$

Hence

$$\theta = 1/\Delta_3 \cdot T$$

$$q_1 = \Delta_1/\Delta_3 \cdot T$$

$$q_2 = \Delta_2/\Delta_3 \cdot T$$

$$q_{12} = q_1 - q_2 = (\Delta_1 - \Delta_2)/\Delta_3 \cdot T.$$

Example 2.1 Determine the torsion shear stress and angle of twist for the two uniform thickness thin-walled box-girder shown in Fig. 2.2.

Solution The shear flow and angle of twist for the thin-walled two cell structure shown in Fig. 2.2 could be determined as follows.

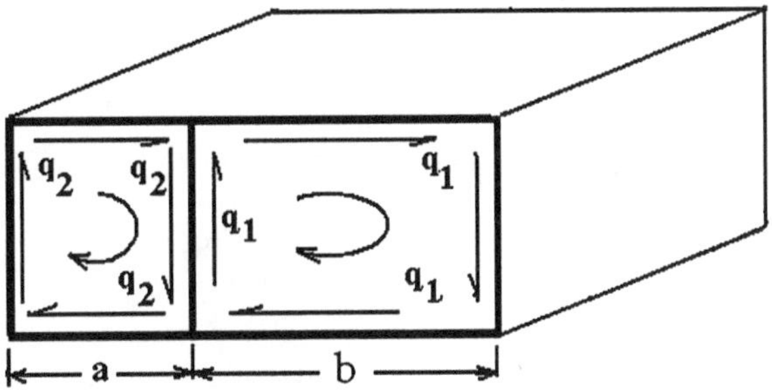

Fig. 2.2 A uniform thin-walled box-girder with two cells

The torsional moment is given by

$$T = 2(q_1A_1 + q_2A_2) \tag{2.3}$$

The angle of twist for cells ***1*** and ***2*** are given by

$$\theta_1 = \frac{1}{2GA_1} \cdot \left(q_1 \oint_1 ds/t - q_2 \int_{12} ds/t \right) \tag{2.4}$$

$$\theta_2 = \frac{1}{2GA_2} \left(q_2 \oint_2 ds/t \oint -q_1 \int_{21} ds/t \right) \tag{2.5}$$

Since the angle of twist is the same for the two cells, then we have

$$\theta_1 = \theta_2 = \theta_3$$

Reformulating equations (2.4) and (2.5), we get

$$\frac{1}{G} \left(q_1 \oint_1 ds/t - q_2 \int_{12} ds/t \right) = 2A_1\theta \tag{2.6}$$

$$\frac{1}{G} \left(q_2 \oint_2 ds/t - q_1 \oint_{21} ds/t \right) = 2A_2\theta \tag{2.7}$$

Let

$$\delta = \text{warping flexibility}$$

$$\delta_{11} = \frac{1}{G} \oint_1 ds/t$$

$$\delta_{22} = \frac{1}{G} \oint_2 ds/t$$

$$\delta_{12} = \delta_{21} = -\frac{1}{G}\int_{12} ds/t$$

Substituting in equations (2.2.4) and (2.2.5), we get

$$\delta_{11}q_1 + \delta_{12}q_2 - 2A_1\theta = 0 \tag{2.8}$$

$$\delta_{12}q_1 + \delta_{22}q_2 - 2A_2\theta = 0 \tag{2.9}$$

Solving equations (2.2.3), (2.2.8) and (2.2.9), q_1 and q_2 could be determined.

Example 2.2 Determine the torsion shear stresses and the rate of twist for the thin-walled ***2***-cell box-girder shown in Fig. 2.3. The girder is subjected to a constant torque T.

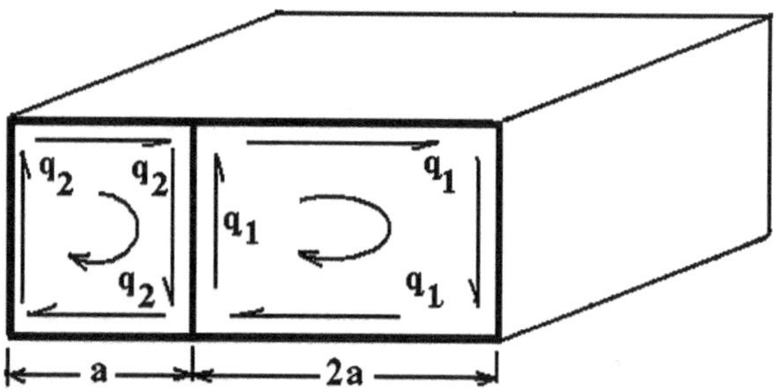

Fig. 2.3 A thin-walled box-girder with two unequal cells

Solution Area of cell (***1***) is given by

$$A_1 = 2a^2$$

Area of cell (***2***) is given by

$$A_2 = a^2$$

Let

$$\delta = \text{warping flexibility}$$

$$\delta_{11} = 1/G \cdot \oint ds/t = 6a/Gt$$

$$\delta_{22} = 4a/Gt$$

$$\delta_{12} = -a/Gt$$

The basic equations are

$$\delta_{11}q_1 + \delta_{12}q_2 - 2A_1\theta = 0 \tag{2.10}$$

$$\delta_{12}q_1 + \delta_{22}q_2 - 2A_2\theta = 0 \tag{2.11}$$

The equilibrium equation gives

$$T = 2(q_1A_1 + q_2A_2) = 2a^2(2q_1 + q_2) \tag{2.12}$$

From equations (2.10) and (2.11), we get

$$\delta_{11}q_1A_2 + \delta_{12}q_2A_2 - \delta_{12}q_1A_1 - \delta_{22}q_2A_1 = 0$$

$$q_1(\delta_{11}A_2 - \delta_{12}A_1) + q_2(\delta_{12}A_2 - \delta_{22}A_1) = 0$$

Hence

$$q_1 = q_2\frac{\delta_{22}A_1 - \delta_{12}A_2}{\delta_{11}A_2 - \delta_{12}A_1} = q_2\frac{2\delta_{22} - \delta_{12}}{\delta_{11} - 2\delta_{12}}$$

From equation (2.12), we get

$$T = 2a^2q_2\left[2\frac{2\delta_{22} - \delta_{12}}{\delta_{11} - 2\delta_{12}} + 1\right]$$

$$T = 2a^2q_2\left[\frac{4\delta_{22} - 4\delta_{12} + \delta_{11}}{\delta_{11} - 2\delta_{12}}\right]$$

From which q_2 is given by

$$q_2 = \frac{T}{2a^2}\left(\frac{\delta_{11} - 2\delta_{12}}{4\delta_{22} - 4\delta_{12} + \delta_{11}}\right)$$

$$q_1 = \frac{T}{2a^2}\left(\frac{2\delta_{22} - \delta_{12}}{4\delta_{22} - 4\delta_{12} + \delta_{11}}\right)$$

Substituting in equation (2.10), we get

$$\begin{aligned}\theta &= \frac{1}{2A_1}\cdot(\delta_{11}q_1 + \delta_{12}q_2)\\ &= \frac{1}{4a^2}\cdot\frac{T}{2a^2}\cdot\left[\frac{2\delta_{11}\delta_{22} - \delta_{11}\delta_{12} + \delta_{12}\delta_{11} - 2\delta_{12}^2}{4\delta_{22} - 4\delta_{12} + \delta_{11}}\right]\\ &= \frac{T}{4a^4}\cdot\left(\frac{\delta_{11}\delta_{22} - \delta_{12}^2}{4\delta_{22} - 4\delta_{12} + \delta_{11}}\right)\end{aligned}$$

Substituting for δ_{12}, δ_{11} and δ_{22}, we get

$$q_1 = \frac{T}{2a^2G}[(8\cdot a/t + a/t)/(16\cdot a/t + 4\cdot a/t + 6\cdot a/t)] = \frac{T}{2a^2G}\times\frac{9}{26}$$

$$q_2 = \frac{T}{2a^2G}\cdot[(6\cdot a/t + 2\cdot a/t)/(26a/t)] = \frac{T}{2a^2G}\times\frac{8}{26}$$

$$\theta = \frac{T}{4a^4G}\left[(24\cdot a^2/t^2 - a^2/t^2)/(26\cdot a/t)\right] = \frac{T}{4a^4G}\cdot 23/26\cdot a/t$$

But

$$\theta = T/GJ$$

Hence

$$J = T/G\theta = 104/23 \cdot a^3 \cdot t.$$

2.2 The General Case of a Uniform Two-Cell Box Girder

This is an indeterminate structural problem and its solution is based on the assumption that the rate of twist for each cell is the same as for the whole section, see Fig. 2.4.

i.e.,

$$\theta_1 = \theta_2 = 0 \quad \text{and} \quad (\theta = d\varphi/dz)$$

$$\theta = T/GJ$$

The torque ***T*** is given by

$$T = 2q_1A_1 + 2q_2A_2$$

$$\theta_1 = \frac{1}{2GA_1}\oint_1 q/t \cdot ds$$

and

$$\theta_2 = \frac{1}{2GA_2}\oint_2 q/t \cdot ds$$

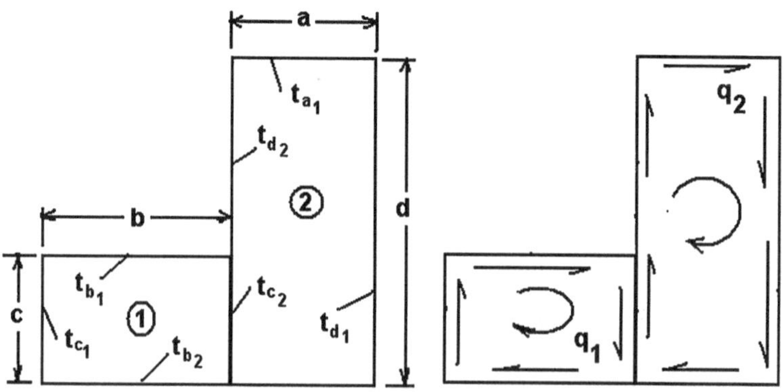

Fig. 2.4 Idealized section and torsion shear flow of a thin-walled two cell structure

i.e.,

$$\frac{1}{G}\left[q_1 \oint_1 q/t \cdot ds - q_2 \cdot (ds/t)_{12}\right] = 2A_1\theta_1 \tag{2.13}$$

$$\frac{1}{G}\left[q_2 \oint_2 ds/t - q_1 (ds/t)_{21}\right] = 2A_2\theta_2 \tag{2.14}$$

Equations (2.13) and (2.14) are simplified to

$$\delta_{11}q_1 + \delta_{12}q_2 = 2A_1\theta$$

$$\delta_{21}q_1 + \delta_{22}q_2 = 2A_2\theta$$

$$\begin{bmatrix} \delta_{11} & \delta_{12} \\ \delta_{21} & \delta_{22} \end{bmatrix} \begin{Bmatrix} q_1 \\ q_2 \end{Bmatrix} = \begin{Bmatrix} A_1 \\ A_2 \end{Bmatrix} 2\theta$$

or

$$[\delta]\{q\} = 2\theta\{A\}$$

The shear flow in each cell is given by

$$\{q\} = [\delta]^{-1} \cdot 2\theta \cdot \{A\}$$

i.e.,

$$\{q\} = 2\theta \cdot [\delta]^{-1} \cdot \{A\}$$

$$\begin{aligned} \delta_{11} &= \frac{1}{G}\oint_1 ds/t \\ \delta_{22} &= \frac{1}{G}\oint_2 ds/t \\ \delta_{12} &= \delta_{21} \\ &= -\frac{1}{G}[ds/t]_{12} \end{aligned} \tag{2.15}$$

The torque is given by

$$T = (2q_1A_1 + 2q_2A_2) \tag{2.16}$$

Solving equations (2.15) and (2.16), we get q_1, q_2 and θ

The torque T is given by

$$T = G \cdot J \cdot \theta$$

Hence J is given by

$$J = T/G\theta.$$

Example 2.3 Determine the torsion shear stresses and angle of twist for the thin-walled box section having uniform thickness t as shown in Fig. 2.5. The section is subjected to a torsion moment ***T***.

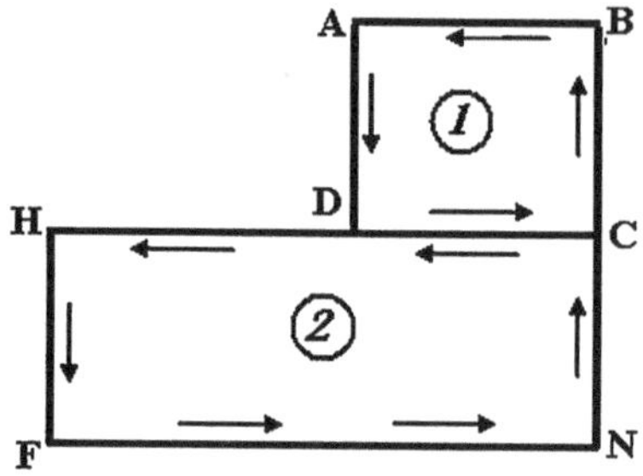

Fig. 2.5 Shear flow due to torsion of a box-girder with two cells

Solution Condition for Compatibility (Consistent Deformation). The warping flexibilities are given by

$$\delta_{11} = 1/Gt \cdot (AB + BC + CD + DA)$$
$$\delta_{22} = 1/Gt \cdot (CH + HF + FN + NC)$$
$$\delta_{12} = -1/G \cdot CD/t\delta_{12} = -1/G \cdot CD/t$$

The basic equations of consistent deformation are given by

$$\delta_{11} \cdot q_1 + \delta_{12} \cdot q_2 = 2A_1 \cdot \theta \tag{2.17}$$

$$\delta_{12} \cdot q_1 + \delta_{22} \cdot q_2 = 2A_2 \cdot \theta \tag{2.18}$$

Solving equations (2.17) and (2.18) we get

$$q_1 = \frac{2\delta_{22} \cdot A_1 - 2\delta_{12} \cdot A_2}{\delta_{11} \cdot \delta_{22} - \delta_{12}^2} \cdot \theta = \Delta_1 \cdot \theta$$
$$q_2 = \frac{2A_1}{\delta_{12}}\theta - \Delta_1 \frac{\delta_{11}}{\delta_{12}}\theta = \Delta_2 \cdot \theta$$

Equation for equilibrium condition is given by

$$T = (2A_1 \cdot q_1 + 2A_2 \cdot q_2) \cdot \theta = \Delta_3 \cdot \theta \tag{2.19}$$

The solution of equations (2.17), (2.18) and (2.19) gives

$$\theta = T/\Delta_3$$

$$q_1 = T \cdot \Delta_1/\Delta_3$$

$$q_2 = T \cdot \Delta_2/\Delta_3$$

$$q_{12} = q_1 - q_2 = T \cdot (\Delta_1 - \Delta_2)/\Delta_3.$$

2.3 Torsion Stresses in a Two Identical Cells Box-Girder

The two identical thin-walled cells box girder, see Fig. 2.6, behaves exactly as a single cell box girder.

Since the two cells are identical we have

$$q_I = q_{II} = q.$$

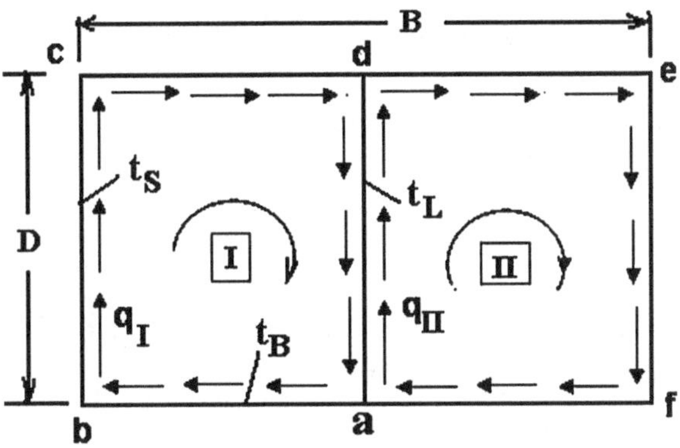

Fig. 2.6 A thin-walled box girder with two identical cells

2.3.1 Shear Flow q

The shear flow q is the same for the two cells and is given by

$$q = T/2A$$

where

$$A = B \cdot D.$$

2.3.2 Shear Stress τ

The shear stress in the sides, top, bottom and the internal partition plating are given by

$$\tau_S = q/t_s, \quad \tau_D = q/t_D, \quad \tau_B = q/t_B, \quad \tau_L = 0.$$

2.3.3 *Rate of Twist θ*

The rate of twist is given by

$$\theta = T/GJ$$

where

$$J = 4A^2 \Big/ \oint ds/t$$

$$\oint ds/t = 2B/t_B + 2D/t_S$$

2.4 Torsion of Three-Cell Box-Girder

Following the same principle that the angle of twist is the same for the three cell box-girder shown in Fig. 2.7.

Then

$$\theta_1 = \theta_2 = \theta_3 = \theta$$

The equations of consistent deformation are given by

$$\delta_{11}q_1 + \delta_{12}q_2 - 2A_1\theta_1 = 0 \quad (2.20)$$

$$\delta_{12}q_1 + \delta_{22}q_2 - 2A_2\theta_2 = 0 \quad (2.21)$$

$$\delta_{32}q_1 + \delta_{33}q_2 - 2A_3\theta_3 = 0 \quad (2.22)$$

The torque is given by

$$T = (2A_1 \cdot q_1 + 2A_2 \cdot q_2 + 2A_3 \cdot q_3)\theta \quad (2.23)$$

Solving equations (2.20)–(2.23), we get q_1, q_2, q_3, q_4 and θ.

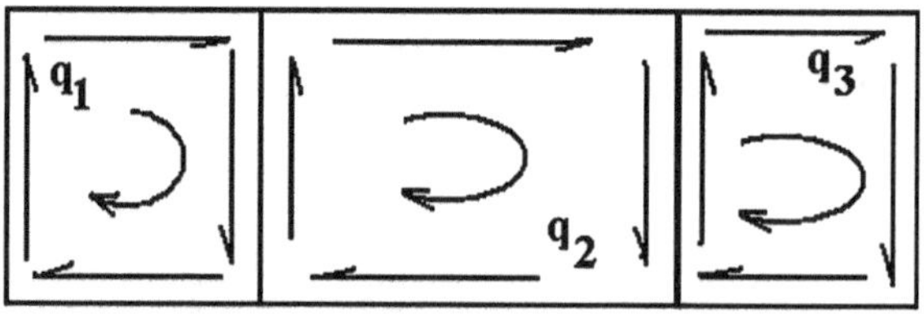

Fig. 2.7 Torsion of a three-cell box-girder

Hence

$$\begin{bmatrix} \delta_{11} & \delta_{12} & 0 \\ \delta_{21} & \delta_{22} & \delta_{23} \\ 0 & \delta_{32} & \delta_{33} \end{bmatrix} \begin{Bmatrix} q_1 \\ q_2 \\ q_3 \end{Bmatrix} = 2\theta \begin{Bmatrix} A_1 \\ A_2 \\ A_3 \end{Bmatrix} \tag{2.24}$$

$$[\delta]\{q\} = 2\theta\{A\}$$

Hence the shear flow in each cell is given by

$$\{q\} = \left[\delta^{-1}\right]\{A\}2\theta$$

and

$$T = 2\theta \cdot \sum A_i q_i$$

where

$$\delta_{11} = \oint_1 ds/t$$

$$\delta_{22} = \oint_2 ds/t$$

$$\delta_{33} = \oint_3 ds/t$$

$$\delta_{12} = \delta_{21} = -[ds/t]_{12}$$

$$\delta_{23} = \delta_{32} = -[ds/t]_{23}.$$

Example 2.4 Determine the shear flow, shear stress and rate of twist for the three-cell box girder shown in Fig. 2.8.

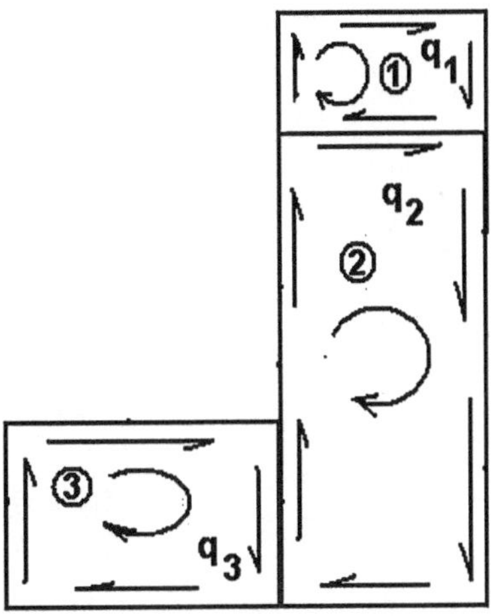

Fig. 2.8 Shear flow due to torsion of a 3-cell thin-walled box girder

Solution Following the same principle that the angle of twist is the same for all cells, see Fig. 2.8, we get

$$\begin{aligned} \delta_{11}q_1 + \delta_{12}q_2 &= 2A_1\theta \\ \delta_{21}q_1 + \delta_{22}q_2 + \delta_{23}q_3 &= 2A_2\theta_2 \\ \delta_{32}q_2 + \delta_{33}q_3 &= 2A\theta_3 \end{aligned} \tag{2.25}$$

but

$$\theta_1 = \theta_2 = \theta_3 = \theta$$

Then

$$\begin{bmatrix} \delta_{11} & \delta_{12} & 0 \\ \delta_{21} & \delta_{22} & \delta_{23} \\ 0 & \delta_{32} & \delta_{33} \end{bmatrix} \begin{Bmatrix} q_1 \\ q_2 \\ q_3 \end{Bmatrix} = 2\theta \begin{Bmatrix} A_1 \\ A_2 \\ A_3 \end{Bmatrix}$$

$$[\delta]\{q\} = 2\theta\{A\} \tag{2.26}$$

$$\{q\} = [\delta^{-1}]\{A\}2\theta \tag{2.27}$$

The torque is given by

$$T = (2q_1A_1 + 2q_2A_2 + 2q_3A_3) \tag{2.28}$$

Solving equations (2.26) and (2.28) we get

$$q_1 = 2\theta\theta^{-1}A_1$$

$$q_2 = 2\theta\theta^{-1}A_2$$

$$q_3 = 2\theta\theta^{-1}A_3$$

Substituting in equation (2.25), we get

$$\theta = 1/2A_1 \cdot (\delta_{11}q_1 + \delta_{12}q_2)$$

where

$$\delta_{11} = 1/G \cdot \oint_1 ds/t$$

$$\delta_{22} = 1/G \cdot \oint_2 ds/t$$

$$\delta_{33} = 1/G \cdot \oint_3 ds/t$$

$$\begin{aligned} \delta_{12} = \delta_{21} &= -1/G \cdot [ds/t]_{12} \\ \delta_{23} = \delta_{32} &= -1/G \cdot [ds/t]_{23}. \end{aligned}$$

2.5 Torsion of Uniform Thin-Walled Multi-Cell Box-Girder

The multi-cell thin-walled structure when subjected to pure torsion is a statically indeterminate problem; see Fig. 2.9.

The torque T is given by

$$T = \sum_{i=1}^{n} 2A_i q_i = GJ\theta J$$

where T = applied uniform torque; A_i = enclosed area of the *i*th cell; J = torsion constant

$$J = 4\sum_{i=1}^{n} A_i \delta^{-1} A_i$$

The angle of twist per unit length

$$\theta = d\varphi/dz$$
$$\theta_i = \theta_j = \theta_{ij} = \theta_{jn}$$

where

$$\theta_i = 1/2GA_i \cdot \left(q_i \oint ds/t\right)$$

The angle of twist for cell ***i*** is given by

$$\theta_i = 1/2GA_i \cdot \left(q_i \oint ds/t - q_{i-1} \int ds/t - q_{i+1} \int ds/t\right) \tag{2.29}$$

Equation (2.29) represents a series of simultaneous equations which gives $q_1, q_2, q_3, \ldots, q_n$.

The set of equations of consistent deformation is given by

$$\delta_{11} q_1 + \delta_{12} q_2 - 2A_1\theta = 0 \tag{2.30}$$

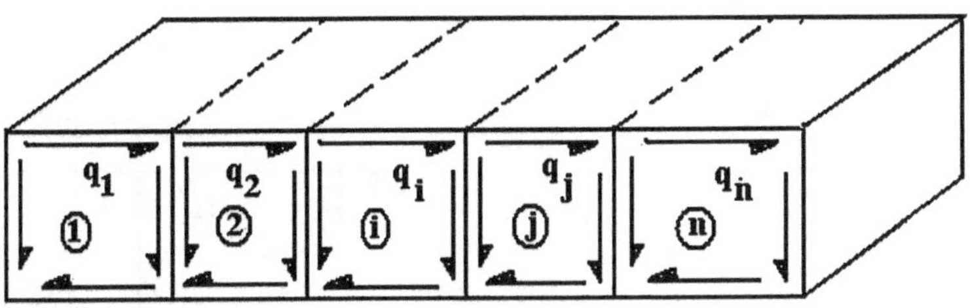

Fig. 2.9 Torsion of a multi-cell thin-walled box-girder

$$\delta_{12}q_1 + \delta_{22}q_2 - 2A_2\theta = 0 \tag{2.31}$$

$$\delta_{32}q_1 + \delta_{33}q_2 - 2A_3\theta = 0 \tag{2.32}$$

This set of equations could be put in the following form

$$[\delta]\{q\} = 2\theta\{A\}$$

Hence, for a multi cell box girder, the shear flow in each cell is given by

$$\{q\} = [\delta]^{-1}\cdot 2\theta \cdot \{A\} \qquad (i = 1, 2, \ldots, n)$$

where

$$\delta = \begin{pmatrix} \delta_{11} & \delta_{12} & 0 & 0 & 0 & 0 \\ \delta_{21} & \delta_{22} & \delta_{23} & 0 & 0 & 0 \\ 0 & \delta_{32} & \delta_{33} & \delta_{34} & 0 & 0 \\ 0 & 0 & \delta_{43} & \delta_{44} & \delta_{45} & 0 \\ 0 & 0 & 0 & \delta_{54} & \delta_{55} & \delta_{56} \\ 0 & 0 & 0 & 0 & \delta_{65} & \delta_{66} \end{pmatrix}$$

The torsion shear stresses are given by

$$\tau_1 = q_1/t_1, \quad \tau_2 = q_2/t_2, \quad \tau_3 = q_3/t_3.$$

2.6 Combined Open and Closed Thin-Walled Sections

For the combined open and closed section, sec Fig. 3.1, the angle of twist is the same for the whole section whether it is an open or closed section.

2.6.1 Combined Open Section with One Closed Cell

The total torque T for the thin-walled section shown in Fig. 2.10 is given by

$$T = \sum_{l=1}^{2} T_i = G \cdot J \cdot \theta$$

Fig. 2.10 Combined open and closed one-cell thin-walled section

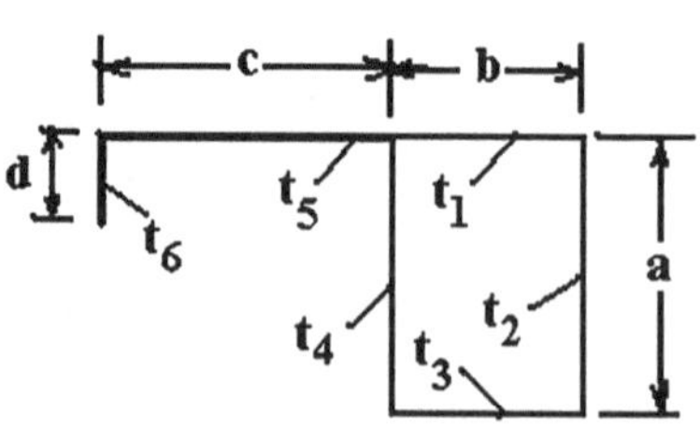

Hence

$$\theta = T/GJ$$

where

$$T_1 = GJ_1 \cdot \theta$$

J_1 = torsion constant of the open section; J_2 = Torsion constant of the closed section.

For the open part of the structure, the shear flow q_1 is given by

$$q_1 = T_1 \cdot t_1^2/J_1$$

$$T_1 = G \cdot J_1 \cdot \theta$$

For the closed section of the structure, the shear flow q_2 is given by

$$q_1 = T_2/2A_2$$
$$T_2 = 2A_2 \cdot q_2 = G \cdot J_2 \cdot \theta.$$

2.6.2 Combined Open Section with Two Closed Cells

The applied torque T for the thin-walled structure shown in Fig. 2.11 is given by

$$T = \sum_{j=1}^{n} 2q_j A_j + \sum_{i=1}^{m} GJ_i\theta$$

In the above particular example

$$T = 2q_1 A_1 + 2q_2 A_2 + GJ_3\theta$$

where J_3 = the torsion constant of the open section part of the structure and is given by $\mathbf{J_3}$

$$T = 1/3 \sum_{i=1}^{n} b_i t_i^3$$

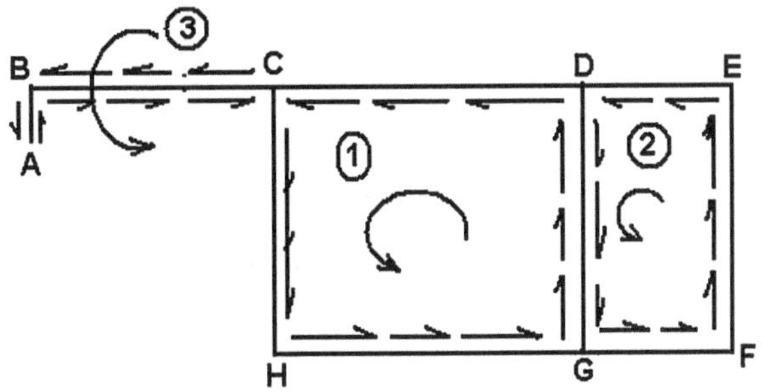

Fig. 2.11 Combined thin-walled open and closed two-cell structure

The shear flow in the two cells is given by

$$\delta_{11}q_1 + \delta_{12}q_2 = 2A_1\theta$$

$$\delta_{21}q_1 + \delta_{22}q_2 = 2A_2\theta$$

$$\delta_{11} = 1/G \cdot \oint_1 ds/t$$

$$\delta_{22} = 1/G \cdot \oint_2 ds/t$$

$$\delta_{12} = \delta_{21} = -1/G \cdot [ds/t]_{12}$$

In the general case, for combined open and closed sections, the shear flow in each cell is given by

$$q_i = \delta^{-1} 2\theta\theta_i \quad (i = 1, 2, \ldots, n)$$

And in each open member the shear flow is given by

$$q_i = T/J \cdot t_i^2$$

And the angle of twist is given by

$$\theta = T/GJ$$

where T = the torque and is given by

$$T = 4\sum_{i=1}^{n} A_i' \delta^{-1} A_i + \frac{1}{3}\sum_{j=1}^{m} b_j t_j^3.$$

Example 2.5 Determine the shear flow distribution and rate of twist for the idealized ship section shown in Fig. 2.12. The ship section is subjected to a torque T.

Solution The torque T is distributed among the thin-walled structural members of the ship section as follows

$$T = \sum_{i=1}^{4} T_i$$

where

$$T_1 = 2A_1 \cdot q_1, \quad T_2 = 2A_2 \cdot q_2, \quad T_3 = 2A_3 \cdot q_3,$$

$$T_4 = GJ_4 \cdot \theta = \frac{q_4 \cdot J_4}{t_4^2}, \qquad J_4 = 1/3 \cdot St_4^3$$

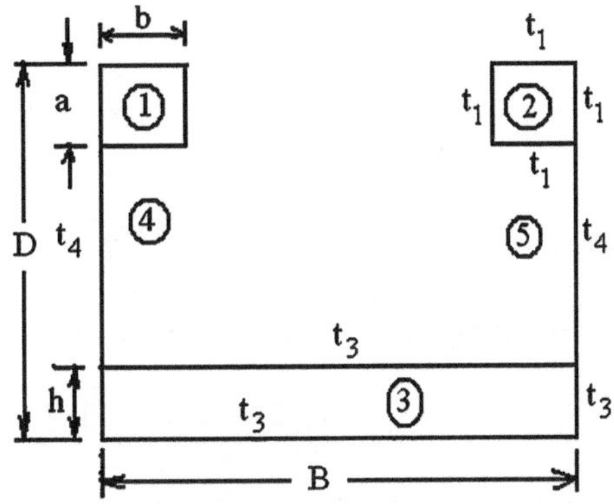

Fig. 2.12 Idealized ship section

because of symmetry of the ship section, we have

$$T_1 = T_2, \quad T_4 = T_5$$

Hence

$$T_1 = T_2 = 2A_1 \cdot q_1 = G \cdot J_1 \cdot \theta$$

Thus

$$T = 2GJ_1 \cdot \theta + 2GJ_4 \cdot \theta + GJ_3 \cdot \theta = \sum_{j=1}^{3} G \cdot J \cdot \theta$$

where

$$G = \frac{E}{2(1+\upsilon)} = \frac{E}{2 \cdot 6}$$

$$J = 2\,J_1 + 2\,J_4 + J_3$$

where

$$J_1 = 4A_1^2 \Big/ \oint \frac{ds}{t} = 4(a \cdot b)^2/(2b/t + 2a/t)$$

$$J_3 = A(B \cdot h)^2/(2B/t_3 + 2h/t_3)$$

$$J_4 = 1/3 \cdot S_4 \cdot t_4^3 = 1/3 \cdot (D - a - h) \cdot t_4^3$$

Hence

$$\theta = T \Big/ \sum GJ$$

Substituting, we get the torque carried by each structural element.

Hence

$$T_1 = GJ_1 \cdot \theta$$

$$T_3 = GJ_3 \cdot \theta$$

$$T_4 = GJ_4 \cdot \theta$$

Substituting, we get the shear flow in each structural element as follows

$$q_1 = T_1/2A_1, \quad q_3 = T_3/2A_3, \quad \text{and} \quad q_4 = T_4/2A_4.$$

Example 2.6 Determine the shear flow and rate of twist for the ship section of bulk carrier shown in Fig. 2.13.

Solution The torque T is given by

$$T = 2[2A_1q_1 + 2q_2A_2 + GJ_6\theta + 2A_3q_3 + 2A_4q_4 + 2A_5q_5]$$

The torsion constant J is given by

$$J = 2\left\{\sum_{i=1}^{5} 4A_i'\delta^{-1}A_i + 1/3 \cdot \lambda_6 \cdot t_6^3\right\}$$

The rate of twist θ is given by

$$\theta = T/GJ$$

The set of equations of consistent deformation for cells (*1*) and (*2*) is given by

$$\delta_{11}q_1 + \delta_{12}q_2 = 2A_1\theta = 2T/J \cdot A_1$$

$$\delta_{21}q_1 + \delta_{22}q_2 = 2A_2\theta = 2T/J \cdot A_2$$

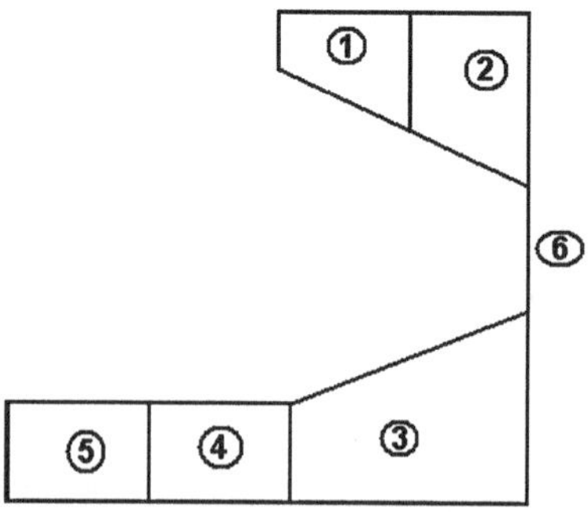

Fig. 2.13 An idealized section of a bulk carrier

This set of equations can be put in the matrix form as follows

$$\begin{bmatrix} \delta_{11} & \delta_{12} \\ \delta_{21} & \delta_{22} \end{bmatrix} \begin{Bmatrix} q_1 \\ q_2 \end{Bmatrix} = 2\theta \cdot \begin{Bmatrix} A_1 \\ A_2 \end{Bmatrix}$$

i.e.,

$$(\delta)\{q\} = 2T/J \cdot \{A\}$$

Hence, the torsion shear flow in cells (*1*) and (*2*) are given by

$$q_1 = \delta^{-1} \cdot 2T/J \cdot A_1$$

$$q_2 = \delta^{-1} \cdot 2T/J \cdot A_2$$

Similarly, the torsion shear flow in cells (*3*), (*4*) and (*5*) are given by

$$q_3 = \delta^{-1} \cdot 2T/J \cdot A_3$$

$$q_4 = \delta^{-1} \cdot 2T/J \cdot A_4$$

$$q_5 = \delta^{-1} \cdot 2T/J \cdot A_5$$

where

$$\delta = \begin{bmatrix} \delta_{33} & \delta_{34} & 0 \\ \delta_{43} & \delta_{44} & \delta_{45} \\ 0 & \delta_{54} & \delta_{55} \end{bmatrix}$$

$$\delta_{ii} = \left[1/G \cdot \sum_{J=1}^{m} \lambda_j/t_j\right], \quad i = 1, 2, \ldots, n = \text{N of cells}$$

$$\delta_{rj} = 1/G \cdot \left[\lambda_{rj}/t_{rj}\right]_i$$

where r, and j are cells, having a common boundary; i = cell No. i.

Chapter 3
Torsion Warping Deformations and Stresses

3.1 Torsion of Thin-Walled Variable Section Beams

In this case, plane sections no longer remain plane and torsion deformation is accompanied with warping deformation. The degree of warping of the section depends on the type of end constraints. In general, there are two cases for the calculation of warping

1. Free warping,
2. Constrained warping.

3.1.1 Free Warping

A thin-walled open section free at both ends and subjected to a torsional moment experiences rotational deformation as well as linear warping deformation, see Fig. 3.1. The magnitude of the angle of twist and the linear warping deformations depends on the geometrical configuration, dimensions and scantlings of the thin-walled section as well as on the magnitude of the torsional moment.

As sections are free to warp, no in-plane stresses are induced and the sections are distorted as shown in Figs. 3.1 and 3.2.

3.1.2 Constrained Warping

When warping is constrained, normal stresses are developed in the thin-walled member, see Fig. 3.3. The magnitudes of these stresses depend on the degree of constraint and the geometry of the section, see Figs. 3.4 and 3.5.

M. Shama, *Torsion and Shear Stresses in Ships*,
DOI: 10.1007/978-3-642-14633-6_3, © Springer-Verlag Berlin Heidelberg 2010

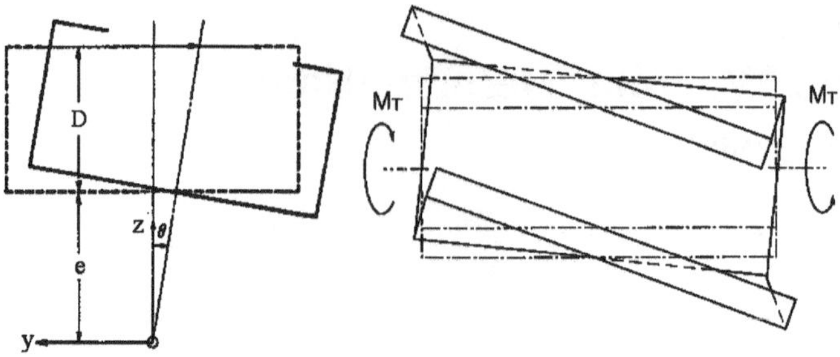

Fig. 3.1 Free warping of open thin-walled sections

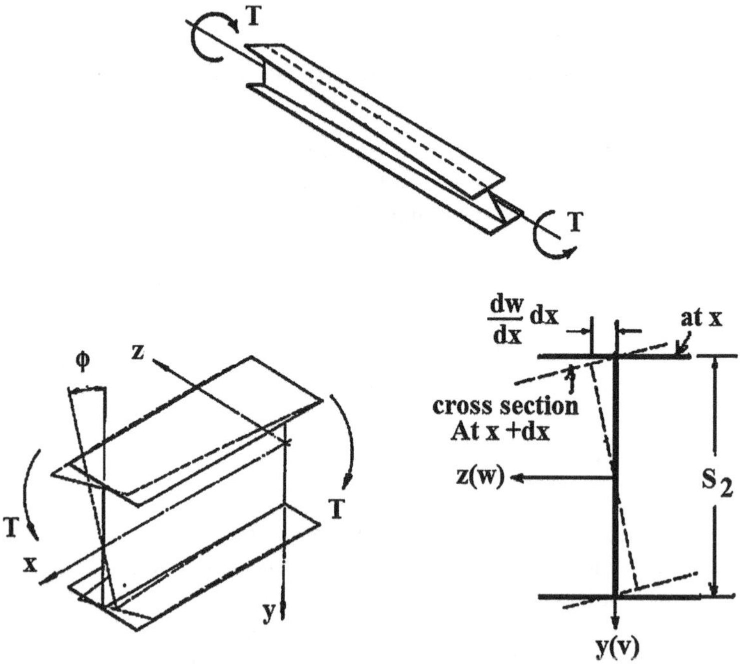

Fig. 3.2 Free warping deformation of an open thin-walled I-section

Fig. 3.3 Constrained warping of a thin-walled box-girder

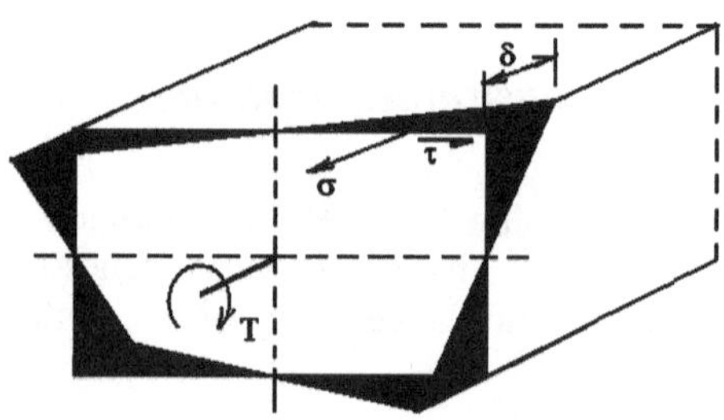

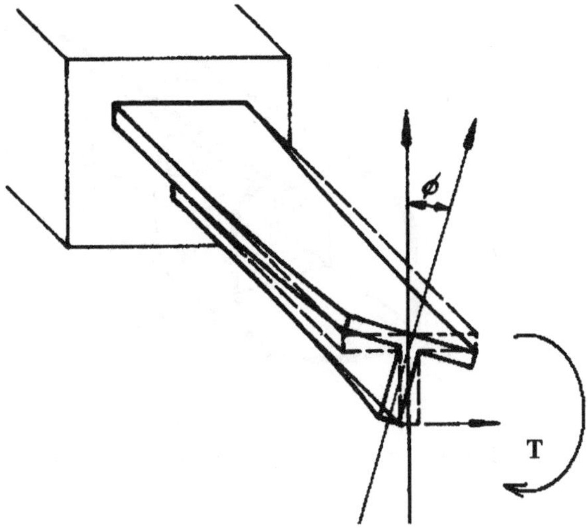

Fig. 3.4 Constrained warping deformation of an open thin-walled T-section

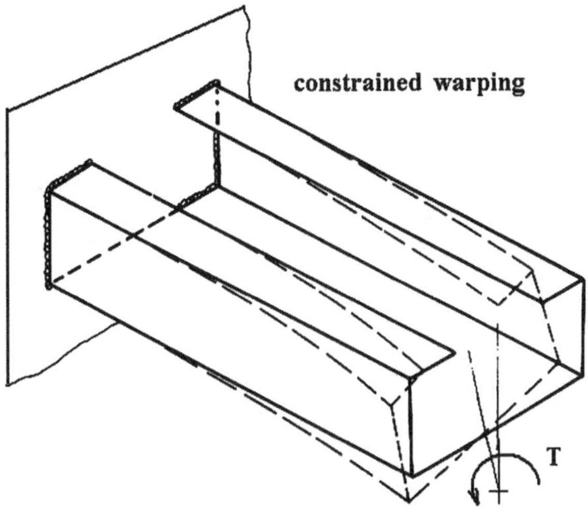

Fig. 3.5 Constrained warping if a thin-walled open section

3.1.3 Warping of Thin-Walled Sections

Assume that a beam of constant cross-section is loaded by a torque *T* at both ends as shown in Fig. 3.6.

Assume that the x-axis coincides with the twisting centre of the cross-section. Let the displacements in x, y, and z directions be u, v, and w respectively.

Assume that the contour of the cross-section maintains its shape when the member is twisted.

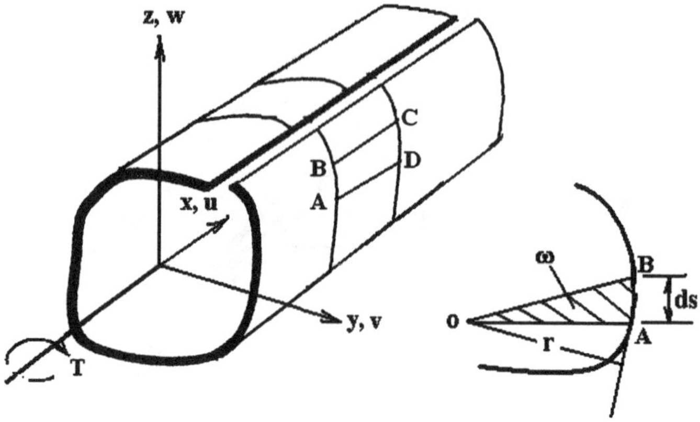

Fig. 3.6 Torsion and warping deformations of an open thin-walled section

Fig. 3.7 Torsional deformation of an element

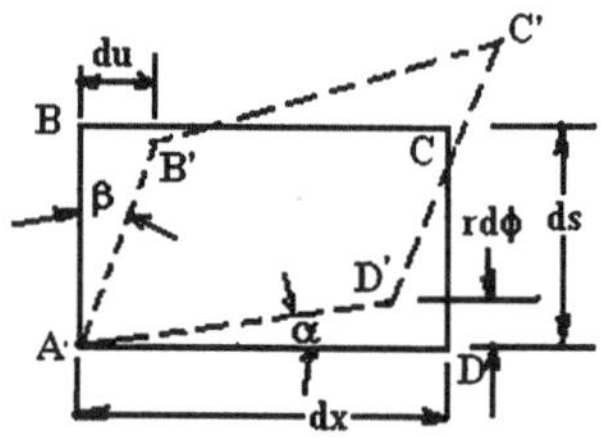

Consider the element ***ABCD***, see Fig. 3.7. The shear angle is given by

$$\gamma = \alpha + \beta$$

where

$$\alpha = \mathrm{r} \cdot \mathrm{d}\varphi/\mathrm{dx} = \mathrm{r}\theta$$

and

$$\beta = \mathrm{du}/\mathrm{ds}$$

Hence

$$\gamma = \mathrm{r} \cdot \theta + \mathrm{du}/\mathrm{ds} = \tau/\mathrm{G}$$

or

$$\mathrm{du} = (\tau/\mathrm{G} - \mathrm{r} \cdot \theta)\mathrm{ds}$$

On the median line of an open contour, $\tau = 0$, see Fig. 3.8.

Thus

$$\mathrm{du} = -\theta \cdot \mathrm{r} \cdot \mathrm{ds}$$

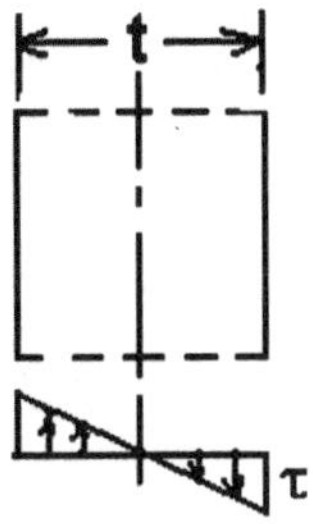

Fig. 3.8 Distribution of shear stress due to torsion over the plate thickness

and

$$u = -\theta \int r \cdot ds + u_0$$

For a closed thin-walled member,

$$du = (q/G \cdot ds/t - r \cdot \theta \cdot ds)$$

i.e.,

$$u = 1/G \int q ds/t - \theta \int r \cdot ds + u_0$$

where

$$u_o = \text{displacement at point S} = 0 \text{ in the } x \text{ direction.}$$

In the case of pure torsion of a thin-walled member with no warping restraints, the choice of the axis of rotation is completely arbitrary.

Thus, the relative warping displacement is given by

$$\bar{u} = 1/G \int \tau_{xs} ds - 2\theta \cdot \omega_s$$

where

$$2\omega_s = \int_0^s r \cdot ds$$

This equation is valid for both open and closed sections.

For open sections

$$\bar{u} = -2\theta\theta_s$$

For closed sections

$$\bar{u} = \frac{T}{G}\left(\frac{1}{2A}\int ds/t - 2\omega_s/J_t\right)$$

If the integrations is taken over the whole periphery of the section

$$\bar{u} = \frac{T}{G}\left(\frac{1}{2A}\oint ds/t - 2A/J_t\right) = 0$$

This gives

$$J_t = 4A^2 \Big/ \oint ds/t.$$

3.1.4 Flexural Warping Stresses

With the warping varying along the *x*-axis, normal stresses are induced over the bar sections.

For a segment AB of length dx, see Fig. 3.9, we have

$$\varepsilon = du/dx = -2\omega \cdot d\theta/dx$$

Hence

$$\sigma = -2E\omega\frac{d\theta}{dx}$$

The development of normal stresses varying along the *x*-axis inevitably entails the appearance of secondary or warping shear stresses τ_2 at the cross-sections of the bar, see Fig. 3.10. The warping shear stresses τ_2 are determined in the same way as in the case of transverse bending.

$$\tau_2 \cdot t = -\int_A d\sigma/dx \cdot dA$$

$$\tau_2 \cdot t = 2E \cdot d^2\theta/dx^2 \cdot \int_A \omega \cdot dA$$

The secondary shearing stresses τ_2 are uniformly distributed over the thickness of the section and do not vanish on the median line as the primary shear stress τ.

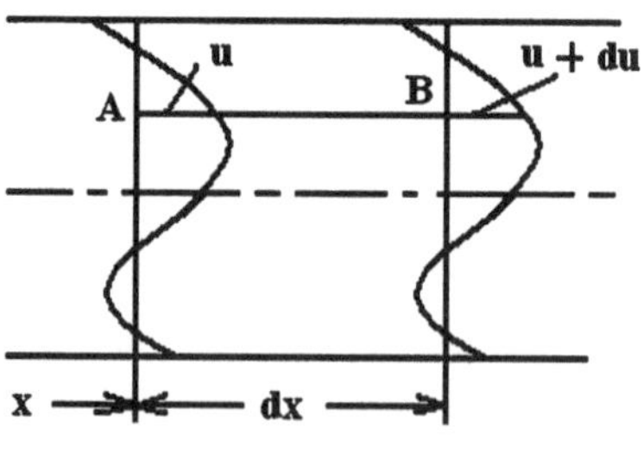

Fig. 3.9 Flexural warping deformation

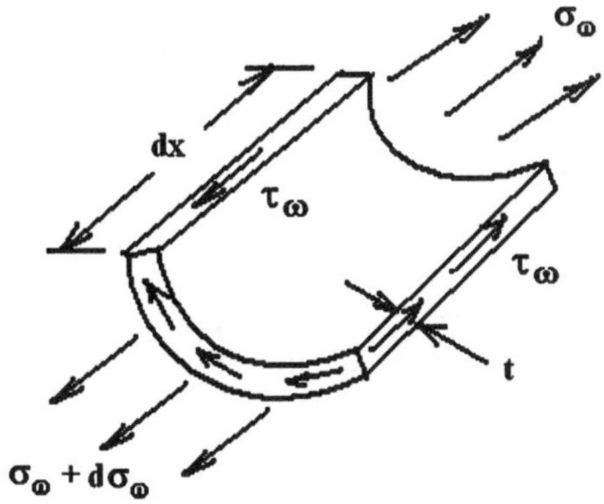

Fig. 3.10 Flexural and shear stresses over a thin-walled open section

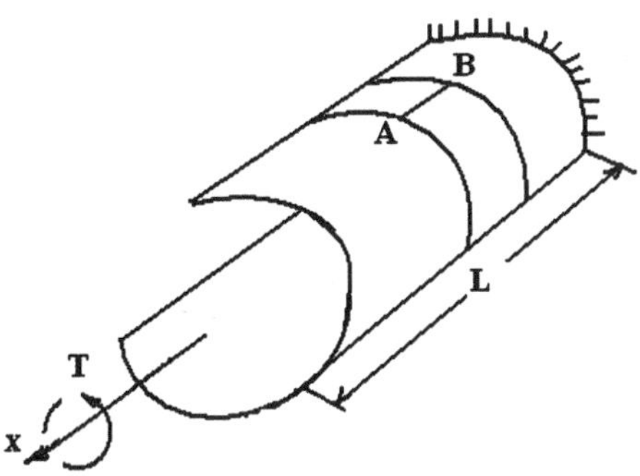

Fig. 3.11 An open thin-walled structure built-in at one end and loaded by a torque T at the free end

3.1.5 Development of the General Equation of Torsion

3.1.5.1 General Case of Warping of Thin-Walled Sections

Assume a bar built-in at one end subjected to a torque $\boldsymbol{T}$ at the free end, see Fig. 3.11.

The warping and rate of twist are ***zero*** at the fixed end and increase towards the free end. On the median line of the open contour the shear stress equals ***zero***.

$$\tau = 0$$

The linear warping deformation is given by

$$\mu = -2 \cdot \theta \cdot \omega$$

Since $\boldsymbol{\theta}$ varies along the length of the member, $\boldsymbol{\mu}$ will also very along the length as shown in Fig. 3.9. The valuation in $\boldsymbol{\mu}$ induces normal stresses given by

$$\sigma_\omega = \mathrm{E} \cdot \varepsilon$$

but

$$\varepsilon = du/dx = -2 \cdot \omega \cdot d\theta/dx$$

hence

$$\sigma_{\omega} = -2 \cdot E \cdot \omega \cdot d\theta/dx$$

The development of $\boldsymbol{\sigma_{\omega}}$ entails the appearance of warping shear stresses $\boldsymbol{\tau_{\omega}}$ which could be determined from the equilibrium conditions of an element, see Fig. 3.10 as follows

$$\tau_{\omega} \cdot t = -\int_{A} d\sigma_{\omega}/dx \cdot dA$$

hence

$$\tau_{\omega} \cdot t = 2 \cdot E \cdot d^2\theta/dZ^2 \cdot \int_{A} \omega \cdot dA$$

The warping shear stress $\boldsymbol{\tau_{\omega}}$ is uniformly distributed across the thickness of the section. In this connection, it should be realized that shear stresses are *zero* on the contour line. Consequently, the actual law of variation of $\boldsymbol{\mu}$ over the section differs from the law of sectorial area as given by

$$\mu = -2 \cdot \theta \cdot \omega$$

where θ = rate of twist.

When the rate of twist $\boldsymbol{\theta}$ varies along the length of member, the derived expressions still give sufficiently accurate values for $\boldsymbol{\sigma_{\omega}}$ and $\boldsymbol{\tau_{\omega}}$.

Since the twisting moment $\boldsymbol{T}$ is the sum of the *Saint–Venant* torque and the warping torque i.e.,

$$T = T_{\omega} + T_s$$

The shearing stresses on the section are the sum of the *Saint–venant* shear stress $\boldsymbol{\tau_s}$ and the warping shear stress $\boldsymbol{\tau_{\omega}}$.

The *Saint–Venant* torque is given by

$$T_s = G \cdot J_t \cdot \theta$$

The warping torque is determined as follows, see Fig. 3.12

$$T_{\omega} = \int_{S} \tau_{\omega} \cdot t \cdot r \cdot ds = E \cdot d^2\theta/dx^2 \int_{A} \left(\int_{A} \omega \cdot dA \right) d\omega$$

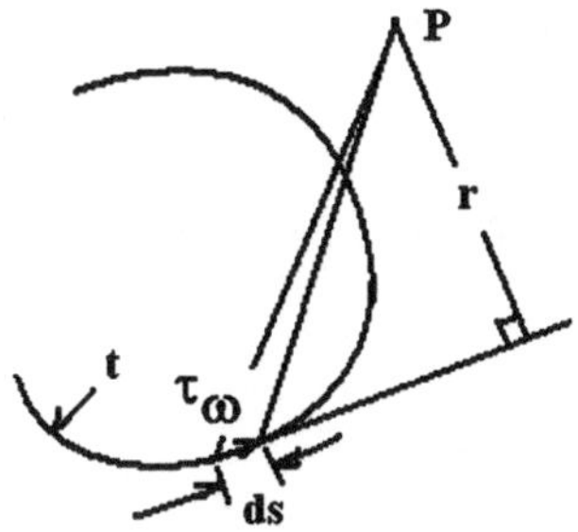

Fig. 3.12 Warping shear stressing a thin-walled open section

Integrating by parts, we get

$$\int_A \left(\int_A \omega \cdot dA \right) d\omega = \left[\omega \int \omega \cdot dA \right]_{s_1}^{s_2} - \int_A \omega^2 dA$$

Since ω is the principal sectorial area diagram,

$$\int_A \omega \cdot dA = 0$$

Thus

$$T_\omega = -E \cdot J_\omega \cdot d^2\theta/dx^2$$

Therefore, the torsion equation is given by

$$G \cdot J_t \cdot \theta - E \cdot J_\omega \cdot d^2\theta/dx^2 = T$$

This general equation could be written in the following simplified general form

$$C \cdot d\varphi/dx - C_1 \cdot d^3\varphi/dx^3 = T$$

where

$$C = G \cdot J_t$$

$$C_1 = E \cdot J_\omega$$

Example 3.1 Calculate the angle of twist, maximum shear and flexural warping stresses, maximum torsion stress for the *I*-beam section shown in Fig. 3.13 when the beam length is *1.0 m*. Assume, $\nu = 0.3$, $E = 2 \times 10^6$ kg/cm^2.

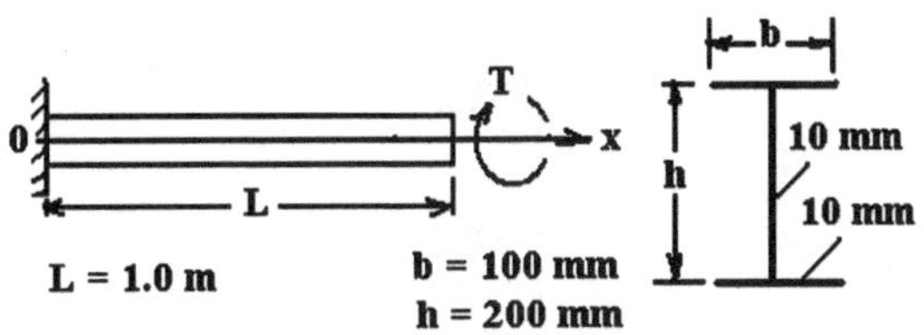

Fig. 3.13 A thin-walled I-section beam subjected to a torque at its free-end

Solution The general torsion equation is given by

$$d^2\theta/dx^2 - k^2\theta = -k^2T/GJ_t$$

The particular solution of this equation is

$$\theta = T/GJ_t$$

Hence

$$\theta = A_1 \cdot \sinh(kx) + A_2 \cdot \cosh(kx) + T/GJ_t$$

The constants A_1 and A_2 are determined from the following boundary conditions

$$\text{at } x = 0, \quad u = 0$$

Hence

$$\text{at } x = 0, \quad \theta = 0$$

Thus

$$0 = A_2 + T/GJ_t$$

i.e.,

$$A_2 = -T/GJ_t$$

$$\text{at } x = L, \quad \sigma_\omega = 0$$

Hence

$$\text{at } x = L, \quad d\theta/dx = 0$$

Thus,

$$0 = A_1 \cdot k \cdot \cosh(kL) + A_2 \cdot k \cdot \sinh(kL)$$

Hence

$$A_1 = T/GJ_t \cdot \tanh(kL)$$

Therefore,

$$\theta = T/GJ_t \cdot [1 + \tanh(kL) \cdot \sinh(kx) - \cosh(kx)]$$

The maximum angle of twist is given by

$$\begin{aligned}\varphi_{max} &= \int_0^L \theta dx \\ &= T/GJ_t \cdot [x + \tanh(kL)/k \cdot \cosh(kx) - \sinh(kx)/k]_0^L \\ &= TL/GJ_t \cdot [1 - \tanh(kL)/(kL)].\end{aligned}$$

Note: in pure torsion i.e., no warping, the angle of twist is given by

$$\varphi = TL/GJ_t$$

The maximum flexural warping stress occurs at x = 0. Thus

$$\begin{aligned}(\sigma_\omega)_{max} &= -E \cdot \omega \cdot (d\theta/dx)_{x=0} \\ &= -ET/GJ_t \cdot \omega \cdot k \cdot \tanh(kL)0\end{aligned}$$

The *Saint–Venant* and warping torque components are given by

$$\begin{aligned}T_s &= G \cdot J \cdot \theta \\ &= T \cdot [1 + \tanh(kL) \cdot \sinh(kx) - \cosh(kx)] \\ T_\omega &= -EJ_\omega \cdot d^2\theta/dx^2 \\ &= -T \cdot [\tanh(kL) \cdot \sinh(kx) - \cosh(kx)]\end{aligned}$$

The maximum *Saint–Venant* shear stress is given by

$$(\tau_s)_{max} = 3T_s t \Big/ \sum st^3 = 3T_s/(2b + h)t^2$$

The maximum warping shear stress is given by

$$(\tau_\omega)_{max} = -T_\omega/J_\omega t \cdot \int \omega dA$$

In order to carry out these calculations, the principal sectorial area diagram is required, see Fig. 3.14.

The warping constant J_ω is given by

$$J_\omega = \int_A \omega^2 dA = 1/24 \cdot b^3h^2t$$

The torsion constant is given by

$$J_t = 1/3 \cdot (2b + h)t^3$$

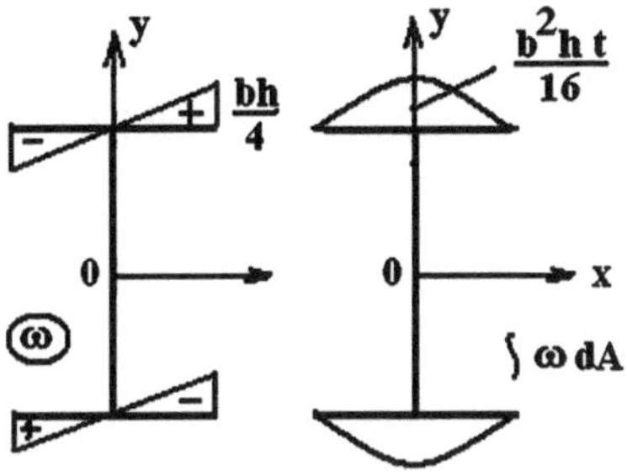

Fig. 3.14 Sectorial properties of a thin-walled I section

Hence

$$k^2 = GJ_t/EJ_\omega$$
$$= 1/2(1+\nu)\cdot\frac{1/3\cdot(2b+h)\cdot t^3}{1/24\cdot b^3\cdot h^2\cdot t} = 3.08\times 10^{-6}\,\text{mm}^{-2}$$

$$k = 1.75\times 10^{-3}\,\text{mm}^{-1}$$

Thus

$$kL = 1.75$$

Maximum φ

$$\varphi_{max} = TL/GJ_t\cdot[1 - 0.9411/1.75] = 0.463\cdot TL/GJ_t$$

The magnitude of the maximum angle of twist for the constrained case is about *46%* of the free warping case.

Maximum flexural warping stress, $(\sigma_\omega)_{max}$

$$(\sigma_\omega)_{max} = \pm ET/GJ_t\cdot bh/4\cdot k\cdot\tanh(kL)$$
$$= 161\times 10^{-4}\cdot T \quad kg/cm^2$$

Maximum torsional shear stress, $(\tau_s)_{max}$

$$(\tau_S)_{max} = 3T_s/(2b+h)t^2$$

at $x = 0$

$$T_s = 0 \quad \text{and} \quad \tau_s = 0$$

at $x = L$

$$T_s = T\cdot[1 + \tanh(kL)\cdot\sinh(kL) - \cosh(kL)] = 0.6614T$$

Hence

$$(\tau_s)_{max} = 49.6\times 10^{-4}T \quad kg/cm^2$$

Maximum warping shear stress, $(\tau_\omega)_{max}$ at $x = 0$

$$(\tau_\omega)_{max} = T_\omega(S_\omega)_{max}/J_\omega t$$
$$= \pm 3T/2bht = \pm 7.5\times 10^{-4}T \quad kg/cm^2$$

This stress occurs at the junction between the web and flange.

At $x = L$

$$(\tau_\omega) = 0.3386T$$

Hence

$$(\tau_\omega)_{max} = 2.5\times 10^{-4}T \quad kg/cm^2$$

Note: In pure torsion, i.e., no warping constraint, the maximum torsional shear stress is given by

$$(\tau_s)_{max} = 3T/t^2(2b + h) = 75 \times 10^{-4}T \quad kg/cm^2.$$

3.1.5.2 Solution of the General Equation of Torsion for a Uniform Member Subjected to a Non-Uniform Torsion Load

Induced Torsion Load

Consider a bottom longitudinal subjected to lateral pressure, see Fig. 3.15. The load q in *t/m* induced by the lateral pressure p in *t/m²* is given by

$$q = p \times S \quad t/m$$

where p = pressure load t/m²; S = longitudinal spacing.

Due to lack of symmetry, the load $\boldsymbol{q}$ does not pass through the enforced centre of rotation ***C***, see Fig. 3.15. This condition creates torsional loading acting along the length of the member given by

$$m = q \times e \quad t \cdot m/m$$

The torque at any distance $\boldsymbol{x}$ is given by, see Fig. 3.16

$$M_t = M_t^0 + qex \tag{3.1}$$

where M_t^0 = torque at the beam left end.

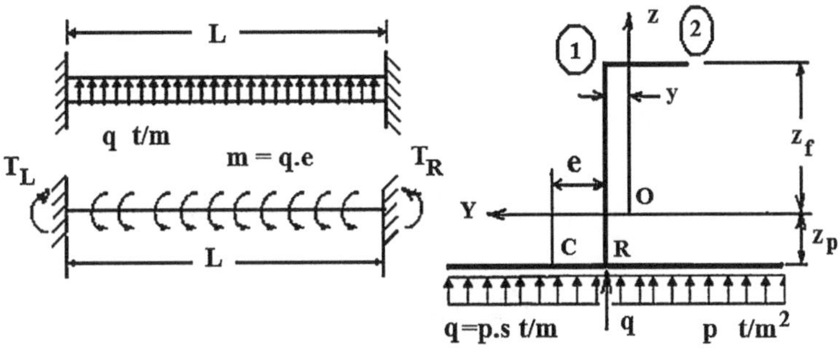

Fig. 3.15 Loading on an asymmetrical thin-walled section

Fig. 3.16 Torque distribution

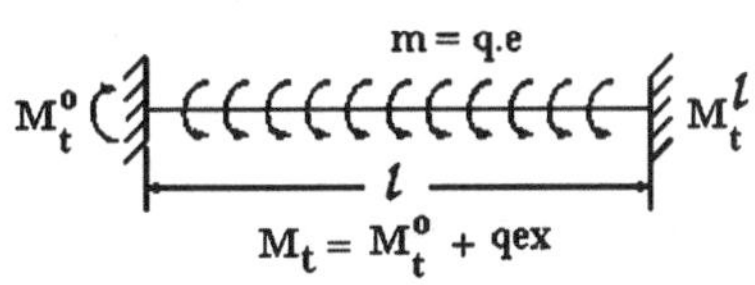

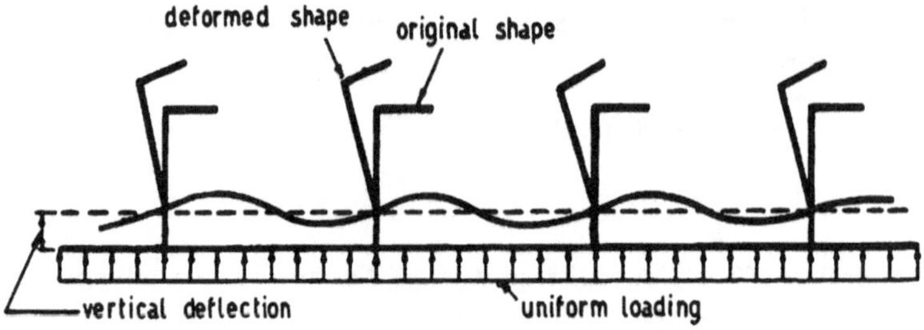

Fig. 3.17 Deformation of asymmetrical sections

Fig. 3.18 Twisting of the asymmetrical section

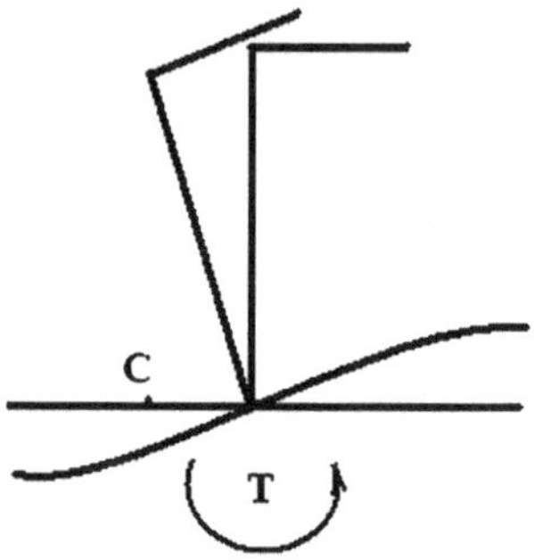

Torsional Deformation of the Section

This uniform torsion loading will cause the asymmetrical section to twist as shown in Figs. 3.17 and 3.18.

3.1.6 Solution of the Torsion Equation

The general equation of a uniform member under a non-uniform torsion is given by

$$M_t = C \cdot d\varphi/dx - C_1 \cdot d^3\varphi/dx^3 \tag{3.2}$$

where

$$C = E \cdot J_t = \text{torsionalrigidity}$$

$$C_1 = E \cdot J_\omega = \text{warping rigidity}$$

Consider the following cases.

3.1.6.1 Uniform Loading

The general solution of Eq. (3.2) for uniform loading is given by

$$\begin{aligned}\varphi = \varphi_0 + \left(\varphi_0^{'}/\mathrm{k}\right) \cdot \sinh(\mathrm{kx}) + (\varphi_0''/\mathrm{k}^2) \cdot [\cosh(\mathrm{kx}) - 1] \\ + \mathrm{M}_\mathrm{t}^0/\mathrm{C} \cdot (\mathrm{x} - \sinh(\mathrm{kx})/\mathrm{k}) + \mathrm{qex}^2/2\mathrm{C}\end{aligned} \tag{3.3}$$

where $\mathrm{k} = \sqrt{\mathrm{C}/\mathrm{C}_1}$, and φ, φ_0', and φ_0'' are constants of integration to be determined from the support boundary condition.

Several cases of boundary conditions are considered here.

Member is Fully Constrained at Both Ends Against Rotation and Warping

At x = 0 and x = λ

$$\varphi = \varphi^{'} = 0$$

Hence

$$\varphi = \frac{\mathrm{qe}\lambda}{\mathrm{kC}} \cdot \begin{bmatrix} \sinh(\mathrm{kx}/2) \cdot \sinh\ \mathrm{k}(\lambda - \mathrm{x}/2)/\sinh(\mathrm{k}\lambda/2) \\ -\mathrm{kx} \cdot (\lambda - \mathrm{x})/2\lambda \end{bmatrix} \tag{3.4}$$

and

$$\varphi'' = \frac{\mathrm{qe}\lambda}{2\mathrm{C}} \cdot [2/\lambda - \mathrm{k} \cdot \cosh(\mathrm{kx} - \mathrm{k}\lambda/2)/\sinh(\mathrm{k}\lambda/2)]. \tag{3.5}$$

Member is Constrained at Both Ends Against Rotation but is Free to Warp

The flexural warping stresses induced by the offset lateral loading are given by

$$\begin{aligned}\sigma_\omega &= -\mathrm{E}\omega \cdot \mathrm{d}^2\varphi/\mathrm{dx}^2 \\ &= -\mathrm{Eqe}\lambda/2\mathrm{C} \cdot [2/\lambda - \mathrm{k} \cdot \cosh(\mathrm{kx} - \mathrm{k}\lambda/2)/\sinh(\mathrm{k}\lambda/2)] \cdot \omega\end{aligned} \tag{3.6}$$

At the fixed ends of the beam, the flexural warping stresses attain the maximum values. Hence at points ***(1)*** and ***(2)***, σ_ω are given by, see Fig. 3.19

$$(\sigma_\omega)_1 = -\frac{\mathrm{Eqe}}{\mathrm{C}} \cdot [1 - (\mathrm{k}\lambda/2)/\tanh(\mathrm{k}\lambda/2)] \cdot \omega_1 \tag{3.7}$$

$$(\sigma_\omega)_2 = -\frac{\mathrm{Eqe}}{\mathrm{C}} \cdot [1 - (\mathrm{k}\lambda/2)/\tanh(\mathrm{k}\lambda/2)] \cdot \omega_2. \tag{3.8}$$

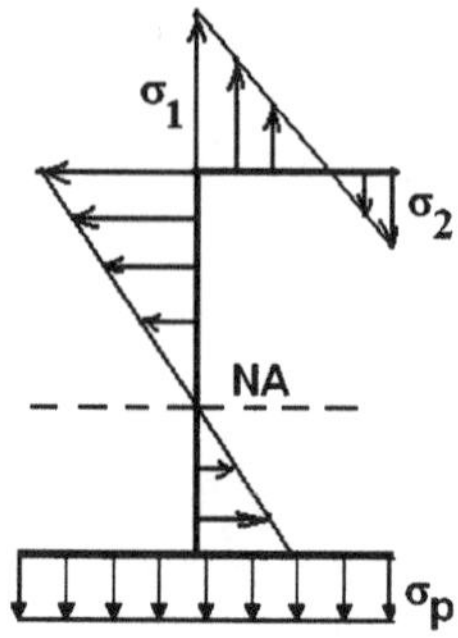

Fig. 3.19 Flexural warping stresses on the flange of the section

Member is Constrained at Both Ends Against Rotation but is Free to Warp

at $x = 0$ and $x = \lambda$

$$\varphi = \varphi' = 0$$

In this case, φ is given by

$$\begin{aligned}\varphi &= \frac{qe}{k^2C} \cdot [\cosh(k\lambda) - 1]/\sinh(k\lambda) \cdot \sinh(kx) - \frac{qe}{k^2C} \cdot [\cosh(kx) - 1] \\ &\quad - \frac{qe\lambda}{2C} \cdot x + \frac{qex^2}{C}\end{aligned} \tag{3.9}$$

and

$$\varphi'' = \frac{qe}{C}\{[\cosh(k\lambda) - 1]/\sinh(k\lambda) \cdot \sinh(kx) - [\cosh(kx) - 1]\} \tag{3.10}$$

At mid-span, σ_ω is given by

$$\begin{aligned}\sigma_\omega &= -E\omega\frac{d^2\varphi}{dx^2} \\ &= -\frac{Eqe}{C} \cdot \left\{ \begin{array}{r} [\cosh(k\lambda) - 1]/\sinh(k\lambda) \cdot \sinh(k\lambda/2) \\ -[\cosh(k\lambda/2) - 1] \end{array} \right\} \cdot \omega.\end{aligned} \tag{3.11}$$

Flexural Warping Stresses

The flexural warping stresses at points *(1)* and *(2)* are given by

$$(\sigma_\omega)_1 = -\frac{Eqe}{C} \cdot \left\{ \begin{array}{r} [\cosh(k\lambda) - 1]/\sinh(k\lambda) \cdot \sinh(k\lambda/2) \\ -[\cosh(k\lambda/2) - 1] \end{array} \right\} \cdot \omega_1 \tag{3.12}$$

and

$$(\sigma_\omega)_2 = -\frac{Eqe}{C} \cdot \left\{ \begin{array}{r} [\cosh(k\lambda) - 1]/\sinh(k\lambda) \cdot \sinh(k\lambda/2) \\ -[\cosh(k\lambda/2) - 1] \end{array} \right\} \cdot \omega_2. \qquad (3.13)$$

3.1.6.2 Linearly Varying Torque

The general solution of a uniform member loaded by a linearly varying torsional moment is given by, see Fig. 3.20

$$\varphi''' - k^2\varphi' = -M_t^o/C_1 \qquad (3.14)$$

where

$$M_t = M_t^o + qex^2/2\lambda \qquad (3.15)$$

The complementary function is given by

$$\varphi = A + A_1 \cdot \cosh(kx) + A_2 \cdot \sinh(kx)$$

In order to get the particular integral, assume a solution in the form

$$\varphi = a_0 + a_1x + a_2x^2 + a_3x^3 + a_4x^4 + a_5x^5 \qquad (3.16)$$

Solving (3.14), (3.15) and (3.16), we get

$$\begin{aligned} \varphi = {} & A + A_1 \cdot \cosh(kx) + A_2 \cdot \sinh(kx) \\ & + \left(qe/\lambda ck^2 + M_0/C\right) \cdot x + qex^3/6\lambda C \end{aligned} \qquad (3.17)$$

Member is Fully Constrained at Both Ends Against Rotation and Warping

At $x = 0$ $x = \lambda$

$$\varphi = \varphi' = 0$$

Hence, φ is given by

$$\varphi = \frac{qe\lambda}{6kc} \left\{ \begin{array}{r} [3/\sinh(k\lambda) - \beta \cdot \tanh(k\lambda/2)] \cdot [1 - \cosh(kx)] \\ \cdot[kx - \sinh(kx)] \cdot \beta + \left(kx^3/\lambda^2\right) \end{array} \right\}$$

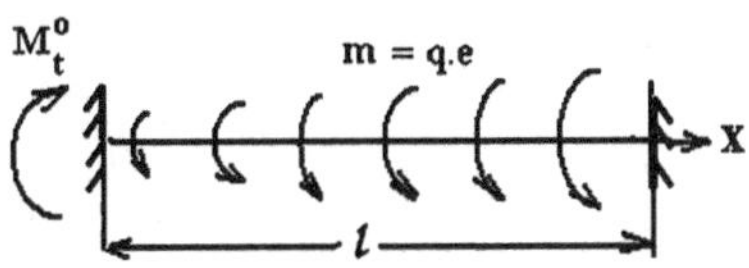

Fig. 3.20 Linearly varying torsional moment

and

$$\varphi'' = -\frac{qe\lambda}{6C}\left\{\begin{array}{r}[3/\sinh(k\lambda) - \beta\cdot\tanh(k\lambda/2)]k\cdot\cosh(kx)\\ +k\cdot\beta\cdot\sinh(kx) - (6x/\lambda^2)\end{array}\right\}$$

where

$$\beta = \frac{3\times[\cosh(k\lambda)-1] - (k\lambda)\cdot\sinh(k\lambda)}{(k\lambda)\cdot\sinh(k\lambda) + 2[1-\cosh(k\lambda)]}$$

at $x = 0$

$$\varphi'' = -\frac{qek\lambda}{6C}\cdot[3/\sinh(k\lambda) - \beta\cdot\tanh(k\lambda/2)]$$

at $x = \lambda$

$$\varphi'' = -\frac{qe\lambda}{6C}\cdot\left\{\begin{array}{r}[3/\sinh(k\lambda) - \beta\cdot\tanh(k\lambda/2)]k\cdot\cosh(k\lambda)\\ +k\cdot\beta\cdot\sinh(k\lambda) - (6/\lambda)\end{array}\right\}.$$

Member is Constrained at Both Ends Against Rotation but is Free to Warp

The flexural warping stresses at points *(1)* and *(2)* are given by

$$(\sigma_\omega)_1 = -E\omega_1\varphi''$$
$$(\sigma_\omega)_2 = -E\omega_2\varphi''.$$

Member is Constrained at Both Ends Against Rotation but is Free to Warp

At $x = 0$ and $x = \lambda$

$$\varphi = \varphi' = 0$$

In this case, φ is given by

$$\varphi = -\frac{qe}{k^2C}\cdot[\sinh(kx)/\sinh(k\lambda)] + \frac{qe\lambda}{6C}\cdot\left[6/k^2\lambda^2 - 1\right]\cdot x + \frac{qex^3}{6\lambda C}$$

and

$$\varphi'' = \frac{-qe}{C} \cdot [\sinh(kx)/\sinh(k\lambda)] + \frac{qex}{\lambda C}$$

The flexural warping stresses attain their maximum values at

$$x = 1/k \cdot \cosh^{-1}[\sinh(k\lambda)/k\lambda].$$

Flexural Warping Stresses

In this case σ_ω is given by

$$\sigma_\omega = -\frac{Eqe}{C} \cdot \{(x_m/\lambda) - \sinh(k) \cdot \cosh^{-1}[\sinh(k\lambda)/(k\lambda)]/\sinh(k\lambda)\}$$

At points *(1)* and *(2)*, σ_ω is given by

$$(\sigma_\omega)_1 = -E\omega_1\varphi''$$

$$(\sigma_\omega)_2 = -E\omega_2\varphi''.$$

Example 3.2 Calculate the flexural warping stresses in the flange of the fabricated asymmetrical section with attached plating shown in Fig. 3.21 for the two cases of web plate thickness $t_w = 15\ mm$.

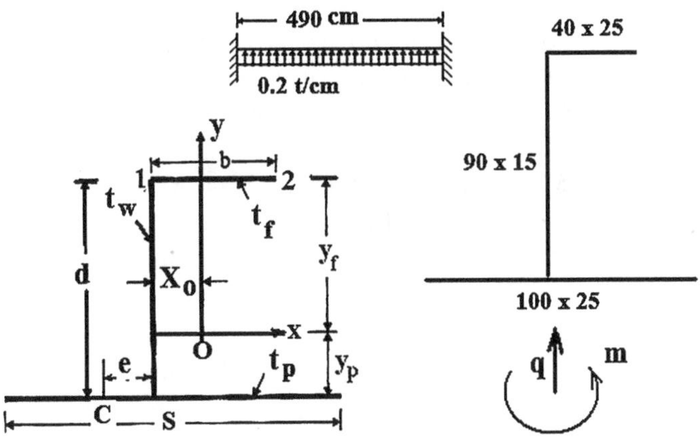

Fig. 3.21 A fixed ended beam having an asymmetrical thin-walled section

Solution Using flexural properties

$$L = 490\ \text{cm} \quad \text{and} \quad q = 0.20\ \text{t/cm}$$

$$A_f = b \times t_f = 40 \times 2.50 = 100.0\ \text{cm}^2$$

$$A_w = d \times t_w = 90 \times 1.50 = 135.0\ \text{cm}^2$$

$$A_p = S \times t_p = 100 \times 2.50 = 250.0\ \text{cm}^2$$

$$A_t = 485\ \text{cm}^2$$

$$y_f = 45 \times [2.0 - (2 \times 100 + 135)/486] = 58.95\ \text{cm}$$

$$y_p = 90 - 58.95 = 31.05\ \text{cm}$$

$$x_o = (40 \times 100)/(2 \times 485) = 4.21\ \text{cm}$$

$$I_x = \left[485 \times (58.95)^2\right] + \left[(90)^2 \times (250 + 153/3)\right]$$
$$- 0 \times 58.95 \times (135 + 500) + \left[250 \times (2.5)^2/3 + 100 \times (2.5)^2/12\right]$$
$$= 70.52 \times 10^4\ \text{cm}^4$$

$$I_y = 250 \times (100)^2/12 + 100 \times (40)^2/3 + (40)^2 \times (100)^2/4$$
$$= 25.355 \times 10^4\ \text{cm}^4$$

$$I_{xy} = [100 \times 58.95 \times (4.12 - 20)] + [135 \times 4.12 \times (58.59 - 45)]$$
$$+ 250 \times 4.12 \times (58.95 - 90.00) = -117.70 \times 10^3\ \text{cm}^4$$

The flexural bending stresses are given by, see Fig. 3.21

$$(\sigma_b)_2 = \frac{M_x \cdot (I_y \cdot y_2 - I_{xy} \cdot x_2)}{I_x \cdot I_y - I_{xy}^2}$$
$$= \frac{(25.355 \times 10^4 \times 58.95) - (117.7 \times 10^3[40 - 4.1])}{(70.52 \times 10^4 \times 25.355 \times 10^4) - (117.7 \times 10^3)^2} \cdot M_x$$
$$= 0.65 \times 10^{-4} M_x$$

$$(\sigma_b)_1 = \frac{(25.355 \times 10^4 \times 58.95) + (117.7 \times 10^3 \times 4.1)}{1651 \times 10^8} \cdot M_x$$
$$= 0.935 \times 10^{-4} M_x$$

$$M_x = \frac{ql^2}{12}$$
$$= \frac{0.2 \times (490)^2}{12} = 0.4 \times 10^4\text{t.m.}$$

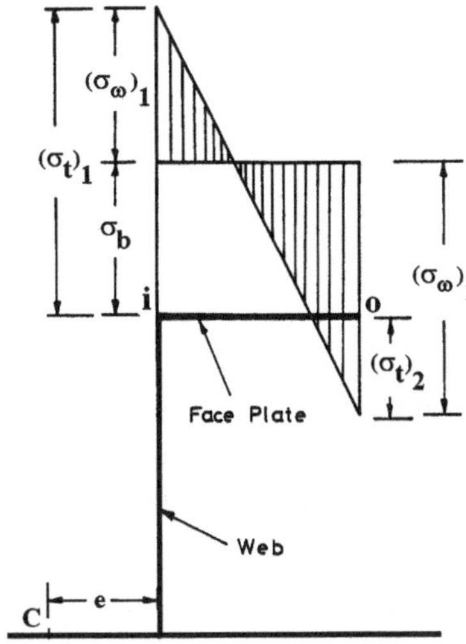

Fig. 3.22 Flexural warping stresses

Using the simple beam theory The bending stress at the flange is given by

$$(\sigma_b)' = \frac{M_x \cdot y_f}{I_x}$$
$$= \frac{58.95}{70.52 \times 10^4} \times 0.4 \times 10^4 = 0.334 \quad t/cm^4.$$

Using sectorial properties The flexural warping stresses are calculated as follows, see Fig. 3.22

Position of the shear center

$$e = -I_p \times I_{xy} \cdot d \Big/ I_y \times I_x - I_{xy}^2$$
$$= -\frac{2.5 \times 100^3/12 \times (-117.7 \times 10^3) \times 90}{1651 \times 10^8} = +13.38 \qquad cm$$

e $\approx$ 14 cm from *C.L.* of web.

Torsion constant

$$J_t = \frac{1}{3}\left(b \cdot t_f^3 + d \cdot t_w^3 + S \cdot t_p^3\right) = 831.2 \quad cm^4.$$

Sectorial properties

$$\omega_c = 45/485 \times [14 \times 135 - 1200] = 64$$
$$\omega_c = 45/485 \times [14 \times 135 - 1200] = 64$$
$$\omega_i = -(90 \times 14) + 64 = 1196$$
$$\omega_o = -1196 + 90 \times 40 = 2404$$
$$J_\omega = (90)^2 \times \left\{135 \times (14)^2/3 + 100 \times \left[(14)^2 + (40)^2 \Big/ 3 - 14 \times 40\right]\right\}$$
$$- (45)^2 \Big/ 485 \times (14 \times 135 + (28 - 40) \times 100)^2$$
$$= 2.07 \times 10^8 \quad cm^6.$$

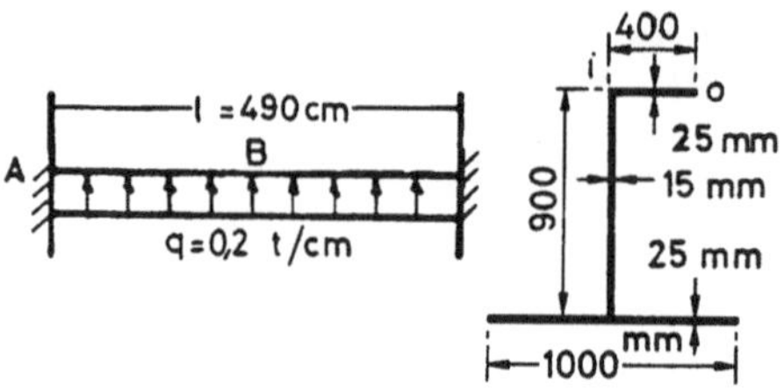

Fig. 3.23 Configuration and loading of the offset flange member

Flexural warping stresses

$$k^2 = GJ_t/EJ_\omega = \frac{831.2}{2.07 \times 10^8} \times \frac{1}{2.6}$$
$$k = 1.245 \times 10^{-3}$$
$$kL/2 = 1.245 \times 10^{-3} \times 490/2 = 0.3045$$
$$\sinh(kL/2) = 0.309$$
$$1 - (kL/2)/\tanh(kL/2) = 1 - 0.3045/0.2954 = -0.03$$

The flexural warping stresses at the inner and outer points of the face plate are given by, see Fig. 3.22

$$(\sigma_\omega)_1 = -\frac{2.6 \times 0.2 \times 14}{831.2} \times 0.03 \times (-1196) = +0.314 \quad t/cm^2$$
$$(\sigma_\omega)_2 = -\frac{2.6 \times 0.2 \times 14 \times 0.03 \times 2404}{831.2} = -0.63 \quad t/cm^2$$

The total bending and flexural stresses on the inner and outer points of the face plate are given by

$$\sigma_t = 0.314 + 0.334 = 0.648 \quad t/cm^2$$
$$\sigma_o = -0.63 + 0.334 = -0.296 \quad t/cm^2.$$

Using FEM 2-D and *3-D FEA* modeling and analysis were made on the same thin-walled offset flange structural member. For the *2-D* idealization, the face plate is idealized by a bar element with different percentages from the original area, see Figs. 3.23 and 3.24.

It is shown that the results obtained from the *FEM* is not far from the results obtained using sectorial properties of the thin-walled section.

Concerning the results of the *3-D FEA*, the stress at the inner point of the face plate is given by, see Fig. 3.25

$$\sigma_o = -0.296 \quad t/cm^2$$

The results of the *2-D FEA* show that the effectiveness of the face plate is only *31.5%* of the total sectional area of the face plate.

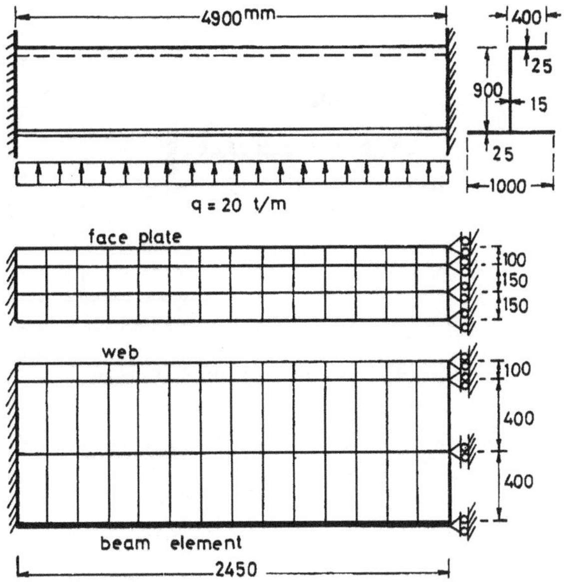

Fig. 3.24 3-D FE idealization of the thin-walled offset flange structure

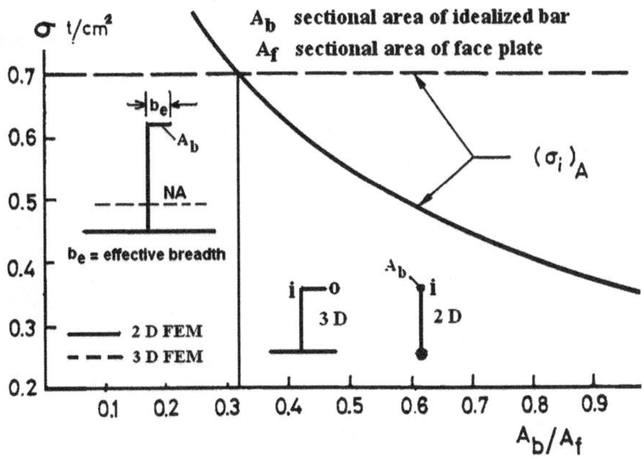

Fig. 3.25 Results of the 2-D and 3-D FEA

Example 3.3 Calculate the flexural and warping stresses at the inner and outer edges of the flange of the thin-walled section of the beam shown in Fig. 3.26. The scantlings of the beam section are given by L = 520 cm, q = 20t/tm, b = 250 mm, t_f = 25 mm, d = 950 mm, t_w = 20 mm, S = 1155 mm, t_p = 37.5 mm.

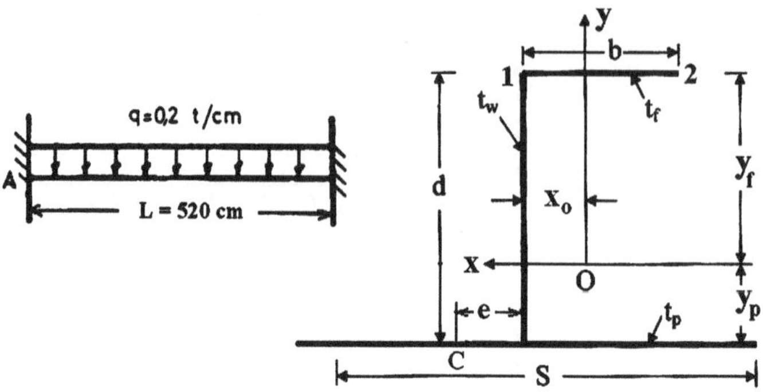

Fig. 3.26 Offset flange section with attached plating

Solution Geometrical, torsional and sectorial characteristics of the section

$$
\begin{aligned}
I_x &= 7.42968 \times 10^5 \quad cm^4 \\
I_y &= 3.6505 \times 10^5 \quad cm^4 \\
I_{xy} &= -0.53715 \times 10^5 \quad cm^4 \\
J_t &= 0.011842 \times 10^5 \quad cm^4 \\
J(\omega) &= 7.28196 \times 10^7 \quad cm^6 \\
e &= 6.715 \quad cm \\
k &= \sqrt{C/C_1} = 0.0025 \quad cm^{-1} \\
kL &= 1.3 \\
\omega_i &= -546 \quad cm^2 \\
\omega_o &= 1829 \quad cm^2
\end{aligned}
$$

Flexural and warping stresses

$$
\begin{aligned}
(\sigma_b)' &= \frac{qL^2 \cdot y_f}{12 I_x} = 0.417 \quad t/cm^2 \\
(\sigma_\omega)_1 &= 0.2164 \quad t/cm^2 \\
(\sigma_\omega)_2 &= -0.723 \quad t/cm^2 \\
(\sigma_t)_1 &= 0.6334 \quad t/cm^2.
\end{aligned}
$$

Chapter 4
Torsion of Container Ships

4.1 Torsion Loading on Ships

Ship hull girder is subjected to shear loading, bending moments, torsional moments, transverse and local loading. The distribution of torsional moments along ship length depends on the distribution of the cargo load over ship length and across the ship breadth. The magnitude and distribution of the torsional moments depend also on the direction of ship advance relative to the encountered waves.

The torsional moments are in general composed of *St. Venant* torsional moment T_S and warping moment T_ω. These two moments constitute the torsional loading on a ship hull girder. Hence the torsional loading is given by

$$T = T_S + T_\omega$$

In most practical cases, the effect of one component may be neglected as compared to the effect of the other. Warping torsion is usually negligible (or may best be considered by local corrections) in slender members with compact, solid or hollow, closed cross sections.

St. Venant torsion is valid only for beams of circular sections where plane sections remain plane and may be neglected in thin-walled open cross sections. In thin-walled sections, plane sections no longer remain plane and warping takes place.

The structural mechanic aspect of the analysis however is independent of whether T_S or T_ω occur separately or together. Both theories of bending and warping torsion assume that the cross sections of a loaded member maintain their shape. There are of course structural members which do not satisfy this condition.

Hull girder torsion loading on container ships represent a major loading element particularly when combined with hull girder shear and bending.

M. Shama, *Torsion and Shear Stresses in Ships*,
DOI: 10.1007/978-3-642-14633-6_4,

4.2 Torsion Loading of Open-Decked Ships

The demand on container ships is increasing very rapidly as a result of the expansion in world trade and the general increase in the use of containerization systems.

Figure 4.1 shows a picture of a container ship. Figure 4.2 shows a longitudinal section of the vessel and Fig. 4.3 shows profile and deck plans of a feeder container ship.

Container ships are characterized by exceptionally wide hatch openings, see Figs. 4.4 and 4.5. These types of ships, therefore, are sometimes called open ships or open deck ships. The use of wide hatches has significant effects on the torsional strength and rigidity of ship hull girder.

The torsional strength and rigidity of open ships depend mainly on the structural arrangement in the parallel middle body of the cargo space, the constraints induced by the structural arrangements of both ship ends and by the distribution of torsion loading along ship length.

All ships are subjected to torsional moments which tend to twist the hull girder along its length, see Fig. 4.6. Normally, torsional stiffness is more than adequate to

Fig. 4.1 A container ship

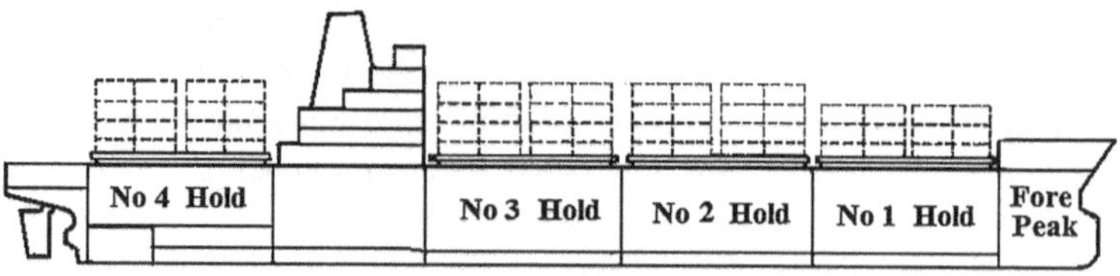

Fig. 4.2 A structural profile of a container ship

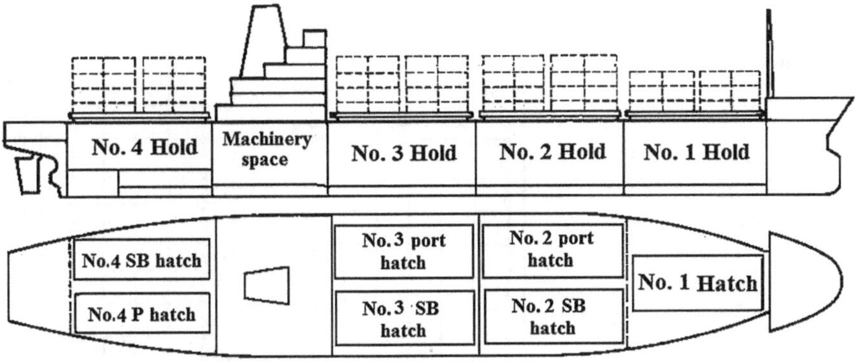

Fig. 4.3 Profile and deck plans of a feeder container ship

Fig. 4.4 A typical section of a conventional feeder container ship

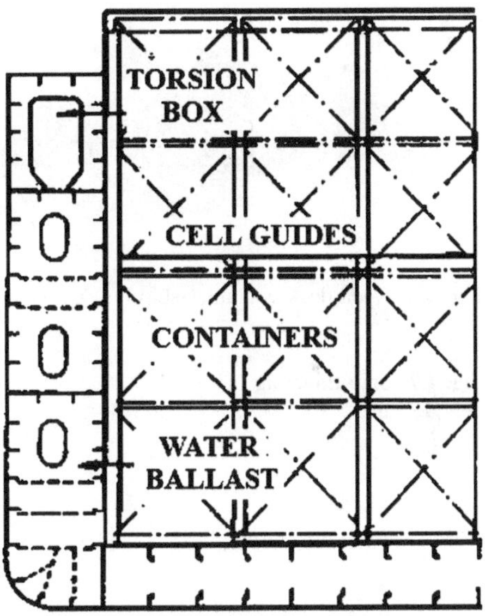

prevent undue distortion of the structure. Torsional loading induces additional stresses, normally called warping stresses, near hatch corners.

Wave induced torsion loading results from the motion of a ship in oblique waves, see Fig. 4.7. When the ship is sailing obliquely into the predominant waves (as compared with the ahead condition) the vertical wave bending moments are reduced, but the horizontal bending moments and torsional moments are increased. In a sea-way, torsional moments are set up by both hydrostatic and hydrodynamic forces including slamming, and by the mass-acceleration forces as a result of the ship's motions. Model experiments have shown the large dynamic influence of rolling on the induced torsional loading.

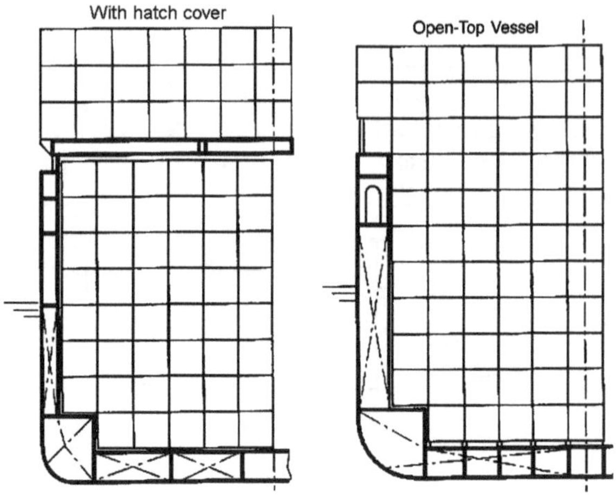

Fig. 4.5 Sections of container ships with and without hatch covers

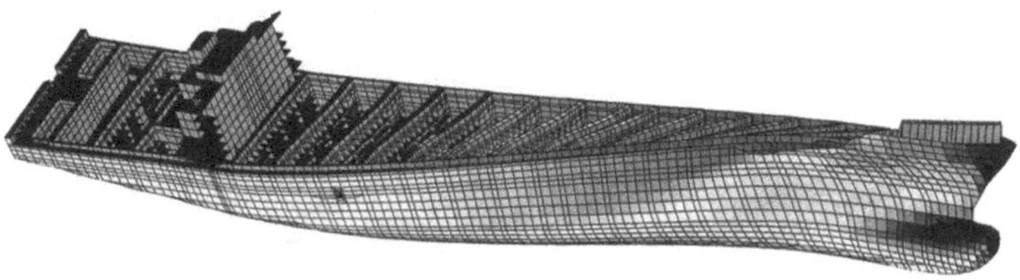

Fig. 4.6 Open deck container ship hull girder torsional deformation

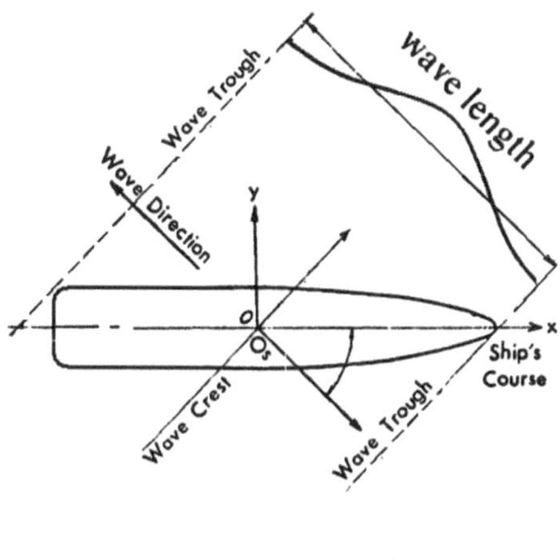

Fig. 4.7 Ship traveling in oblique waves

All external forces acting on the hull girder, and which do not pass through the axis of shear center, will produce torsion. Open ships, therefore, are also subjected to additional torsional loading induced by the horizontal component of the shearing force. The latter acts at a distance from the shear center of the ship section see Figs. 4.8, 4.9 and 4.10.

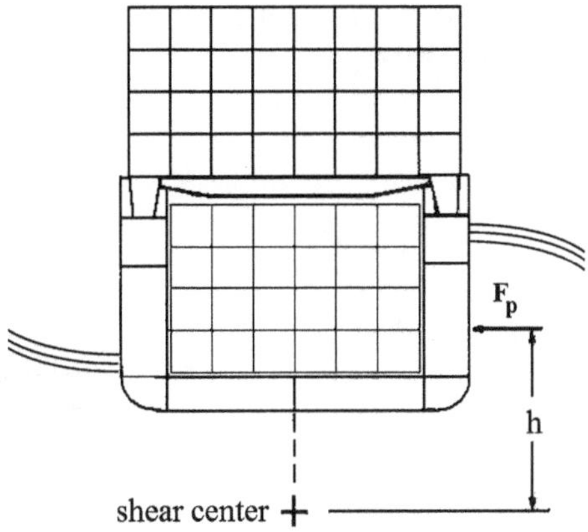

Fig. 4.8 Torsion load resulting from the effect of wave forces, cargo loading and the offset location of the shear center

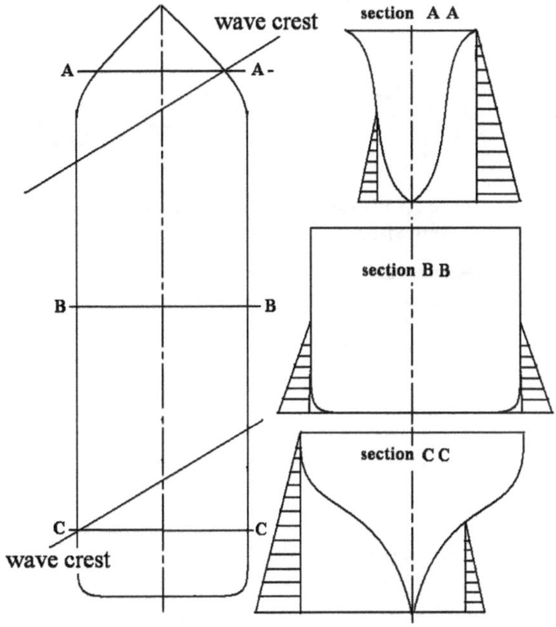

Fig. 4.9 Hull girder torsion loading at several ship sections along her length for a ship traveling obliquely among waves

Improper distribution of cargo loading and fuel also induces still-water torsional loading. Stillwater torsional moments could be determined from the published Rulers of Classification Societies.

An approximate value of the Stillwater torque is given by

$$T_S = k \cdot B \cdot W_T \text{ kN} \cdot \text{m}$$

where K = 0.004; B = breadth of the vessel, m; W_T = Maximum total container weight, kN.

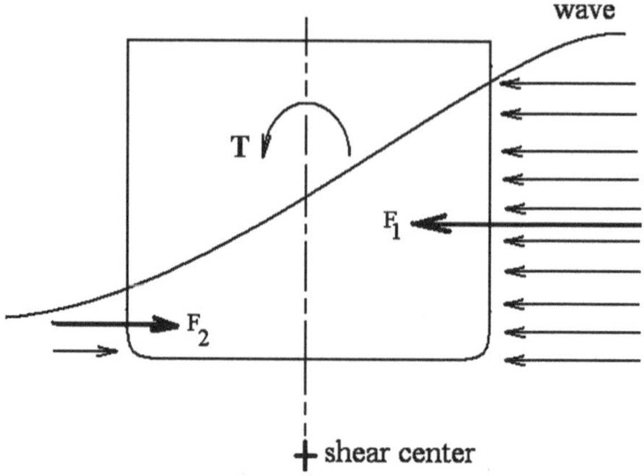

Fig. 4.10 Asymmetric loading on ship section

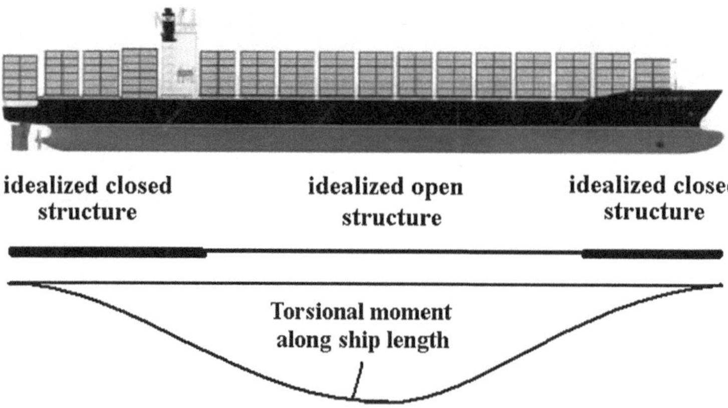

Fig. 4.11 Torsional loading distribution along ship length

The distribution of torsional loading along ship length and the extent of ship sections having closed and open deck structure are shown in Fig. 4.11.

The torsional response of open ships is a function of the geometry of the transverse sections, the longitudinal extent of ship length having open deck structure and the degree of torsional rigidity of the ship structure at both ends of the ship.

4.3 Torsion Loading on Catamaran Vessels

Catamaran vessels are high speed crafts constructed of thin-walled steel or aluminum structure. A typical configuration of a catamaran vessel is shown in Fig. 4.12. In order to simplify the torional analysis of the complex structure of a catamaran vessel, the vessel could be idealized by a *3-D* simplified configuration, see Fig. 4.13.

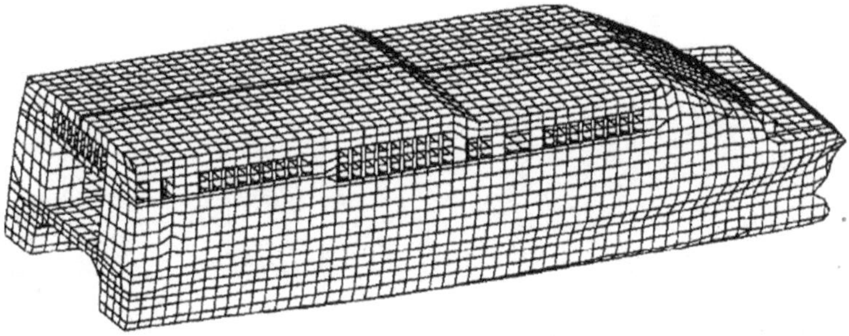

Fig. 4.12 A typical view of a conventional catamaran vessel

Fig. 4.13 3-D Idealization of a catamaran vessel

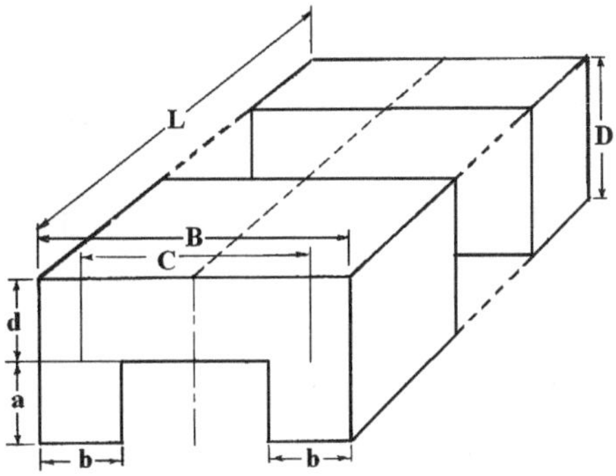

Fig. 4.14 A catamaran vessel encountering different wave lengths

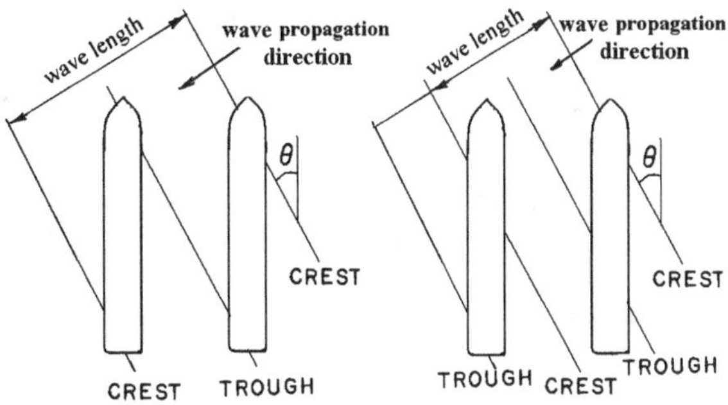

As catamaran vessels travel at high speed among waves, wave bending moments, torsional moments are lateral forces are induced. Therefore, high shear and bending stresses could be developed. Figure 4.14 shows a catamaran vessel traveling at high speed against the encountering waves. Two conditions of encountering wave lengths are considered. Figure 4.15 shows the torsion induced loading due to the hydrodynamic forces induced by the obliquely encountered waves. Figure 4.16 shows the lateral forces induced by the encountered waves. Because of the thin-walled structural configuration of catamaran vessels, the shear center does not coincide with the centroid of the transverse section of the vessel.

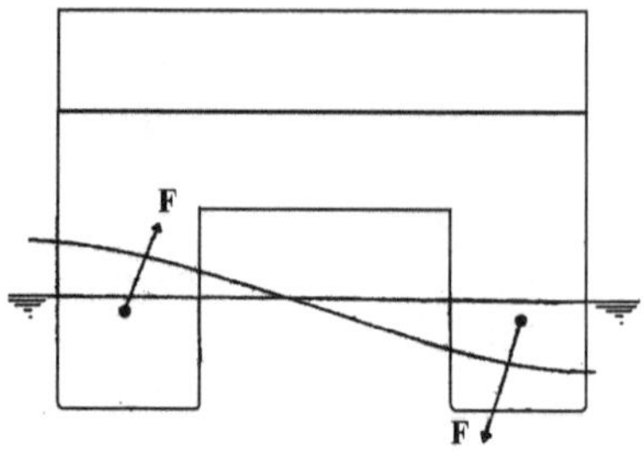

Fig. 4.15 Torsion loading on a catamaran vessel due to oblique waves

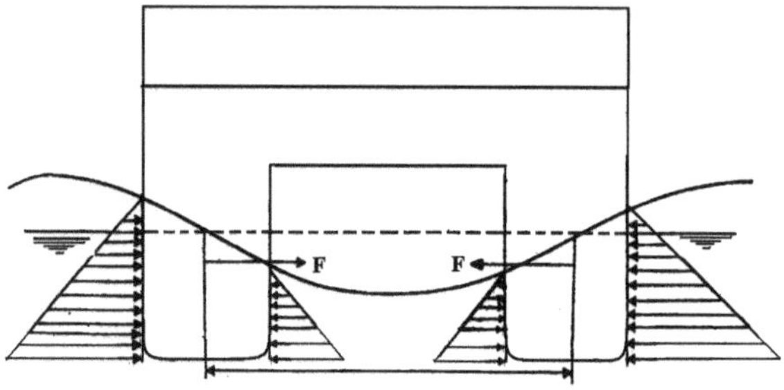

Fig. 4.16 Transverse loading on a catamaran section

Additional torsion loading are therefore induced by virtue of the offset location of the shear center from the location of the lateral wave forces.

4.4 Warping Deformations and Stresses in the Deck Structure of Container Ships

The open deck thin-walled structure of container ships is constrained at both ends by virtue of the high stiffness and rigidity of the closed structure at both ends of the ship. When this constrained open structure of the cargo space is subjected to torsional moments, warping deformations and stresses are developed in the deck structure and the lateral ligaments. The magnitude of these deformations and stresses depend on the degree of constraint at both ends of the open deck provided by the closed structure at both ends of the vessel and the degree of flexibility of the structural ligaments between holds, see Fig. 4.17. The local loading and deformation of the structural connection between the deck plating and a lateral ligament is shown in Fig. 4.18.

4.5 Torsional Deformation of Ship Hull Girder

The torsional deformation of the hull girder of an open deck ship depends on the magnitude and distribution of the torsional loading and the magnitude of the

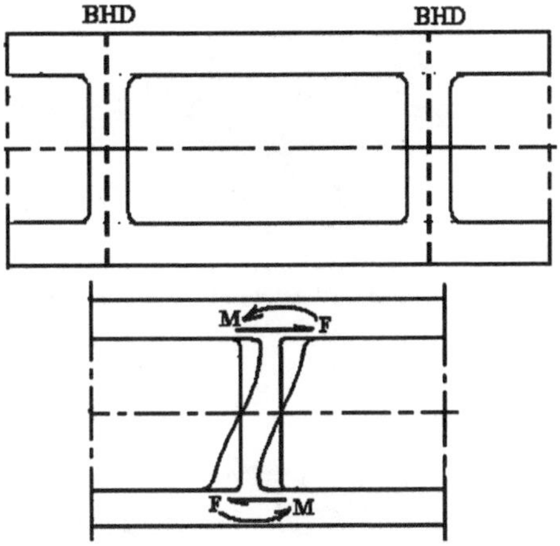

Fig. 4.17 Warping deformation of a deck strut due to torsion of ship hull

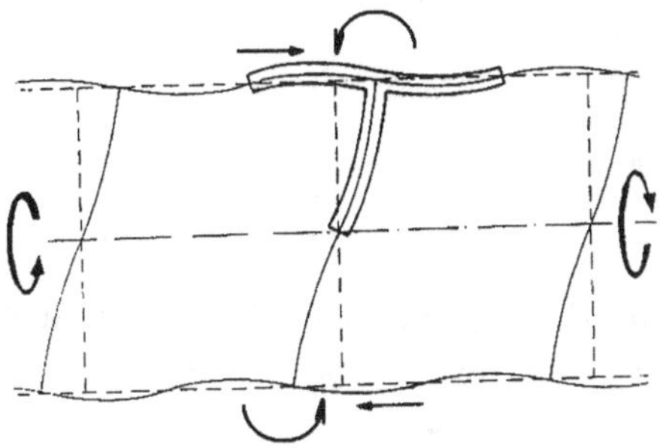

Fig. 4.18 Distortion of deck strut

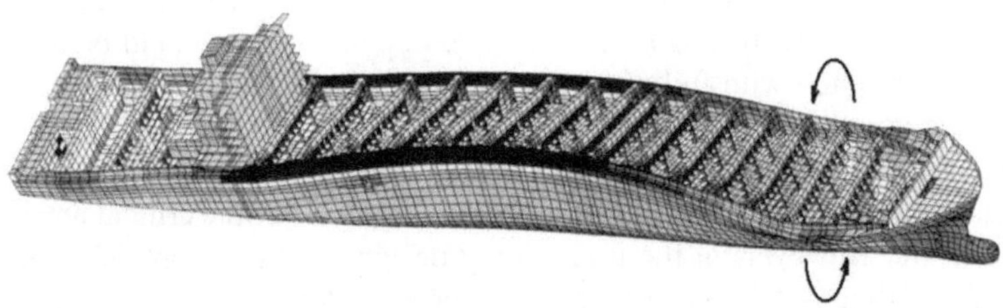

Fig. 4.19 Torsional loading at the fore end of the ship

torsional rigidity of the hull girder structure. Figure 4.19 shows the deformed shape of the hull girder of an open deck ship subjected to a torsion loading at the fore end of the vessel.

Figure 4.20 shows the deformed shape of the hull girder of an open deck ship subjected to a torsion loading at the aft end of the vessel.

Figure 4.21 shows the deformed shape of the hull girder of an open deck ship subjected to torsion loading at both ends of the vessel.

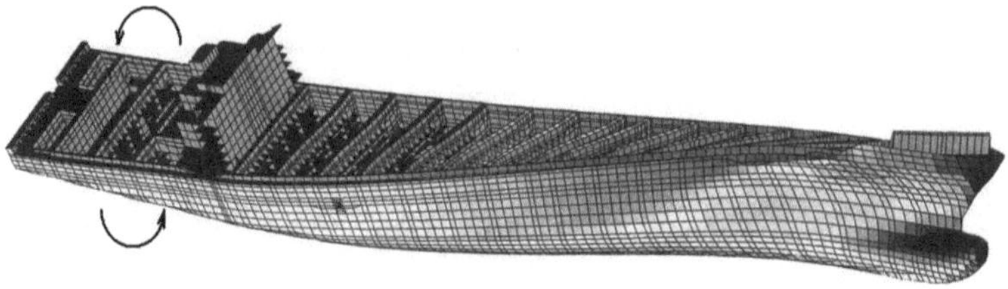

Fig. 4.20 Torsional loading at the aft end of the ship

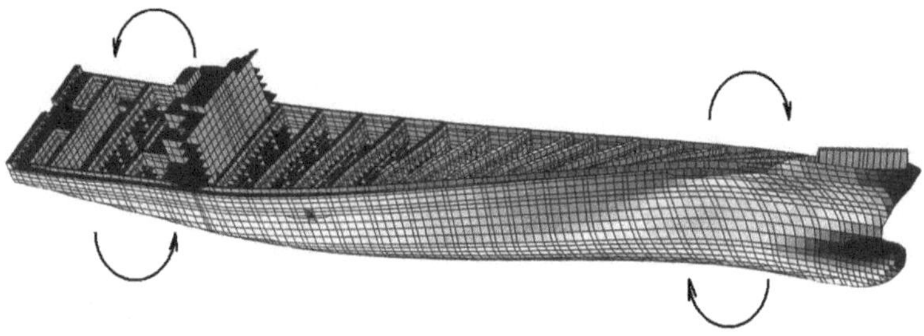

Fig. 4.21 Torsional loading at both ends of the ship

4.6 An Approximate Method for Torsion Analysis of Open Deck Vessels

The torsional behavior of open ships could be studied theoretically, using the finite element method or experimentally, using model testing. A simplified approach, based on the torsional behavior of thin-walled open sections, could be also used. However correlation with full-scale measurements becomes necessary in order to confirm the validity of the method of analysis.

Since the finite element method and model testing are costly and time consuming, their use should be confined to the final analysis of the critical areas of the ship structure. However, in the preliminary design stages of container ships, for comparing different designs and when the torsional loading cannot be accurately determined, a simplified procedure for structural design and analysis of open ships should be very useful. The simplified method gives a general procedure for calculating the shear and flexural warping stresses as well as the torsional deformations induced in open ships by torsional loading. The method is based on the sectorial properties of ship section. The latter are obtained using a simplified idealization of the ship section configuration.

The solution of the torsion equation is given for the general case of a polynomial torque distribution as well as for the particular cases of linear and parabolic torque distributions. The solution takes into account the degree of fixity provided by the closed structures at both ends of the open length of the ship.

4.7 Calculation of the Shear and Flexural Warping Stresses

The shear and flexural warping stresses are calculated as follows

$$\tau(\omega)=T(\omega)\cdot S(\omega)/J(\omega)t \tag{4.1}$$

$$\sigma(\omega)=\omega\cdot M(\omega)/J(\omega) \tag{4.2}$$

where $M(\omega)$ = bi-moment $= -EJ(\omega)d^2\phi/dx^2$; $T(\omega)$ = warping torsional moment $= -EJ(\omega)d^3\phi/dx^3$; $J(\omega)$ = warping constant of ship section = sectorial moment of inertia; $S(\omega)$ = sectorial static moment; ω = sectorial coordinate; E = modulus of elasticity; ϕ = angle of twist; t = thickness.

Therefore, in order to calculate the warping stresses, the following information is required

(i) The sectorial properties of the ship section i.e., ω, $S(\omega)$ and $J(\omega)$.
(ii) The solution of the torsion equation so as to calculate $M(\omega)$ and $T(\omega)$.

4.8 Solution of the Torsion Equation

The differential equation of the combined bending and torsion is given by

$$EJ(\omega)d^3\phi/dx^3 - GJ_t d\phi/dx = -T_x \tag{4.3}$$

where T_x = torsional moment; x = length coordinate; GJ_t = torsional rigidity; J_t = *Saint-Venant* torsional constant; $EJ(\omega)$ = warping rigidity.

The solution of Eq. 4.3 depends on the torque distribution, T_x, and the boundary conditions.

4.8.1 Boundary Conditions

There are three general cases of boundary conditions

4.8.1.1 Fixed End i.e., No Warping

This is satisfied by

$$\phi = d\phi/dx = 0$$

4.8.1.2 Free End i.e., Free Warping

This is satisfied by $d^2\phi/dx^2 = 0$

4.8.1.3 Constrained End i.e., Constrained Warping

$$d\phi/dx = \frac{T_e}{GJ_t}(1 - f)$$

where T_e = torsional moment at the end of the member, f = a factor representing the degree of constrained against warping, and $0 < f < 1.0$.

The magnitude of f, for any particular ship, depends on the length, rigidity and configuration of the end structure.

4.8.2 Distribution of Torsional Loading

The magnitude and distribution of the torsional loading acting on a ship hull girder depend on several factors; among them are the sea condition, shape of ship form, particularly the shape and size of flare of the bow sections and the location of the shear center. Different formulations for the magnitude and distribution of the hull girder torsion loading are given in the published Rules of Classification Societies.

It should be indicated here that any shape of torque distribution along ship length could be represented by a polynomial equation. A typical polynomial torque distribution is given by

$$T_x = \sum_{i=0}^{n} b_i x^i$$

The general solution of the torsion equation could be obtained by summing the solution for the various terms of the polynomial, as given in Chap. 6.

However, in order to present the simplified procedure of calculations, two simple torque distributions are considered, see Fig. 4.22.

4.8.2.1 Linear Torque Distribution

An assumed linear torque distribution could be presented by the following equation

$$T_x = T_0[1 - (1 - \alpha)x/a]$$

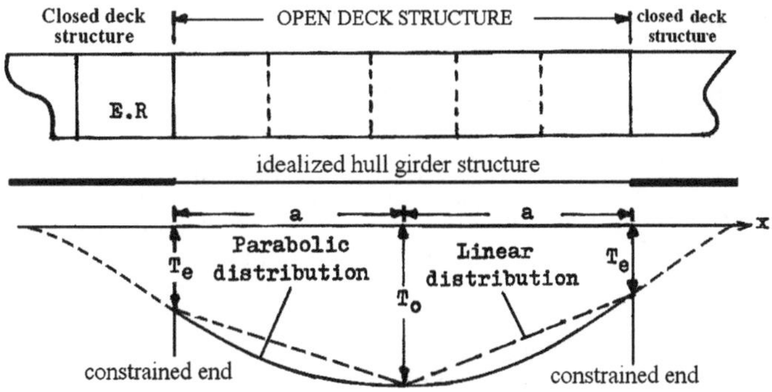

Fig. 4.22 Assumed torsion loading over the cargo hold length

4.8.2.2 Parabolic Torque Distribution

An assumed parabolic torque distribution could be given by the following equation

$$T_x = T_0[1 - (1 - \alpha)(x/a)^2]$$

In both cases, the origin of x is at mid-length, see Fig. 4.22, and T_o is the maximum value of the torque at x = 0.0.

The magnitude of T_o could be obtained from the published Rules of Classification Societies.

4.8.3 Solution of the Torsion Equation for Constrained Warping

Assuming constrained warping at both ends of the open length, see Fig. 4.22, the boundary conditions are

At x = 0

$$\phi = d^2\phi/dx^2 = 0$$

At x = a

$$d\phi/dx = \frac{T_e}{GJ_t}(1 - f)$$

Using these boundary conditions, the solution of Eq. 4.3 is obtained for the linear and parabolic torque distributions as follows.

4.8.3.1 Linear Torque Distribution

The solution of the torsion equation is given by

$$\phi = \frac{T_0 a}{GJ_t}\left[\begin{array}{r} X/a - 1/2(1-\alpha)(X/a)^2 - f\alpha\dfrac{\sin(kX)}{ka\cosh(kX)} \\ -(1-\alpha)/(ka)^2\eta \end{array}\right]$$

$$M(\omega) = T_o a\left[f\alpha\frac{\sinh(kx)}{ka\cosh(ka)} + (1-\alpha)/(ka)^2\eta\right]$$

$$T(\omega) = T_o\left\{\begin{array}{r} f\,\alpha\sinh(kx)/\cosh(ka) + (1-\alpha)/ka \\ [\tanh(ka)\cosh(kx) - \sinh(kx)] \end{array}\right\}$$

where

$$k^2 = GJ_t/EJ(\omega)$$

$$\eta = 1 + \tanh(ka)\sinh(kx) - \cosh(kx)$$

4.8.3.2 Parabolic Torque Distribution

The solution of Eq. 4.3 gives

$$\phi = \frac{T_o a}{GJ_t}\left\{\begin{array}{r} x/a - 1/3(1-\alpha)(x/a)^3 - f\alpha\dfrac{\sinh(kx)}{ka\ \cosh(ka)} \\ -2(1-\alpha)/(Ka)^2[x/a - \sinh(kx)/ka\cosh(ka)] \end{array}\right\}$$

$$M(\omega) = T_o a\left\{\begin{array}{r} f\alpha\sinh(kx)/ka\cosh(ka) + 2(1-\alpha)\Big/(ka)^2 \\ [x/a - \sinh(kx)/ka\cosh(ka)] \end{array}\right\}$$

$$T(\omega) = T_o\left\{f\alpha\frac{\cosh(kx)}{\cosh(ka)} + \frac{2(1-\alpha)}{(ka)^2}\left[1 - \frac{\cosh(kx)}{\cosh(ka)}\right]\right\}$$

4.8.4 Calculation of the Sectorial Properties of Ship Section

The sectorial properties of a thin-walled section of an open ship could be calculated using the following general procedure as follows.

4.8.4.1 Idealization of a Ship Section

A section of the double-skin structure of an open ship, see Fig. 4.23, could be idealized by a simplified open section, as shown in Fig. 4.24. The dimensions and scantlings of the idealized section are obtained from the dimensions and scantlings of the original configuration of the ship section.

The geometrical and flexural properties and scantlings of the idealized ship section are given by

$$2b = (B_i + B_o)/2$$

Let t_w = effective thickness of the side structure; t_b = effective thickness of the bottom structure; A_i = sectional area of girder i; A_b, A_w = sectional area of the bottom and side respectively; 2 m, 2n = number of girders in the side and bottom structures respectively

$$I_Y = 2A_w(b/2)^2 + A_b b^2/12 + 2\sum_{i=1}^{n} A_i Z_i^2 + 2\sum_{j=1}^{m} A_j\left(Z_j^2 + p_j^2/12\right)$$

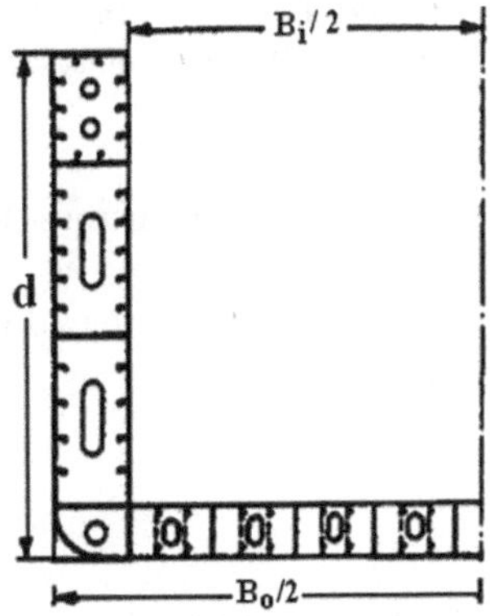

Fig. 4.23 Ship section of a container ship

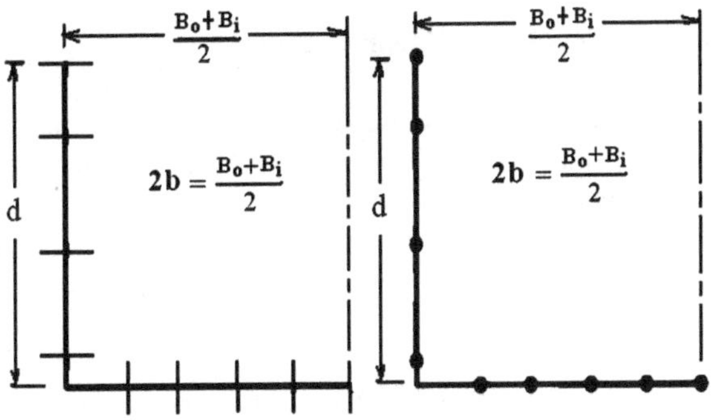

Fig. 4.24 Simplified idealization options of ship section

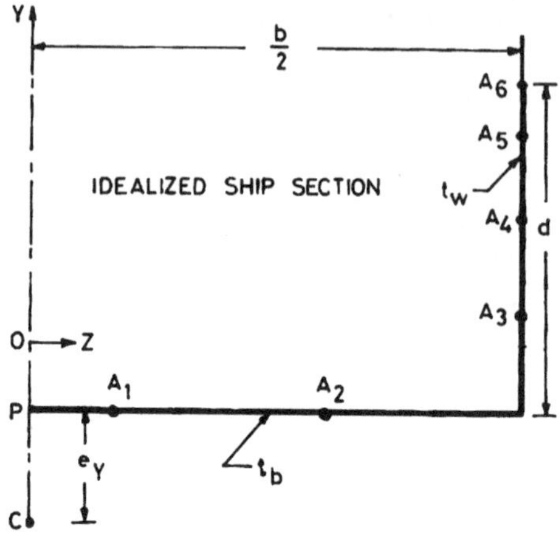

Fig. 4.25 Idealized ship section

There are two options for idealization of the ship section, see Fig. 4.24.

In the first option, the scantlings of the horizontal and vertical girders in the side and double bottom structure are used directly in the idealized ship section. In the second option, the sectional areas of the horizontal and vertical girders in the side shell and double bottom structure are idealized by lumped areas.

In the following analysis, the ship section is idealized as shown in Fig. 4.25.

4.8.4.2 Torsion Constant J_t

The torsion constant for a double-skin structure is calculated from the original ship section as follows

$$J_t = \sum_{i=1}^{r} 4A_i^2 \Big/ \oint ds/t$$

where r = number of cells; A_i = enclosed area of cell i.

4.8.4.3 Position of Shear Center

Due to symmetry about the *Y*-axis, the *Y*-axis is a principal axis and the shear center is located on the *Y*-axis at a distance e_Y from *P*, see Fig. 4.25. The distance e_Y is given by

$$e_Y = 1/I_Y \int_A z\omega' dA$$

ω' = sectorial coordinate based on an assumed pole at *P*, see Fig. 4.25.

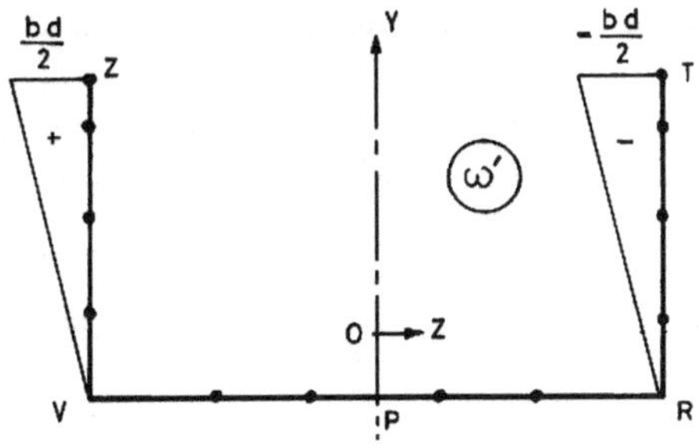

Fig. 4.26 Assumed ω' diagram

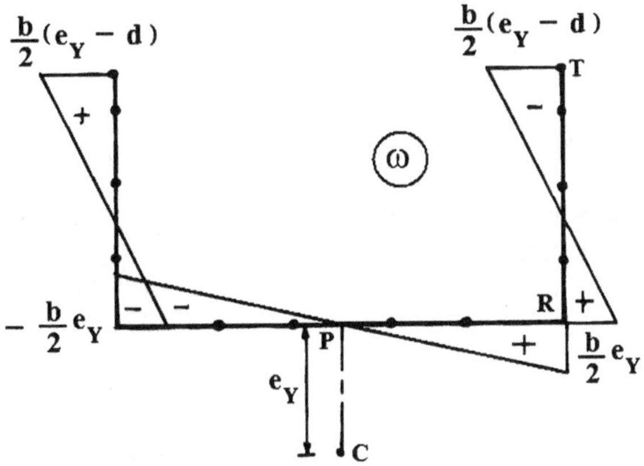

Fig. 4.27 Principal sectorial area diagram

Substituting we get, see Fig. 4.26.

$$\int_A z\omega' dA = b^2 d^2 t_w/4 + b\sum_{i=1}^{m} \omega'_i A_i$$

Hence

$$e_Y = \left(b^2 d^2 t_w/4 + b\sum_{i=1}^{m} \omega'_i A_i\right)/I_Y$$

where A_i = sectional area of girder *i*; m = half the total number of girders in the side structure (Fig. 4.27)

$$\begin{aligned} I_Y &= 2A_S\,(b/2)^2 + A_p b^2/12 \\ &= b^2(A_p + 6A_S)/12 \end{aligned}$$

4.8.4.4 Principal Sectorial Area Diagram

Having determined the location of the shear center, the principal sectorial area diagram could be determined as follows

$$\omega = \int_0^{b/2} e_Y ds$$

Substituting we get

$$\omega = -e_Y b/2 + \int_0^{b/2} b/2 ds$$

4.8.4.5 Warping Constant *J*(ω)

The warping constant *J*(ω) is calculated as follows using the principal sectorial area diagram shown in Fig. 4.28

$$J(\omega) = \int_A \omega^2 dA$$
$$= 2\sum_{i=1}^{m+n} \omega_i^2 A_i + A_b b^2 e_Y^2/12 + b^2 t_w/6\left[e_Y^3 - (e_Y - d)^3\right]$$

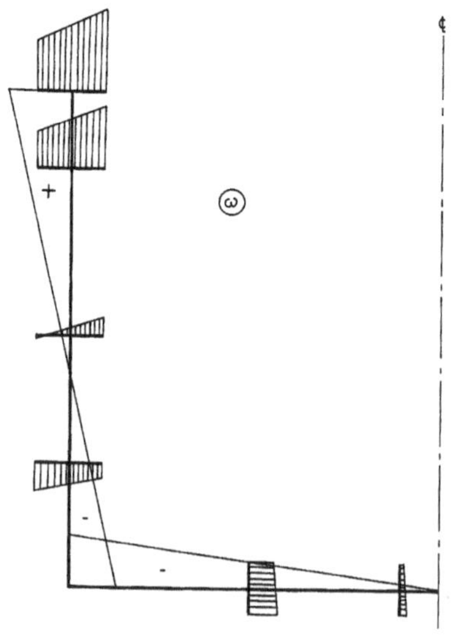

Fig. 4.28 Principal sectorial area diagram

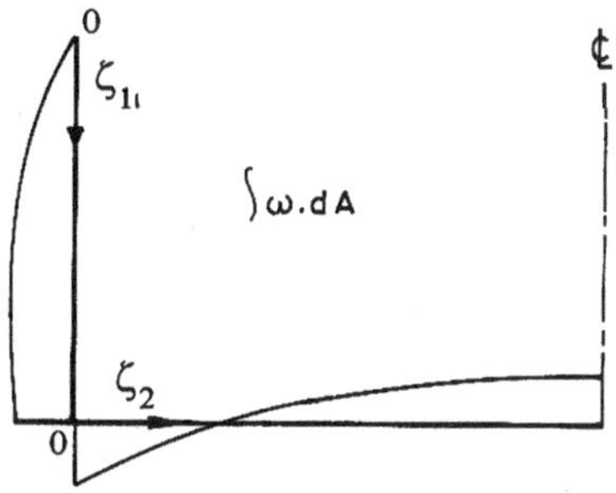

Fig. 4.29 Sectorial static moment for the ship section

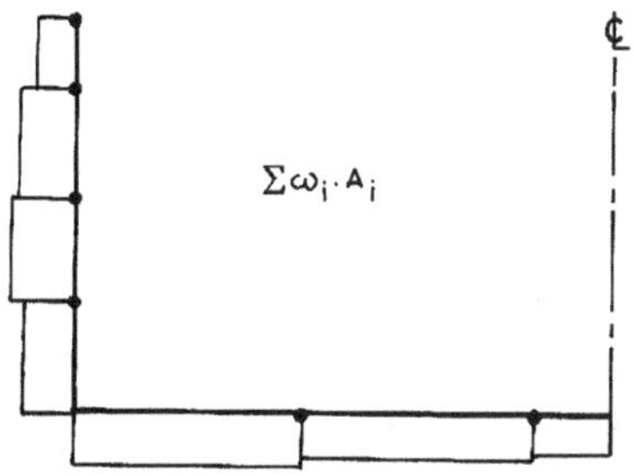

Fig. 4.30 Sectorial static moment for the discrete girder elements

4.8.4.6 Sectorial Static Moment S(ω)

The sectorial static moment is calculated as follows, see Figs. 4.29 and 4.30

$$S(\omega) = \int_A \omega dA + \sum_A \omega_i A_i \quad i = 1, 2, \ldots, (m + n)$$

For the side structure, S(ω) is given by

$$S(\omega) = \frac{bt_w}{2}[\xi_1(d - e_Y) - \xi_1/2]$$

where $0 \leq \xi 1 \leq d$,, see Fig. 4.29.
For the bottom structure, $S(\omega)$ is given by

$$S(\omega) = bdt_w/4(d - 2e_Y) - \left[\xi_2 - \xi_2^2/b\right] bt_b/2e_Y$$

where $0 \leq \xi_2 \leq b/2$.

Example 4.1 Calculate the torsional shear stresses for a container ship having the following main dimensions and is subjected to a torsional moment 8100 tm, see Fig. 4.31. Given that L = 152.9 m, B = 28.0 m, D = 16.2 m, h = 1.6 m, p = 2.2 m, 2a = 120.0 m.

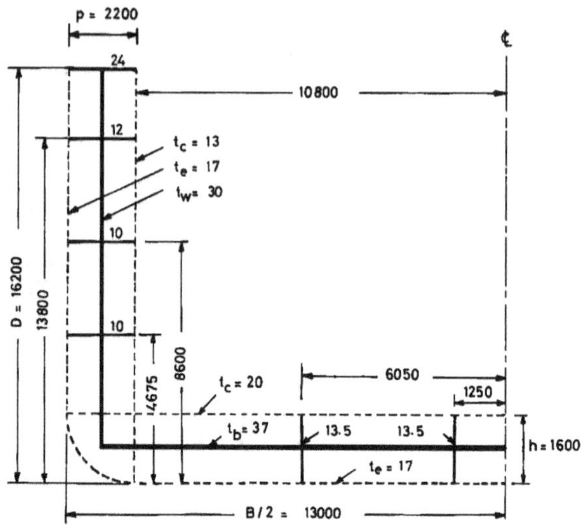

Fig. 4.31 Idealized ship section

Solution The idealized ship section shown in Fig. 4.31 has the following dimensions d = 15.4 m, b = 23.8 m, t_w = 30 mm, t_b = 37 mm, and its sectorial properties are given in Table 4.1.

Table 4.1 Sectorial properties of the idealized ship section

i	1	2	3	4	5	6
A_i, cm^2	216.5	216.5	220	220	264	264
Z_i, m^2	1.25	6.05	11.9	11.9	11.9	11.9
ω'_i, m^2	–	–	46.1	93.0	154.6	183.0
ω_i, m^2	−8.59	−40.7	−34.0	12.75	74.8	103.2
$\omega_i A_i$, m^4	−0.186	−0.882	−0.748	0.281	1.97	5.44
$\int \omega dA$, m^4	0.0	6.405	13.1	12.0	−7.77	−12.1

The results of the various calculations, for the case of the linear torque distribution are shown in Table 4.2 and Figs. 4.31 and 4.32.

Table 4.2 Results calculations for the linear torque distribution

Item	Numerical value
I_Y	208.91 m^4
$J(\omega)$	6150.0 m^6
k	0.02 m^{-1}
$T(\omega)_{max}$	5050.0 t m
$\zeta(\omega)_{max}$	0.133 t/cm^2
e_Y	6.73 m
J_t	6.41 m^4
$M(\omega)_{max}$	1.94 × 10^5 t m^2
$\sigma(\omega)_{max}$	0.40 t/cm^2
φ_{max}	0.214 × 10^{-2} radians

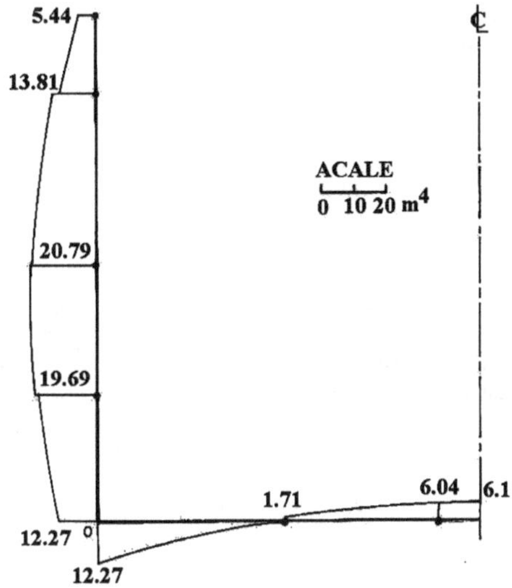

Fig. 4.32 Principal sectorial area diagram

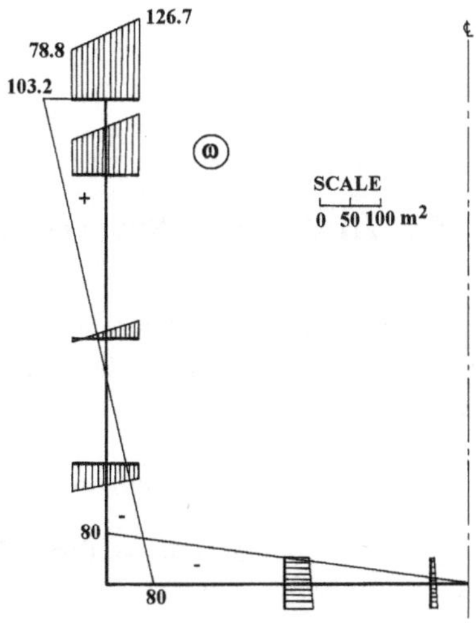

Fig. 4.33 Sectorial static moment

4.9 Total Stress in the Deck Plating of Container Ships due to Hull Girder Bending and Torsional Loading

The deck plating of container ships is subjected to bending stresses induced by hull girder bending and flexural warping stresses induced by the torsional loading of the hull girder (Figs. 4.33, 4.34, 4.35).

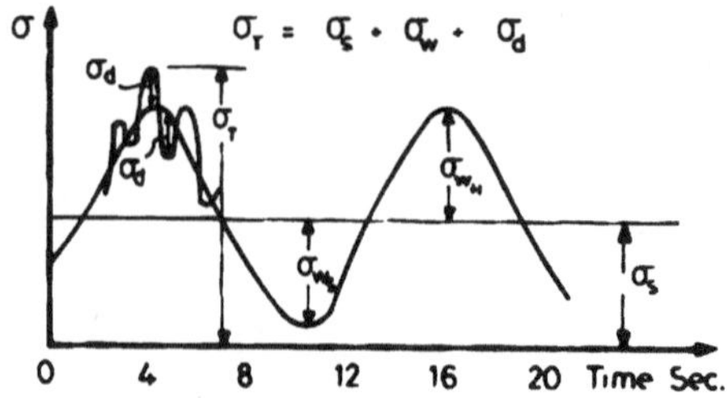

Fig. 4.34 Static and wave induced hull girder bending stresses

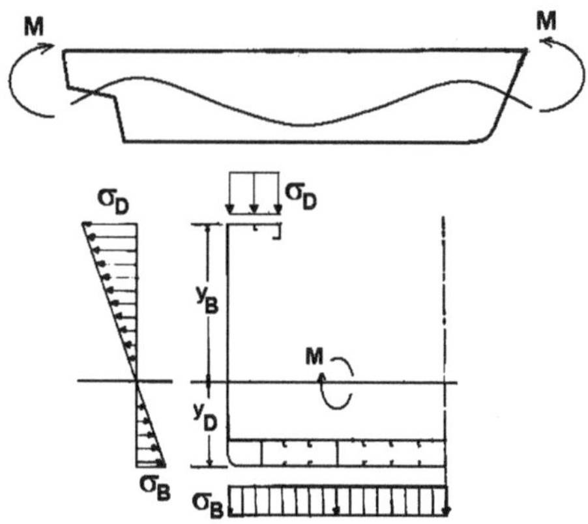

Fig. 4.35 Hull girder primary stress in deck plating

4.9.1 Hull Girder Stresses due to Vertical Bending

An approximate mean value of the still-water bending moment could be calculated using the following formula as given in the published Rules of a Classification Society

$$M_{sw} = +0.015\ C\ L^2\ B(8.167 - C_b)\ \text{kN m} \quad \text{Hog}$$

$$M_{sw} = -0.065\ C\ L^2\ B\left(C^b - 0.70\right)\ \text{kN m} \quad \text{Sag}$$

where

$$
\begin{aligned}
C &= 0.0792\ L & L &\leq 300\ \text{m} \\
&= 10.75 - [(300 - L)/100]^{1.5} & 300\ \text{m} < L &\leq 350\ \text{m} \\
&= 10.75 - [(L - 350)/150]^{1.5} & 350\ \text{m} < L &\leq 500\ \text{m}
\end{aligned}
$$

and L = ship length, m; B = ship breadth, m; C_b = block coefficient at summer load waterline.

4.9.2 Horizontal Hull Girder Bending Stresses

Hull girder bending stresses are also induced by the presence of horizontal wave bending moments. An approximate value of the horizontal wave induced bending moment could be obtained from the published Rules of Classification Societies

$$M_{WH} = 0.32L\ Q_{WHmax}\ C_M\ \text{kN m}$$

where Q_{WHmax} = Maximum value of horizontal shear force.

$$Q_{WHmax} = 2\sqrt{LT}\ B\ C_B C_o\ \text{kN}$$

The distribution of hull girder bending stresses induced by a horizontal bending is shown in Fig. 4.36.

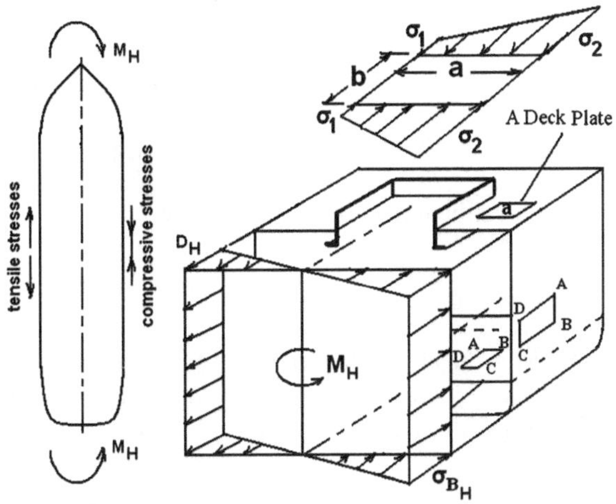

Fig. 4.36 Horizontal bending stresses

The horizontal hull girder bending stress is given by

$$\sigma_2 = M_h \cdot B/2I_H$$

where σ_2 = horizontal hull girder bending stress in deck plating; M_h = horizontal hull girder bending moment; B = ship breadth; I_H = second moment of area of ship section about the centerline vertical axis.

4.9.3 Local Stresses

The configuration of deck structure of a container ship with wide hatch opening is shown in Fig. 4.37. Because of the presence of end constraints, the end moments of deck longitudinals and girders could be calculated as follows, see Fig. 4.38

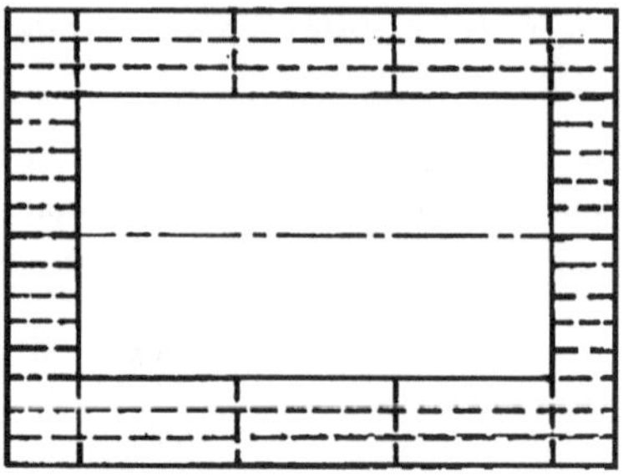

Fig. 4.37 Deck configuration

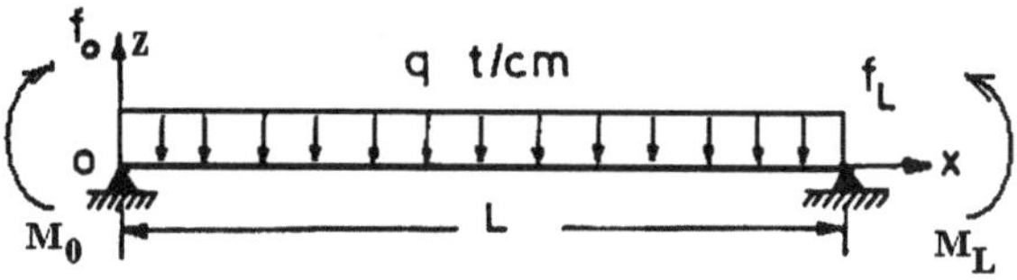

Fig. 4.38 End moments for a constrained member at both ends

$$M_0 = -qL^2/12[(3 - f_L)/2]f_0$$

$$M_L = qL^2/12[(3 - f_0)/2]f_L$$

The local bending stress is given by

$$\sigma_2 = -\frac{My_f}{i}$$

where M = local bending moment; y_p = distance of deck plating from the neutral axis of the section with attached plating, see Fig. 4.39; i = second moment of area of section with attached plating; σ_2 = local stress in deck plating.

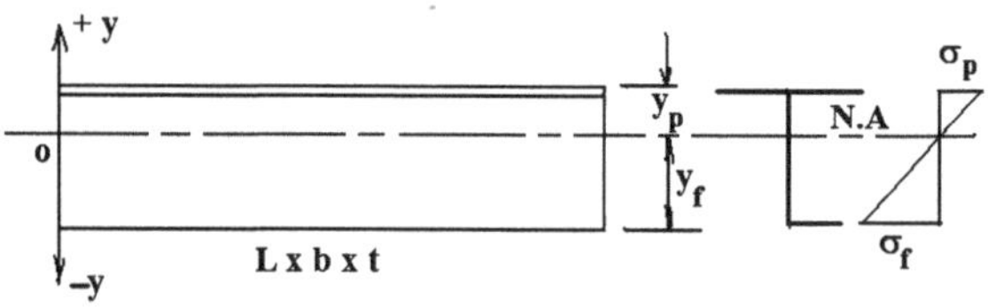

Fig. 4.39 Local stresses in a deck longitudinal

Let M_0 = bending moment at x = 0.0; M_L = bending moment at x = L; f_o = degree of fixity at x = 0.0; f_L = degree of fixity at x = L.

4.9.4 Flexural Warping Stresses

The flexural warping stress $\sigma(\omega)$ is given by, see Fig. 4.40

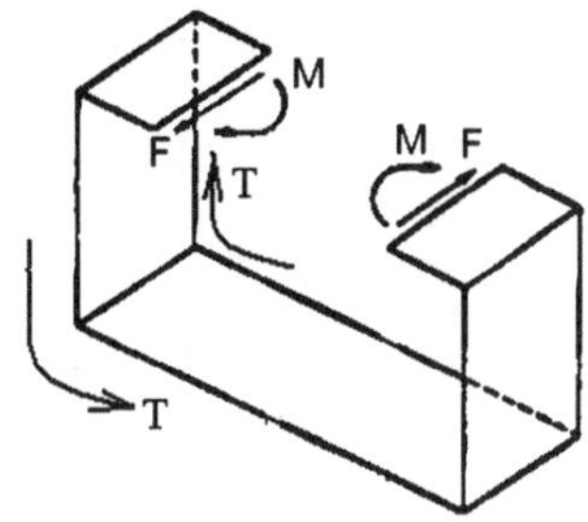

Fig. 4.40 Local bending and shear loading on the deck plating due to warping of the deck structure

$$\sigma(\omega) = \omega M(\omega)/J(\omega)$$

A typical distribution of the flexural warping stresses over the outer edge of the deck plating is shown in Fig. 4.41.

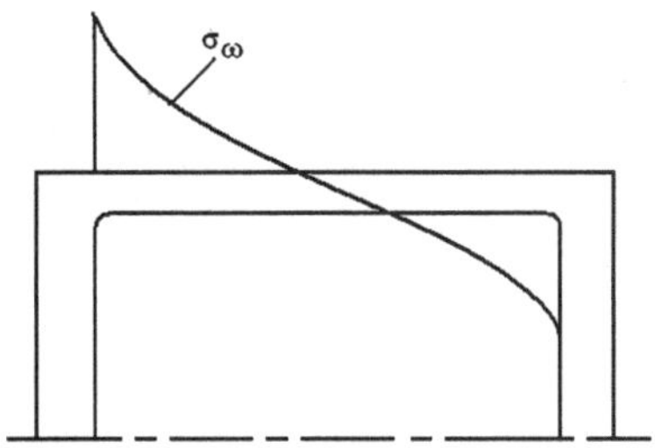

Fig. 4.41 Distribution of flexural warping stress over the inner edge of deck

4.9.5 Total Stress Over the Deck Plating

The various stress components over the deck plating of a container ship are shown in Fig. 4.42 and the total stress is shown in Fig. 4.43.

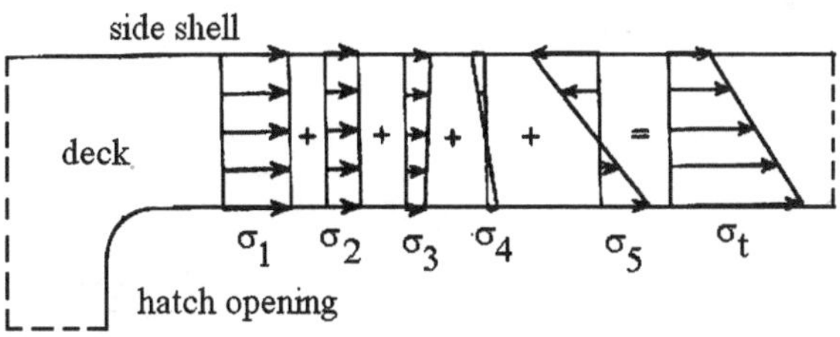

Fig. 4.42 Various stress components in deck plating

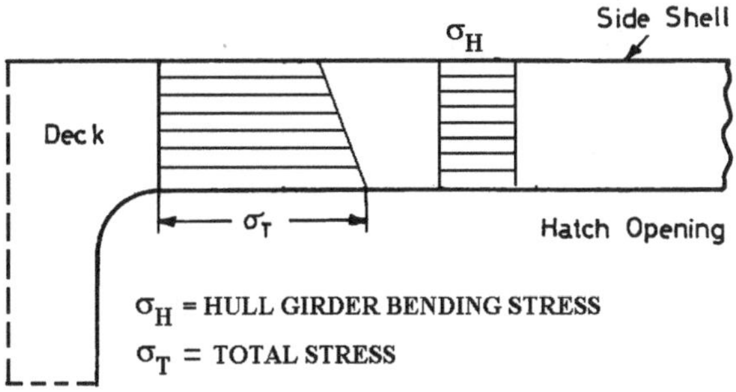

Fig. 4.43 Total stress over the deck plating

The total stress components are given by

$$\sigma_t = \sigma_1 + \sigma_2 + \sigma_3 + \sigma_4 + \sigma_5$$

where σ_1 = hull girder stress in deck plating due to vertical still water and wave bending moments; σ_2 = hull girder stress in deck plating due to horizontal wave induced bending moments; σ_3 = hull girder flexural warping stress in deck plating due to static torsional moments; σ_4 = hull girder flexural warping stress in deck plating due to wave induced torsional moments; σ_5 = local stress in deck plating.

Chapter 5
Sectorial Properties of Thin-Walled Open Sections

5.1 Introduction

Thin-walled sections are widely used in aircraft and ship structures and are becoming very attractive to structural engineers. The economy achieved through the reduced *weight/strength* ratio makes thin-walled sections very desirable.

In structural mechanics of thin-sections, shear, torsion, stability and warping problems become rather significant. The solution of the torsion equation requires the knowledge of the warping constant $J(\omega)$. In structures subjected to torsional buckling, the position of the shear centre and the warping constant $J(\omega)$ must be determined before any solution could he obtained. The calculation of the shear and flexural warping stresses of thin-walled members are also based on the position of the shear centre and the warping constant $J(\omega)$.

The position of the shear centre and the warping constant of thin-walled sections are both determined using the sectorial properties of the section. Examples are given to illustrate the simplicity of using the sectorial properties of thin-wal led sections for calculating the warping constant.

A flexural twist causes a "pair" or "pairs" of bending moments called a "Bi-moment" which is a mathematical function introduced by *Prof. Vlasov*. A bending moment is defined as a pair of forces; a bimoment is a pair of equal and opposite bending moments acting in two parallel planes. The numerical value of a bi-moment = moment × distance between the two moments, see Fig. 5.1.

5.2 Sectorial Properties of Thin-Walled Sections

In addition to the geometrical and flexural properties of sections, i.e., A, S_x, S_y, I_x, I_y, and I_{xy}, there are additional unique characteristics for thin-walled sections which are called "sectorial properties", where A is the sectional area (cm^2), S_x and S_y are the first moment of area about the x and y axes of the section. I_x, I_y and I_{xy}

M. Shama, *Torsion and Shear Stresses in Ships*,
DOI: 10.1007/978-3-642-14633-6_5,

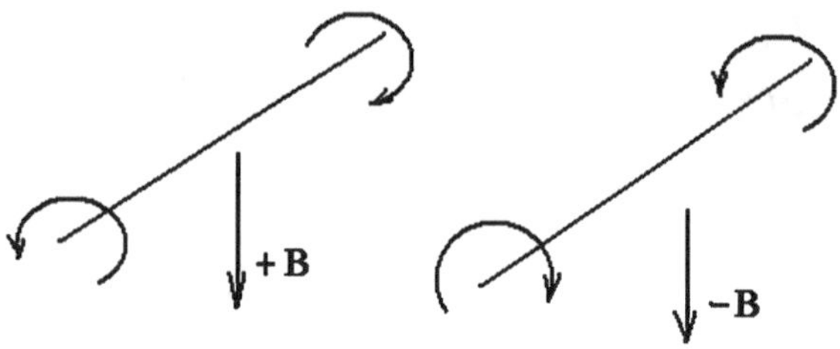

Fig. 5.1 Concept of bi-moment

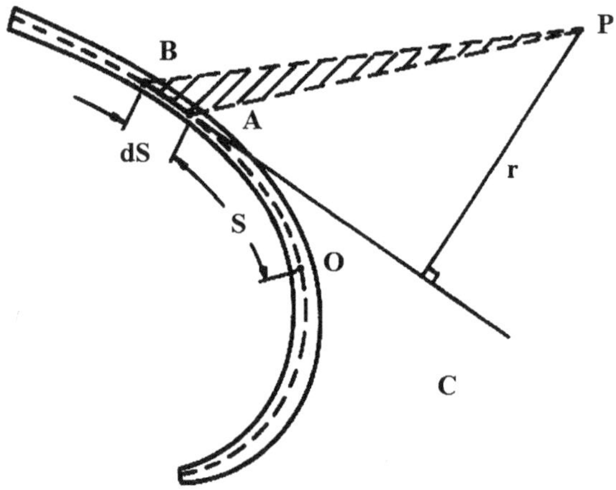

Fig. 5.2 Concept of sectorial coordinates

are the second moment and product moment of area about the x and y axes of the section.

The sectorial properties are also associated with area, moment of area and moment of inertia.

The sectorial properties of a thin-walled open section see Fig. 5.2, having an arbitrary pole P and an arbitrary starting point 0, are as follows

$$\omega = r \cdot ds$$

Sectorial area $= \omega$

$$\mathrm{d}\omega = \int_{s_2}^{s_1} r \cdot \mathrm{d}s \text{ cm}^2$$

Thus, the sectorial area is the double area swept by the radius vector from $s = 0$ to the required point on the contour, where r is the the perpendicular distance from the pole P to the tangent at the point under consideration; see Fig. 5.2. s_1, and s_2 represent end points on the contour of section.

The sectorial area is positive when the radius vector rotates clockwise and is negative when the radius vector rotates anti-clockwise, see Fig. 5.3.

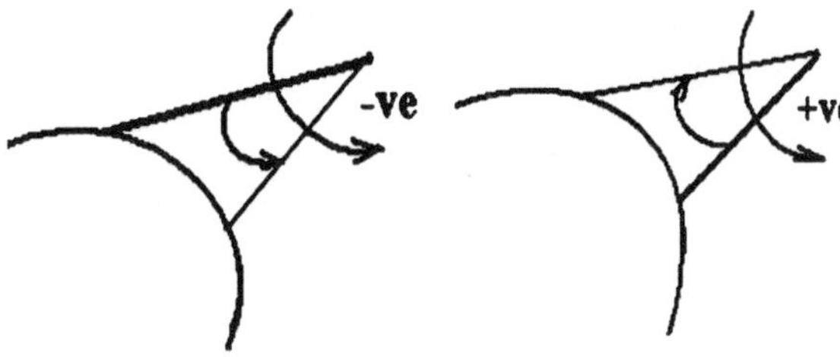

Fig. 5.3 Sign convention for calculating sectorial properties of thin-walled sections

Sectorial static moment $S(\omega)$

$$S(\omega) = \int_A \omega \cdot \mathrm{d}A \text{ cm}^4$$

Sectorial linear moments $S(\omega)_x$, $S(\omega)_y$

$$S(\omega)_x = \int_A \omega \cdot y \cdot \mathrm{d}A \text{ cm}^5$$

$$S(\omega)_y = \int_A \omega \cdot x \cdot \mathrm{d}A \text{ cm}^5$$

Sectorial moment of inertia $J(\omega)$

$$J(\omega) = \int_A \omega^2 \mathrm{d}A \text{ cm}^6$$

For a thin-walled section having n elements: each having a uniform thickness t and a constant r,

$$J(\omega) = \frac{1}{3}\sum_{i=1}^{n} r_i^2 \cdot t_i \cdot \left(s_i^3\right)_{s_1}^{s_2} \text{ cm}^6$$

It is to be noted here that once the sectorial area diagram is determined, the sectorial characteristics of the section could be easily calculated by either direct or numerical integration. In the majority of cases, the integral of the product of two functions is required to be evaluated.

5.2.1 Principal Sectorial Properties of Thin-Walled Sections

In the same way as for the centroidal and principal centroidal axes, see Fig. 5.4, there is also the principal sectorial coordinates. The pole of, the latter system is the

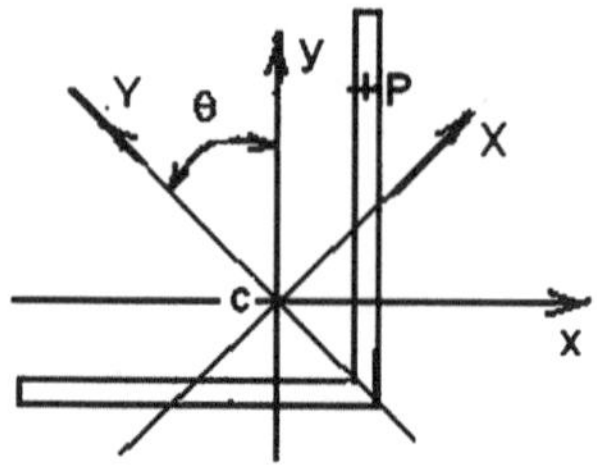

Fig. 5.4 Centroidal and principal axes

"principal pole" or the shear centre of the section. The principal origin is defined by, the "principal radius". The principal sectorial moment of inertia $J(\omega)$ is calculated using the principal sectorial coordinates.

The principal pole and the "principal radius" are determined from the following conditions

$$S(\omega) = \int_A \omega \cdot dA = 0 \tag{5.1}$$

$$S(\omega)_x = \int_A \omega \cdot y \cdot dA = 0 \tag{5.2}$$

$$S(\omega)_y = \int_A \omega \cdot x \cdot dA = 0 \tag{5.3}$$

Condition (5.1) giver the direction of the principal radius, i.e., the location of the 'principal origin, whereas conditions (5.2) and (5.3) give the location of the principal pole, i.e., the shear centre. From these conditions, it is evident that the sectorial linear moments with respect to the principal centroidal axes and a pole coincident with the shear centre are *zero*. The origin of ω is of no importance because if it is shifted, the sectorial area diagram is changed by a constant which does not have any effect on conditions (5.1, 5.2 and 5.3).

5.2.2 Position of the Shear Center

In order to determine the location of the shear centre, a diagram of sectorial area ω' for an arbitrary pole P' is drawn. The location of the shear centre is given by, see Fig. 5.5.

Since the centroidal axes x and y are not the principal centroidal axes, the coordinates of the shear centre with respect to the assumed pole P' are given by

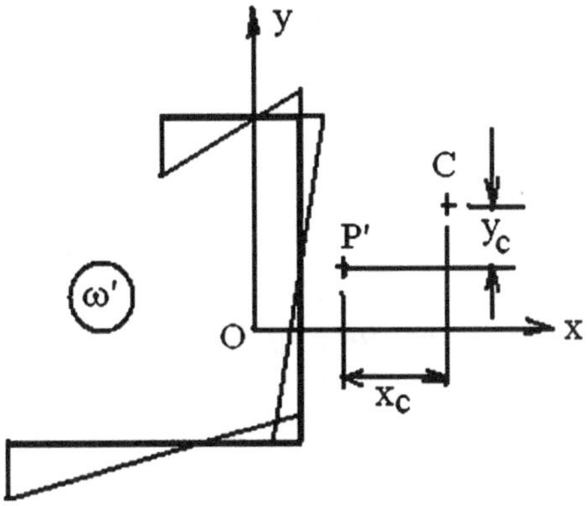

Fig. 5.5 Positions of principal pole and shear center

$$e_x = -\left(I_y \int_A \omega' \cdot y \cdot \mathrm{d}A - I_{xy} \cdot \int_A \omega' \cdot x \cdot \mathrm{d}A\right) \Big/ \left(I_x \cdot I_y - I_{xy}^2\right)$$

$$e_y = -\left(I_x \int_A \omega' \cdot x \cdot \mathrm{d}A - I_{xy} \cdot \int_A \omega' \cdot y \cdot \mathrm{d}A\right) \Big/ \left(I_x \cdot I_y - I_{xy}^2\right),$$

where e_x and e_y are the coordinates of the shear centre with respect to the assumed pole P'. I_x, I_y and I_{xy} are the moments and product of inertia about the centroidal axes x, y.

If the principal centroidal axes X and Y of the section are used, the coordinates of the shear center is given by

$$e_X = -\int_A \omega' \cdot Y \cdot \mathrm{d}A / I_X$$

$$e_Y = -\int_A \omega' \cdot X \cdot \mathrm{d}A / I_Y,$$

where I_X and I_Y are the principal moments of inertia about the X and Y axes, respectively.

5.2.3 Sectorial Area Diagram

In order to determine the principal sectorial area diagram, the shear centre is used as the principal pole and an arbitrary origin P' is assumed. The sectorial area diagram ω' is calculated based on the arbitrary origin P', see Fig. 5.5. The principal sectorial area at any point is given by

$$\omega = \omega' + \omega_c,$$

where ω_c is a constant given by

$$\omega_c = -\int_A \omega' \cdot \mathrm{d}A/A$$

5.2.4 Procedure of Calculation

The calculation of the principal sectorial area diagram could be summarized as follows

(i) The flexural properties of the section about the principal centroidal axes X, Y or any convenient centroidal axes x, y are determined, i.e., I_X, I_Y or I_x, I_y, and I_{xy} in addition to the sectional area A.
(ii) Choose an arbitrary pole P' and an arbitrary origin O', and then calculate the sectorial area diagram ω'.
(iii) Calculate the coordinates of the shear centre relative to P'.
(iv) Using the calculated shear centre as the principal pole and assuming any arbitrary origin O', the sectorial area diagram ω' is calculated.
(v) The correcting term ω_c is then calculated.
(vi) The principal sectorial area diagram ω is then calculated.

5.3 Applications to Some Typical Sections

5.3.1 Sectorial Properties for Thin-Walled Sections Free to Warp

The position of the sheer centre and the sectorial moment of inertia $J(\omega)$, are calculated for the following sections, using the principal sectorial area diagram.

5.3.1.1 I-Section

Due to symmetry, see Fig. 5.6, the shear centre coincides with the centroid of the section.

Hence

$$e_x = e_y = 0.0$$

$$J(\omega) = \int_A \omega^2 \cdot \mathrm{d}A$$

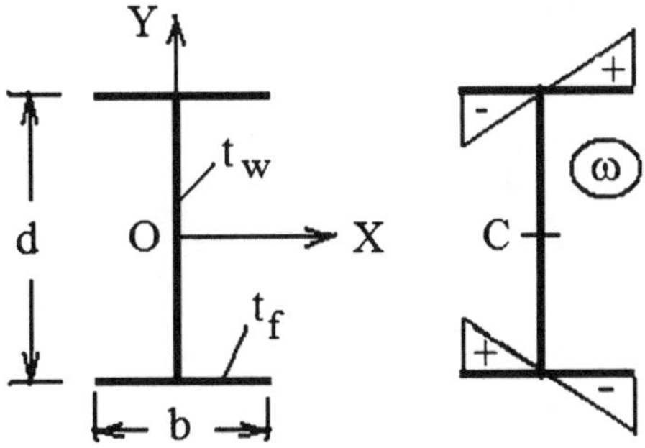

Fig. 5.6 Sectorial area diagram of an I-section

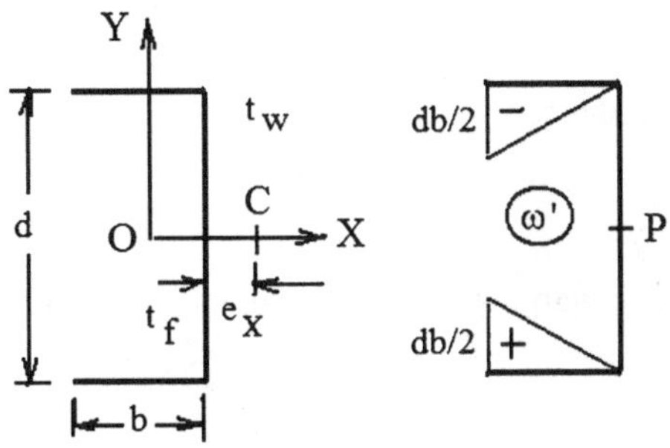

Fig. 5.7 Sectorial area diagram of a thin-walled channel section

$$= 4 \cdot t \cdot f \cdot \int_0^{b/2} (\mathrm{d}s/2)^2 \cdot \mathrm{d}s$$

$$= I_Y \cdot d^2/4$$

5.3.1.2 Channel Section

Due to symmetry about the X axis, see Fig. 5.7, we have

$$e_y = 0.0$$

The shear centre is therefore located along the X axis at a distance e_x given by

$$e_X = -\int_A Y \cdot \omega' \cdot \mathrm{d}A/I_X$$

$$I_X = d^2 \cdot (\mathrm{d}t_w/12 + bt_f/2)$$

$$\int_A \omega' \cdot Y \cdot \mathrm{d}A = t_f \cdot d^2/4 \cdot \int_0^b (-s) \cdot \mathrm{d}s - t_f \cdot d^2/4 \cdot \int_0^b s \cdot \mathrm{d}s = -t_f \cdot b^2 \cdot d^2/4$$

The ω' diagram is shown in Fig. 5.7.

Fig. 5.8 ω Diagram

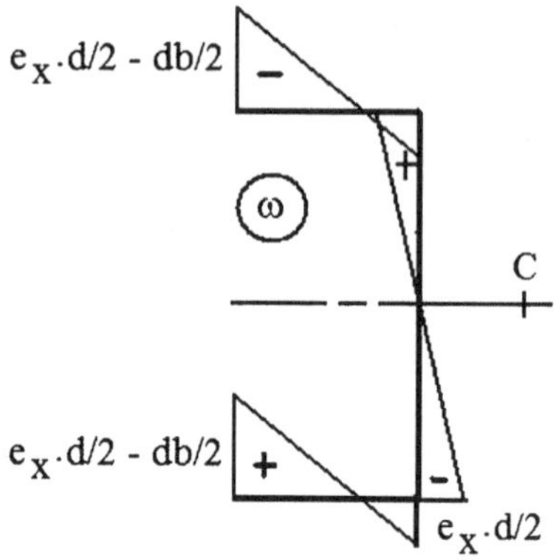

Hence,

$$e_X = b/\left(A_w/3A_f + 2\right).$$

The warping constant is given by, see Fig. 5.8.

$$J(\omega) = \int_A \omega^2 \mathrm{d}A$$

$$= 2t_w \int_0^{d/2} (e_X \cdot s)^2 \cdot \mathrm{d}s + 2t_f \int_0^{b} (e_X \cdot d/2 - d \cdot s/2)^2 \cdot \mathrm{d}s$$

$$= I_X \cdot e_X^2 + t_f \cdot b^2 \cdot d^2 \cdot (b - 3e_X)/6$$

Example 5.1 Determine the torsional and warping rigidities of the thin-walled fabricated offset flange section with attached plating, see Fig. 5.9.

Solution x and y are centroidal axes, X and Y are the principal centroidal axes.

Geometrical and Flexural Properties of the Section

$$A_f = b \cdot t_f$$
$$A_w = b \cdot t_w$$
$$A_p = S \cdot t_p$$
$$A_t = A_f + A_w + A_t,$$

where A_f is the flange area, A_w is the web area, A_p is the attached plate area, A_t = total section area, Ip is the second moment of area of plate and S is the effective breadth of plate.

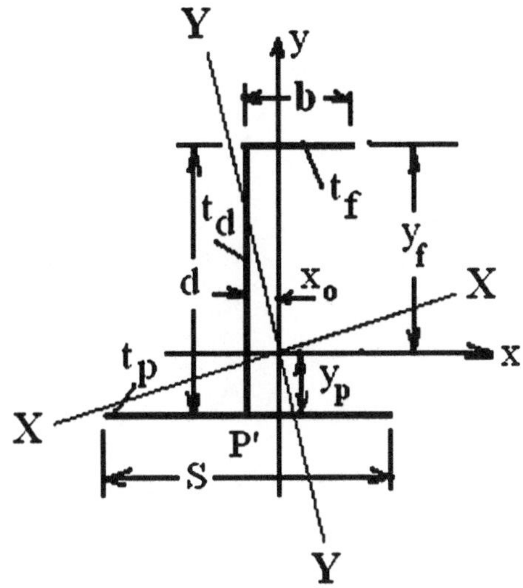

Fig. 5.9 Offset flange thin-walled section with attached plating

$$y_f = d/A_t \cdot (A_p + A_w/2)$$

$$y_p = d/A_t \cdot (A_f + A_w/2)$$

$$I_P = t_p \cdot S^3/12$$

$$x_o = 1/A_t \cdot (A_f \cdot b/2) = bA_f/2A_t$$

$$y_f = d/A_t \cdot (A_f + A_w/2)$$

The moments of inertia about the centroidal axes x, y are given by

$$I_x = A_f\left(y_f^2 + t_f^2/12\right) + A_p\left(y_p^2 + t_p^2/12\right) + A_w\left(\left[d/2 - y_p\right]^2 + d^2/12\right)$$

$$I_y = A_f\left[(b/2 - x_o)^2 + b^2/12\right] + A_p(x_o^2 + S^2/12) + A_w(x_o^2 + t_w^2/12)$$

$$I_{xy} = A_f \cdot y_f(x_o - b/2) + A_w \cdot x_o\left(y_f - d/2\right) + A_p \cdot x_o \cdot y_p$$

The moments of inertia about the principal centroidal axes X, Y are given by

$$I_X = I_X \cdot \cos^2(\alpha) - I_{XY} \cdot \sin(2\alpha) + I_Y \cdot \sin^2(\alpha)$$

$$I_Y = I_X \cdot \sin^2(\alpha) - I_{XY} \cdot \sin(2\alpha) + I_Y \cdot \cos^2(\alpha)$$

$$I_{XY} = \{I_{XY} \cdot \cos(2\alpha) + (I_X - I_Y) \cdot \sin(2\alpha)\}/2$$

$$\tan(2\alpha) = 2I_{XY}/(I_X - I_Y)$$

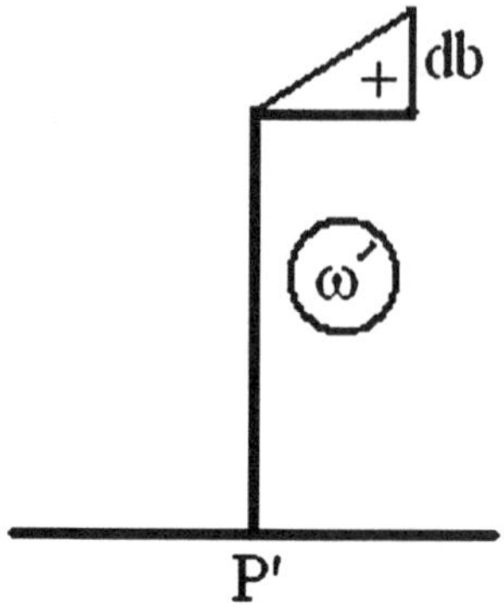

Fig. 5.10 ω' Diagram

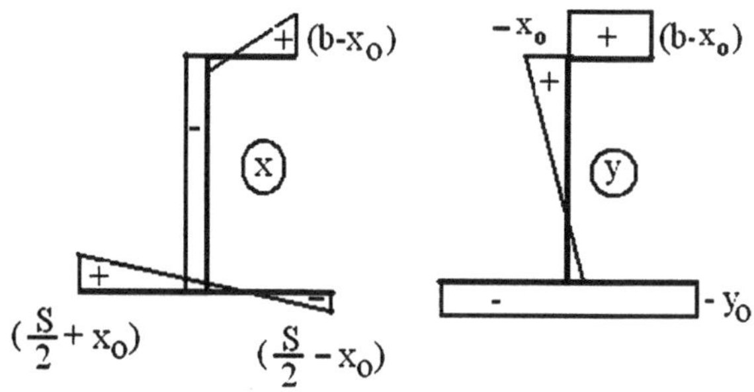

Fig. 5.11 The x and y diagrams

The coordinates of the shear center with respect to the principal centroidal axes are given by

$$Y_C = 1/I_Y \cdot \int_A X \cdot \omega' \cdot dA$$

$$X_C = -1/I_X \cdot \int_A Y \cdot \omega' \cdot dA$$

The sectorial area diagram based on an assumed pole at P' is as shown in Fig. 5.10.

$$\int_A x \cdot \omega' \cdot dA = t_b db^2/2 \cdot (2b/3 - x_o)$$

$$\int_A y\omega' dA = t_b db^2/2 \cdot (d - y_f)$$

The integration can be executed using Fig. 5.11.

The location of the shear center relative to the web of the section is given by

$$e = -I_p \cdot I_{xy} \cdot d \Big/ \left(I_x \cdot I_y - I_{xy}^2 \right),$$

where e = distance of shear center from the web of the section.

Sectorial Properties of the Section

Let ω_c = correcting sectorial area.

$$\omega_c = d/2 \cdot A_t \cdot \left[e \cdot A_w - 2 \cdot A_f (b/2 - e) \right]$$
$$\omega_i = -d \cdot e + \omega_c$$
$$\omega_o = -b \cdot d + \omega_i,$$

ω_i is the sectorial area at point I and ω_o is the sectorial area at point O

Torsion Constant J_t

The torsion inertia J_t is given by

$$J_t = \left(b \cdot t_f^3 + b \cdot t_w^3 + S \cdot t_p^3 \right) \Big/ 3.$$

The warping constant J_ω is given by

$$J_w = d^2 [A_w \cdot e^2/3 + A_f(e^2 + b^2/3 - e \cdot b)] - d^2/4\, A_t \cdot [e\, A_w + (2 \cdot e - b) A_f]^2$$

$$k = \sqrt{C/C_1} = \sqrt{G \cdot J_t / E \cdot J_\omega} = \frac{1}{1.61} \cdot \sqrt{J_t/J_\omega},$$

where C = torsion rigidity $= G \cdot J_t$; C_1 = warping rigidity $= C \cdot E \cdot J_\omega$.

5.4 Sectorial Properties for a Thin-Walled Section with an Enforced Axis of Rotation

5.4.1 A thin-Walled T-Section with an Enforced Axis of Rotation

For the T-section, the Y-axis is an axis of symmetry. In this case the enforced axis of rotation replaces the shear center when calculating the principal sectorial area diagram, see Fig. 5.12. The sectorial moment of inertia (warping constant) is given by

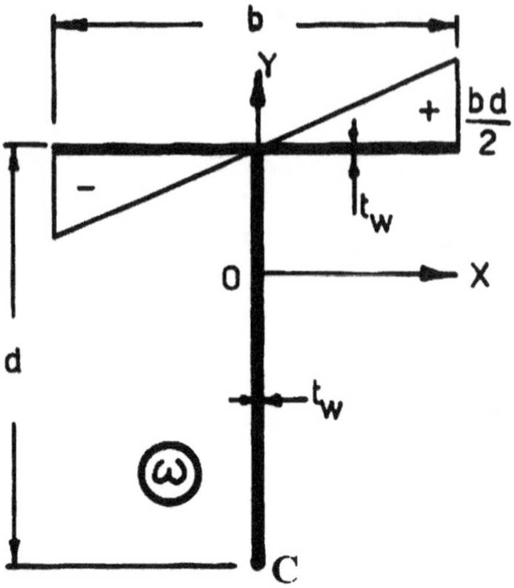

Fig. 5.12 ω Diagram for enforced center of rotation for a thin-walled T-section

$$\begin{aligned}J_\omega &= \int_A \omega^2 \mathrm{d}A \\ &= 2t_f \int_0^{b/2} (\mathrm{d}s)^2 \cdot \mathrm{d}s \\ &= 2 \cdot t_f \cdot d^2 \cdot b^3/24 \\ &= I_Y \cdot d^2\end{aligned}$$

5.4.2 Enforced Center of Rotation for a Thin-Walled Angle Section

The warping constant is given by, see Fig. 5.13.

$$\begin{aligned}J_\omega &= \int_0^b \omega^2 \cdot t_f \cdot \mathrm{d}s = t_f \cdot \int_0^b d \cdot s^2 \cdot \mathrm{d}s \\ &= t_f \cdot d^2 \cdot b^3/3 = I_Y \cdot d^2\end{aligned}$$

5.4.3 Enforced Center of Rotation at a Point C on the Opposite Side of a Thin-Walled Asymmetrical Fabricated Section

The enforced axis of rotation is at a distance e from the web of the section, see Fig. 5.14.

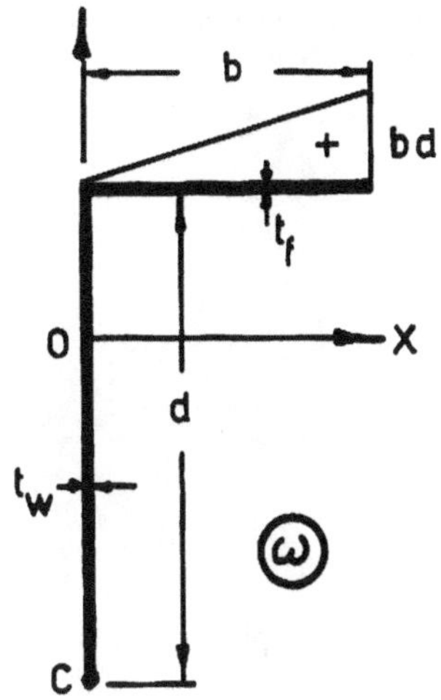

Fig. 5.13 ω diagram for enforced center of rotation for a thin-walled angle section

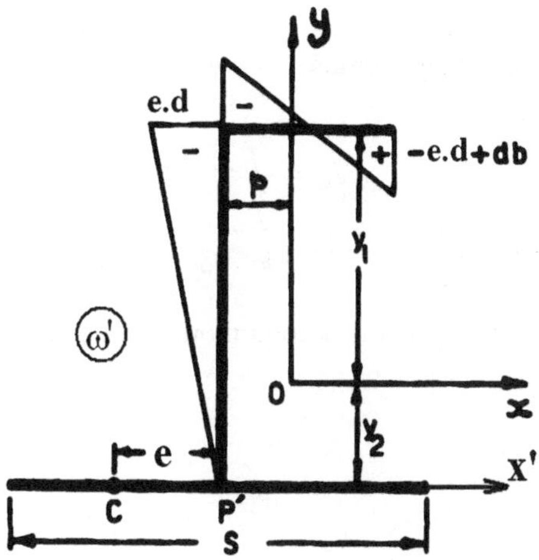

Fig. 5.14 ω' Diagram for enforced center of rotation for a thin-walled fabricated angle section with attached plating

The sectorial coordinates of the section are given by

$$\omega = \omega' + \omega_c,$$

where

$$\omega_c = -1/A_t \cdot \int_A \omega' \mathrm{d}A.$$

Using numerical integration, ω_c is given by

$$\omega_c = -d/2A_t \cdot \left[e \cdot A_w + (2e - b) \cdot A_f\right]$$

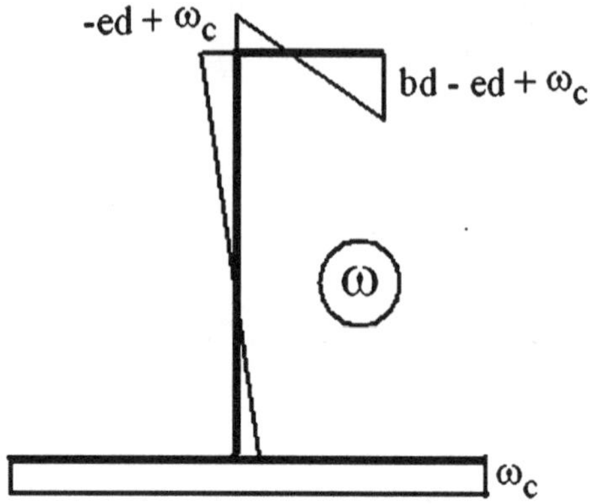

Fig. 5.15 ω Diagram of the offset section

The warping constant is given by

$$J_\omega = \int_A \omega^2 \mathrm{d}A = \int_A (\omega_c + \omega')^2 \cdot \mathrm{d}A$$

The ω diagram is shown in Fig. 5.15. Using numerical integration we have

$$\begin{aligned} J_\omega &= \int_0^d (e \cdot s)^2 \cdot t_w \cdot \mathrm{d}s + \int_0^b (-e \cdot d + d \cdot s)^2 \cdot t_f \cdot \mathrm{d}s \\ &= e^2 \cdot I_{x'} + t_f \cdot b^2 \cdot d^2(b/3 - e), \end{aligned}$$

where $I_{x'}$ is the second moment of area about x' axis, see Fig. 5.14.

Chapter 6
General solution of the torsion equation

The general torsion equation is given by

$$\varphi''' - k^2.\varphi' = -T_X/C_1 \tag{6.1}$$

where T_X = external torque; C = torsional rigidity; C_1 = warping rigidity $k = \sqrt{C/C_1}$.

The solution is given in the form

$$\varphi(x) = \varphi_1(x) + \varphi_2(x) \tag{6.2}$$

where $\varphi_1(x)$ = complementary function, and is given by

$$\varphi_1(x) = A_0 + A_1 \cdot \cosh(kx) + A_2 \cdot \sinh(kx) \tag{6.3}$$

where A_i, (i = 0, 1, 2), are constants to be determined from the support boundary conditions; $\varphi_2(x)$ = particular integral, P.I.

The P.I. depends on the function of the torque T_x.

If T_x is given in a polynomial form, i.e.,

$$T_x = \sum_{i=0}^{n} b_i \cdot X^i \tag{6.4}$$

The P.I. could be obtained for any individual term of the polynomial. The required $\varphi_2(x)$ is obtained by summing up the P.I. of each term in the polynomial.

Assuming the general term of the torque equation to be

$$T_r = b_r \cdot x^r \tag{6.5}$$

And substituting in Eq. 6.1 we get

$$(D^3 - k^2 \cdot D) \cdot \varphi = -b_r/C_1 \cdot X^r$$

M. Shama, *Torsion and Shear Stresses in Ships*,
DOI: 10.1007/978-3-642-14633-6_6, © Springer-Verlag Berlin Heidelberg 2010

Table 6.1 Particular Integral for $T_r = b_r\, x^r$

r	T_r	$\varphi_2(x)$
0	b_0	$b_0/C_1 \cdot x$
1	$b_1 \cdot x$	$b_1/C_1 \cdot (x^2 + 1/k^2)$
2	$b_2 \cdot x^2$	$b_2/C_1 \cdot (x^3/3 + 2x/k^2)$
3	$b_3 \cdot x^3$	$b_3/C_1 \cdot (x^4/4 + 3x^2/k^2 + 6/k^6)$
4	$b_4 \cdot x^4$	$b_4/C_1 \cdot (x^5/5 + 4x^3/k^2 + 24x^2/k^4)$
5	$b_5 \cdot x^5$	$b_5/C_1 \cdot (x^6/6 + 5x^4/k^2 + 60x^2/k^4 + 120/k^6)$

Hence

$$
\begin{aligned}
\varphi_2(x) &= D^{-1}/k^2 \cdot \left[1 - (D/k)^2\right]^{-1} \cdot B_r \cdot x^r \\
&= \left[D^{-1}/k^2 + D/k^4 + D^3/k^6 + D^5/k^8\right] B_r \cdot x^r \\
&= B_r \left[\frac{x^{r+1}}{k(r+1)} + \frac{r \cdot x^{r-1}}{k^4} + \frac{r(r-1)(r-2) \cdot x^{r-3}}{k^6}\right] \\
&\quad + B_r\left[r(r-1)(r-2)(r-3)(r-4) \cdot x^{r-5}/k^8 + \cdots\right]
\end{aligned}
\tag{6.6}
$$

where

$$B_r = b_r/C_1$$

The P.I. for r_i (I = 0, 1, 2, 3, 4, 5) is given in Table 6.1, from which $\varphi_2(x)$ could be obtained by summing the appropriate terms of the polynomial. However, if the torque distribution is given in any other form, a polynomial equation could be fitted and the P.I. could then be obtained from Table 6.1.

Part II
Shear Loading and Stresses in Ships

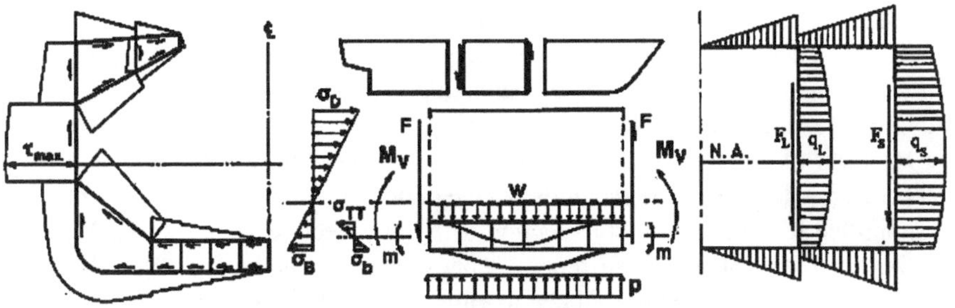

Chapter 7 introduces the basic concept of shear stress, average shear stress, shear stresses due to bending, shear flow and stress in thin-walled symmetrical and asymmetrical sections, shear center, shear deflection, shear deformation and shear lag.

Chapter 8 introduces methods of calculating shear flow and stresses in thin-walled symmetrical and asymmetrical single and multi-box girders.

Chapter 9 introduces methods of calculating shear flow and stresses for different types of ships. The methods are based on using the concept of effective thickness and a simplified ship section idealization.

Chapter 10 introduces methods of calculation of shear flow and stresses in tankers subjected to longitudinal vertical shear forces. Tankers having one and two longitudinal bulkheads are considered. A method is given to calculate the shear load carried by the longitudinal bulkheads and the side shell plating.

Methods for calculating shear flow and stress distribution over a ship section of an oil tanker experiencing a local damage at different locations on the side shell plating and longitudinal bulkhead addressed. Both coastal and sea-going oil tankers are considered.

Chapter 11 introduces shear loading and stresses in ship hull girder of bulk carriers due to longitudinal vertical shearing forces. The Stillwater, wave-induced and dynamic shear force components are presented. Methods for calculating the shear flow distribution over the top wing tanks and over the hopper tanks in a ship section of a bulk carrier are given. The methods are based on using the concept of effective thickness and a simplified structure idealization of the ship section. Shear loading on local structural elements are also presented.

Introduction

This book presents the basic principles and concepts of shear flow, shear stress, shear deformation and the application of these principles to ship structure. The shear lag effect in thin-walled structures resulting from the effect of shear stress on bending stress is presented.

Methods of calculating the distribution of shear flow and stresses over symmetrical and asymmetrical thin-walled open sections are given.

Shear centre for symmetrical and asymmetrical thin-walled sections is explained. The distribution of shear stresses over thinwalled single and multi-box girders is given. The methods of calculation are explained and supported by numerical examples.

Methods of calculation of the distribution of shear flow and stress over ship sections are addressed. The methods of calculation are based on the introduction of a simplified idealization of ship section using an effective thickness for the shell plating and the attached stiffeners. For ship sections having closed boxes, a correcting shear flow is introduced to eliminate any torsional distortions induced by the assumed shear flow distribution.

The method is used to calculate the shear flow distribution over ship sections of single and double deck cargo ships and oil tankers with one and twin longitudinal bulkheads. A method for calculating shear load carried by the side shell plating and longitudinal bulkheads is given.

A damage occurring in any part of the ship structure will cause redistribution of the shear and bending stresses over the remaining intact structural members. Some structural members will be over stressed and others may be lightly stressed. The shear stress distribution over ship sections experiencing local damages is examined so as to ensure adequate safety of the overloaded structural members.

The importance of calculating the distribution of shear stresses over ship sections of the hull girder is emphasized so as to determine the maximum allowable shearing force for a given ship section.

Shear loading on ship hull girder is given together with shear force distribution for alternate hold loading in bulk carriers. Bulk carriers experience unique problems which result from the particular structural configuration and loading of these ships (alternate hold loading system).

In bulk carriers, the longitudinal vertical shearing force is carried by the side shell plating, top wing tanks, and hopper tanks. The side shell, therefore, may carry a high proportion of the longitudinal vertical shearing force. Consequently, shear stresses in the side shell plating may reach unfavorable values. Shear buckling, or high values of combined stresses, may occur in some panels in the side shell plating. Adequate measures should be taken, therefore, to prevent the initiation of instability and high stresses.

In order to calculate the shear stress distribution over a typical ship section of a bulk carrier, the ship section is idealized by a simplified configuration so as to reduce the laborious calculations associated with shear flow distribution. The idealized structure should affect neither the magnitude nor the distribution of shear flow distribution around the top wing tanks, hopper tanks, and side shell. Procedures for calculating shear flow and stresses in bulk carriers are given in derail.

Chapter 7
Shear Stresses in Thin-Walled Structures

7.1 Basic Principles

Consider an element ABCD of a member subjected to a shear force F, see Fig. 7.1.

The force F produces shear stresses on AB and CD. For equilibrium of the element ABCD, complementary shear stresses are produced on AD and BC. The average shear stress is given by

$$\tau_{\text{mean}} = F/A$$

where A is the area subjected to shear stresses, and F is shear force.

The shear strain is given by

$$\gamma = \delta/\text{dy} = \tau/G$$

where γ is shear strain, G modulus of rigidity $= E/2(1 + \nu)$, and ν is Poisson ratio.

7.2 Shear Stresses in Beams due to Bending

7.2.1 Solid Beams

The shearing force at any cross-section of a beam, see Fig. 7.2, will set up a shear stress on transverse sections which will vary across the section.

The calculation of the shear stress is based on the following assumptions:

- The shear stress is uniform across the width.
- The presence of the shear stress does not affect the distribution of the bending stress.

M. Shama, *Torsion and Shear Stresses in Ships*,
DOI: 10.1007/978-3-642-14633-6_7,

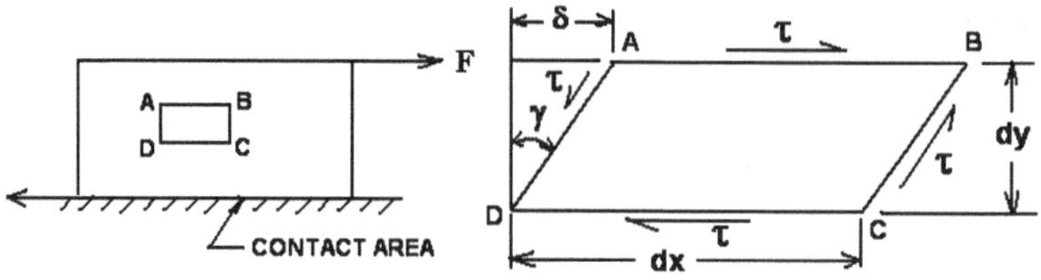

Fig. 7.1 Basic concept of shear stress

Fig. 7.2 Shear force and bending moment diagrams for a simply supported beam

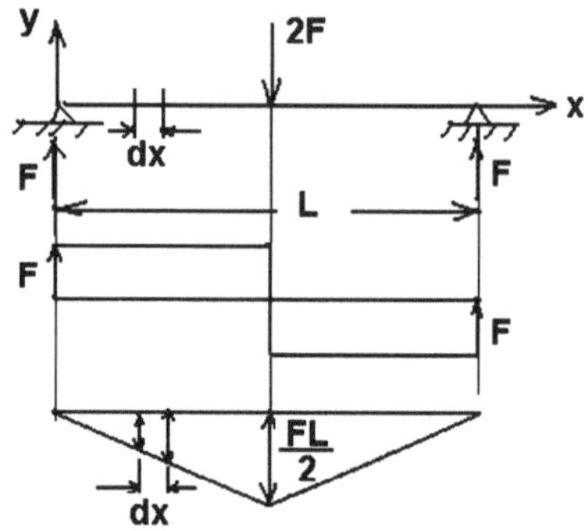

Fig. 7.3 Shear and bending stresses over an element of the beam section

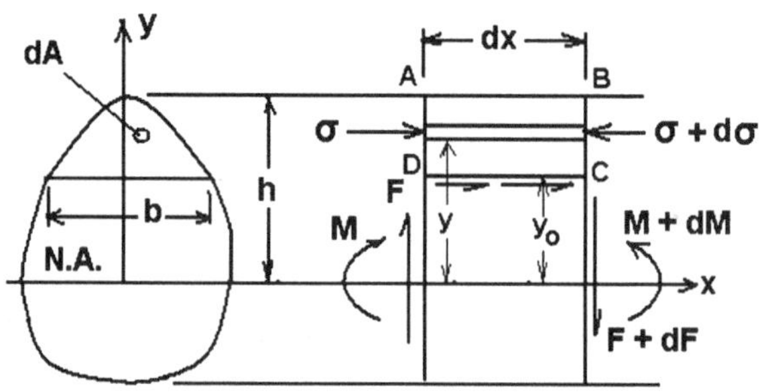

Consider the equilibrium of an element such as ABCD, see Fig. 7.3, we have

$$\tau \cdot b \cdot \mathrm{d}x = \int_{y_0}^{h} \mathrm{d}\sigma \cdot \mathrm{d}A.$$

but

$$\sigma = \frac{M \cdot y}{I}.$$

Thus

$$\mathrm{d}\sigma = \frac{\mathrm{d}M \cdot y}{I}.$$

Substituting in (a), we get

$$\tau \cdot b \cdot \mathrm{d}x = \int_{y_0}^{h} \mathrm{d}M \cdot y/I \cdot \mathrm{d}A.$$

Hence

$$\tau = \mathrm{d}M/\mathrm{d}x \cdot \int_{y_0}^{h} y/Ib \cdot \mathrm{d}A$$

but

$$\mathrm{d}M/\mathrm{d}x = F.$$

Hence

$$\tau = F/Ib \cdot \int_{y_0}^{h} y\mathrm{d}A = F/Ib \cdot A\bar{y}$$

where τ is shear stress at a distance y from N.A., F shear force, A is the sectional area above the point considered, $\bar{\mathbf{y}}$ distance of centroid of A above N.A., B thickness at a distance y from N.A., and I second moment of area of section about N.A.

7.2.1.1 Shear Stresses over a Rectangular Section

The distribution of shear stress over a rectangular section of a beam having depth d and width b and is subjected to a shearing force F is calculated as follows, see Fig. 7.4

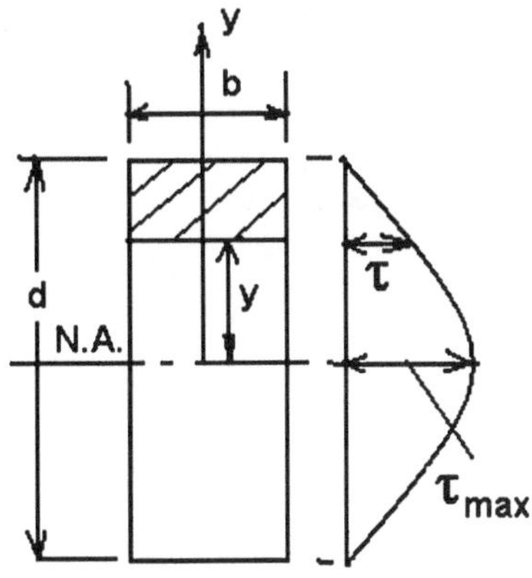

Fig. 7.4 Shear stress distribution over a rectangular section

$$\tau = F/Ib \cdot A\overline{y}$$
$$A\overline{y} = (d/2 - y) \cdot (d/2 + y)/2 \cdot b$$
$$= b/2 \cdot \left(d^2/4 - y^2\right)$$

Thus

$$\tau = F/\left(bd^3/12 \cdot b\right) \cdot b/2 \cdot \left(d^2/4 - y^2\right) = 6F/bd^3 \cdot \left(d^2/4 - y^2\right).$$

This is an equation of a parabola having zero ordinates at the extreme fibers, i.e., a distance given by

$$y = \pm d/2$$

The maximum value of the shear stress τ_{max} occurs at $y = 0$ and is given by

$$\tau_{max} = 6F/4bd = 1.5 \cdot F/A = 1.5 \cdot \tau_{max}.$$

7.2.2 Average Shear Stress

The average shear stress is given for the following three sections.

7.2.2.1 Rectangular Section

The average shear stress for a rectangular section is given by, see Fig. 7.5

$$\tau_{av} = F/A_s = F/ht$$

where τ_{av} average shear stress, A_s shear area, τ_{max} maximum shear stress.

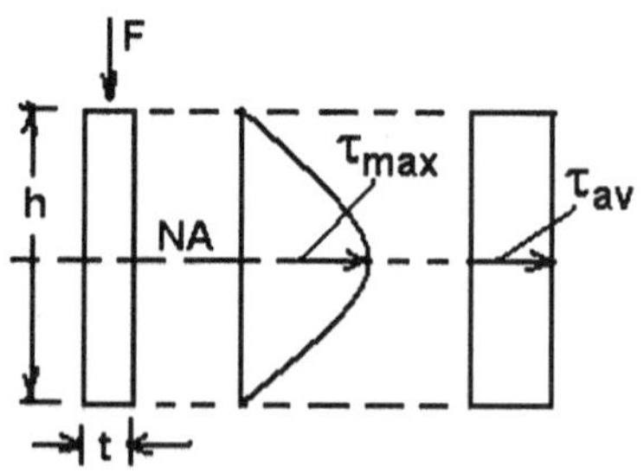

Fig. 7.5 Average shear stress: rectangular section

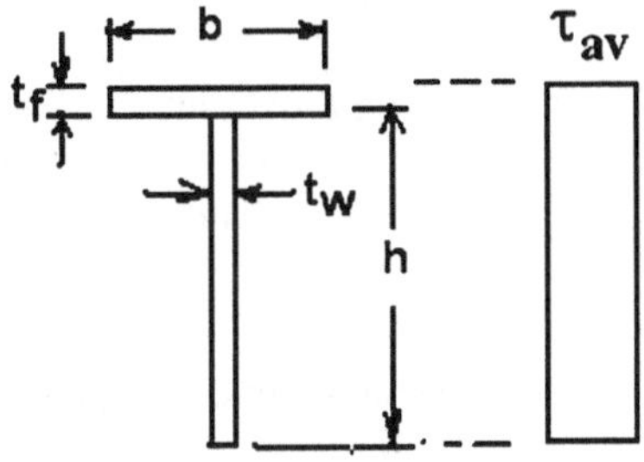

Fig. 7.6 Average shear stress over a T-section

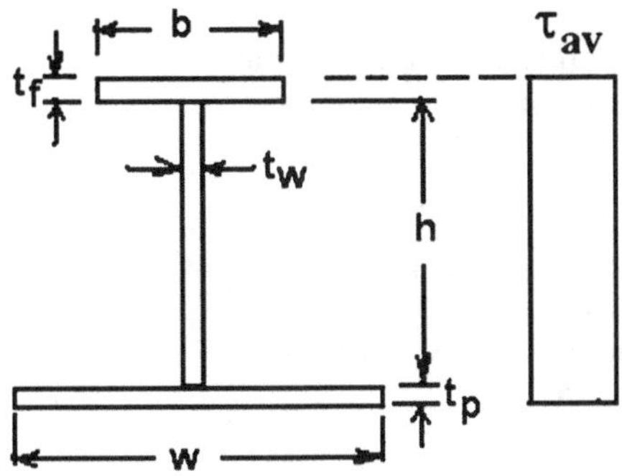

Fig. 7.7 Average shear stress over a fabricated section

7.2.2.2 T-Section

The average shear stress for a T-section is given by, see Fig. 7.6

$$\tau_{av} = F/A_s$$

$$A_s = (h + t_f) \cdot t_W.$$

7.2.2.3 I-Section

The average shear stress for a fabricated section is given by, see Fig. 7.7

$$\tau_{av} = F/A_s$$

$$A_s = (t_W + h + t_f) \cdot t_W.$$

7.2.3 Shear Flow and Stress in Thin-Walled Sections

7.2.3.1 Basic Concepts

The shear flow and stresses over thin-walled open structural sections could be calculated for thin-walled sections having one axis of symmetry using the basic equation given by, see Fig. 7.8

$$q = F/I \cdot A\bar{y}$$

The shear flow distribution over a thin-walled section having two axes of symmetry is shown in Fig. 7.9.

The shear flow over an angle section is shown in Fig. 7.10.

The shear stress distribution over an open section such as a T-section, and I-beam, see Fig. 7.8, is determined using the general equation given by

$$\tau = F/It \cdot A\bar{y}$$

The maximum value occurs at the N.A. of the section, see Fig. 7.9

$$\tau_{\max} \cong (1.1 \sim 1.2) \cdot \tau_{\text{mean}}$$

It should be noted that a vertical shear force will be carried by the web of the section.

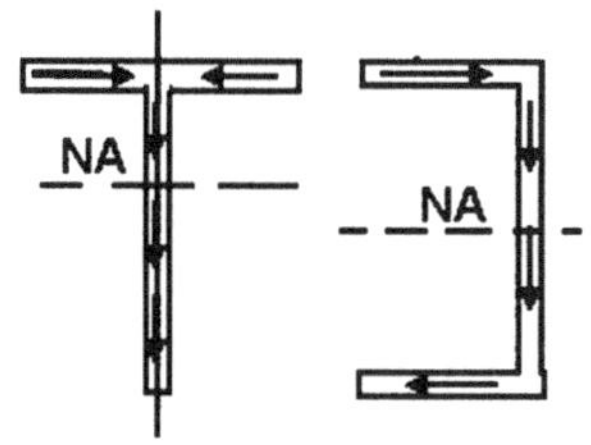

Fig. 7.8 Shear flow distribution over thin-walled structural sections with one axis of symmetry

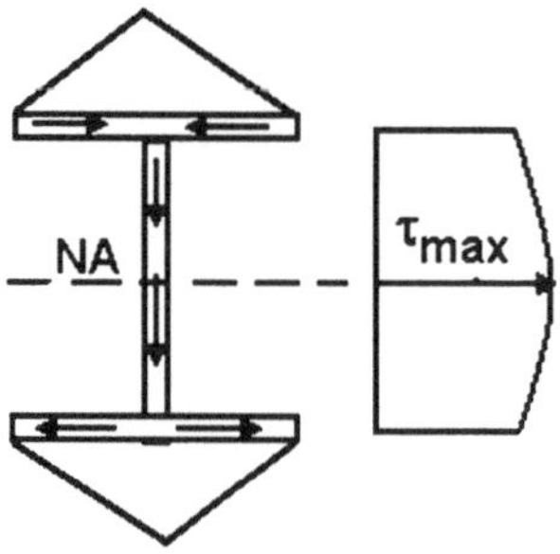

Fig. 7.9 The shear flow distribution over a thin-walled section having two axes of symmetry

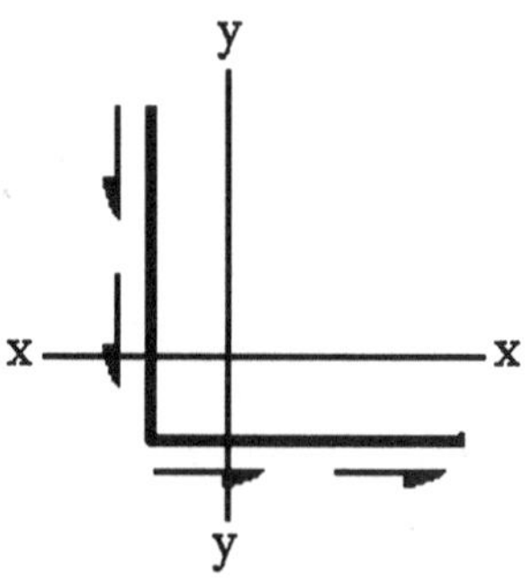

Fig. 7.10 Shear flow distribution over thin-walled structural sections having no axis of symmetry in the xy plane

7.2.3.2 Shear Flow and Stresses in Thin Walled Asymmetrical Open Sections Under Transverse Bending

Shear stresses induced over the plating thickness of the thin-walled cross-sections of a bar are determined in the same manner as for a solid rod, see Fig. 7.11.

Therefore, we have

$$\tau = FQ/I_x t$$

$$Q = \sum_i a\bar{y}$$

where F is shear free, Q first moment of area, t thickens, and I_x second moment of area.

The shearing stress τ is assumed to be uniformly distributed across the thickness "t" of the section. If the direction of the shearing force "F" does not coincide with the principal axes of the section, the shear stress is given by

$$\tau = \frac{F_y \cdot Q_x}{I_x \cdot t} + \frac{F_x \cdot Q_y}{I_y \cdot t}$$

where F_x and F_y are the components of the shear force F along the principal axes X and Y.

The calculation of the magnitude and distribution of shear stresses for thin-walled symmetrical sections could be illustrated by the following examples.

Channel Section

The shear stress distribution over the channel section shown in the Fig. 7.12 is calculated as follows

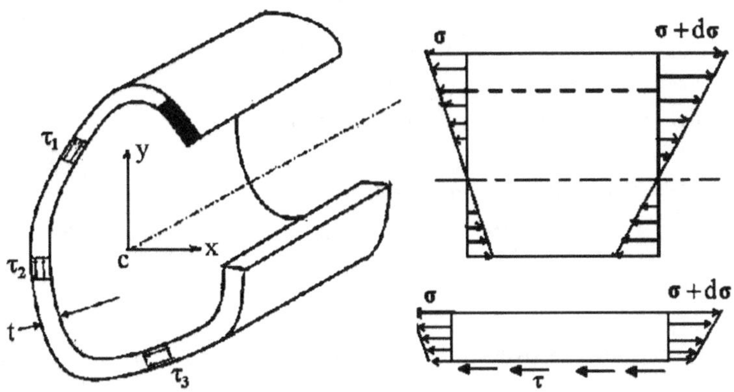

Fig. 7.11 Bending and shear stresses over a thin-walled open-section

Fig. 7.12 Shear stress distribution over a channel section

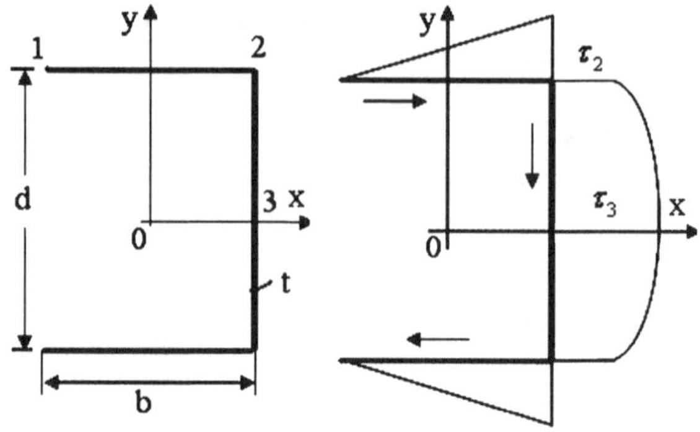

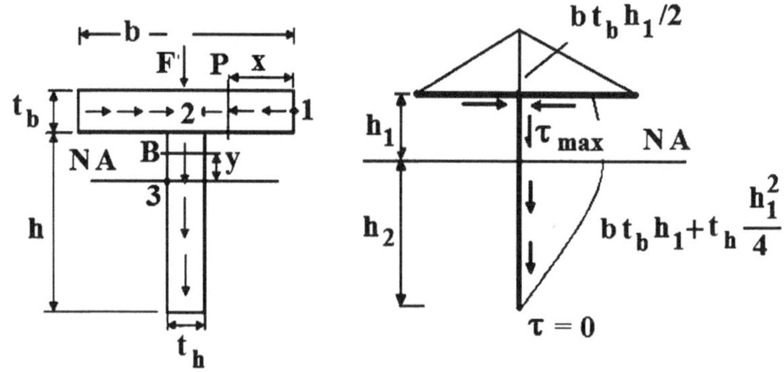

Fig. 7.13 Shear stress distribution over a T-section

$$\tau = \frac{F}{I \cdot t} \cdot A\bar{y}$$

$$(A\bar{y})_1 = 0$$

$$(A\bar{y})_2 = b \cdot t.d/2 = bdt/2$$

$$(A\bar{y})_3 = b \cdot d \cdot t/2 + d/2 \cdot t \cdot d/4 = bdt/2 + d^2t/8$$

$$I = 2bt.(d/2)^2 + td^3/12 = tbd^2/2 + td^3/12$$

Hence the shear stresses are given by: $\tau_1 = 0$

$$\tau_2 = Fbdt/2I.t = F \cdot 6b/\left(6tbd + td^2\right)$$

$$\tau_3 = F.\left(bdt/2 + d^3t/8\right)/It = F \cdot \left(6b + 3/2d^2\right)/\left(6tbd + td^2\right).$$

Shear Flow and Stresses over a Thin-Walled T-Section

The shear flow over the T-section is calculated as follows, see Fig. 7.13

$$\tau = \frac{F}{I.t} \cdot \sum a\bar{y}$$

$$\tau_1 = \tau_2 = \tau_3 = 0.0, \ldots$$

where τ_4 is the maximum value

$$\left(\sum a\bar{y}\right)_1 = 0$$

and

$$\left(\sum a\bar{y}\right)_p = x.t_b.h_1$$

$$\left(\sum a\bar{y}\right)_2 = b/2 \cdot t_b \cdot h_1 = b \cdot t_b \cdot h_1/2$$

$$\left(\sum a\bar{y}\right)_B = [bt_b h_1/2 + t_h(h_1 - y) \times (h_1/2 + y)/2]$$

$$= \left[bt_b h_1/2 + t_h/2 \cdot (h_1^2/2 + h_1 y/2 - y^2)\right]$$

$$\left(\sum a\bar{y}\right)_3 = bt_b h_1/2 + t_h h_1^2/4$$

$$\tau_{5-2} = \tau_{5-3} = \tau_{5-4}/2.$$

Shear Flow and Stresses in an Open Thin-Walled Circular Section

The shear stress distribution over the open thin-walled circular section shown in Fig. 7.14 due to a vertical shear force F is calculated as follows:

$$A\bar{y} = t \int_{\varphi}^{\Pi} R\mathrm{d}\psi \cdot R\sin\psi = R^2 t(1 + \cos\varphi)$$

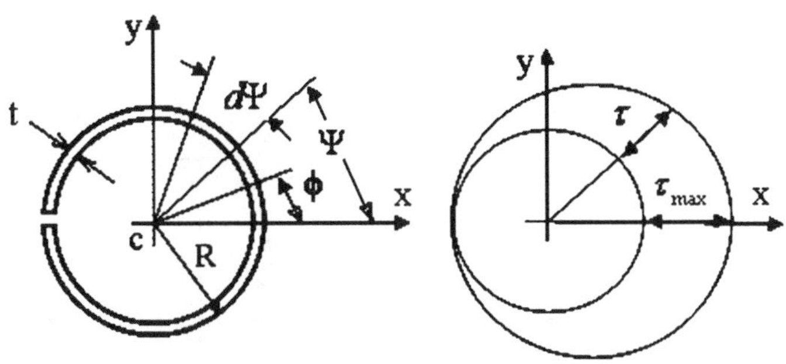

Fig. 7.14 Shear stress distribution over an open thin-walled tube section

$$I = 2tR^3 \int_0^{\pi} \sin^2\psi d\psi$$

$$= 2tR^3 [\Psi/2 - \sin 2\pi/4]_0^{\pi}$$

$$= \pi t R^3$$

Hence

$$\tau = F/\pi t R^3 t \cdot R^2 t (1 + \cos\varphi)$$

$$= F/\pi R t (1 + \cos\varphi)$$

$$\tau_{\max} = 2F/\pi R t.$$

Shear Flow and Stresses in an Angle Section with Attached Plating

The mechanism of shear stress transmission between the bottom plating, the web and the flange of the section is shown in Fig. 7.15.

The equilibrium conditions over the flange of the section are shown in Fig. 7.16. The flange is subjected to normal load N and shear stress τ. This will induce lateral bending of the flange. The resultant stress is the summation of the normal stress resulting from the in- plane load N and the bending stress resulting from the shear stress τ, see Fig. 7.16. The calculation of the maximum stress in the offset flange of the beam could be determined as follows

- The normal stress in the flange is given by

$$\sigma_n = N/bt$$

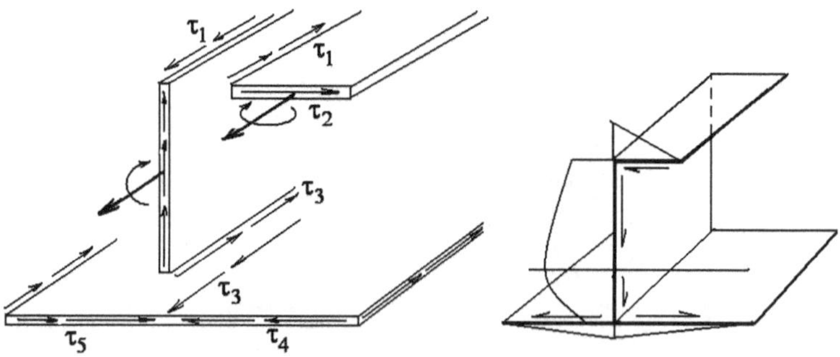

Fig. 7.15 Shear stress distribution over an open thin-walled asymmetrical section

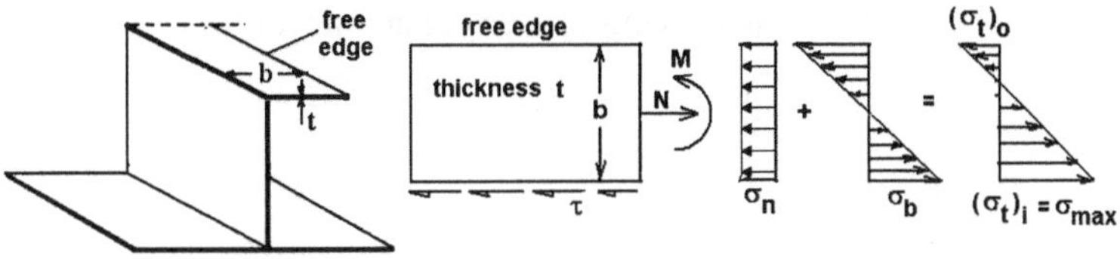

Fig. 7.16 Stresses in the offset flange of a fabricated section

- The bending moment in the flange is given by

$$M = Nb/2$$

- The in plane bending of the offset flange is given by

$$\begin{aligned}\sigma_b &= Mb/2I = Mb/\left(2tb^3/12\right) = 6M/tb^2\\ &= 6/tb^2 \cdot N \cdot b/2 = 3N/tb\end{aligned}$$

- The maximum stress in the flange is given by

$$\begin{aligned}\sigma_{max} &= (\sigma_t)_i = \sigma_n + \sigma_b\\ &= N/bt + 3N/bt = 4N/bt.\end{aligned}$$

7.2.3.3 Shear Flow and Stresses in Thin-Walled Box–Girders

The shear stress distribution over the box-section shown in Fig. 7.17 is calculated as follows

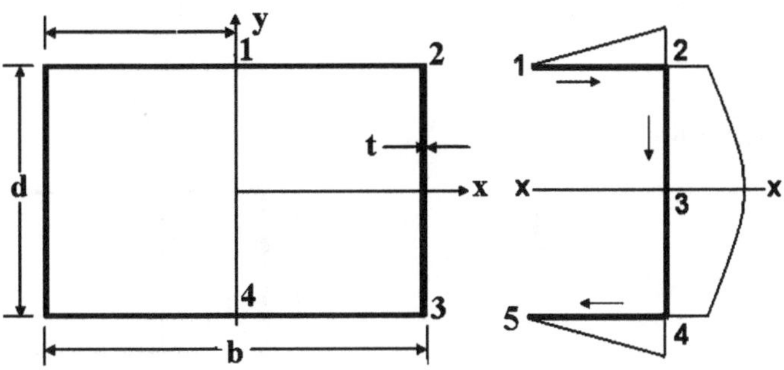

Fig. 7.17 Shear stress distribution over a thin-walled box girder

- Due to symmetry, only half the section could be used as the shear flow is zero at points 1 and 4 on the axis of symmetry.

The shear flow over the channel section is calculated as follows

$$\tau = F/It.\sum a\bar{y}$$
$$\tau_1 = \tau_5 = 0$$
$$\tau_4 = \tau_2$$
$$\sum_2 a\bar{y} = a.t.d/2 = adt/2$$
$$\sum_3 a\bar{y} = \sum_2 a\bar{y} + t \cdot d/2.d/4$$
$$= adt/2 + td^2/8$$

Then

$$\tau_2 = F/It \cdot adt/2$$
$$\tau_3 = \tau_2 + F/It.td^2/8.$$

Example 1.1 Determine the shear stress distribution over the thin-walled asymmetrical box-girder having equal thicknesses for the top and bottom sides and unequal thicknesses for the left and right webs, see Fig. 7.18.

Solution The shear flow distribution is calculated using the basic equation given by

$$q = \frac{F}{I} \cdot \sum a\bar{y}$$

Since the x-axis is an axis of symmetry, the asymmetrical box-girder could be treated as two channel sections, each having one axis of symmetry. The shear flow distribution for both channel sections is shown in Fig. 7.19.

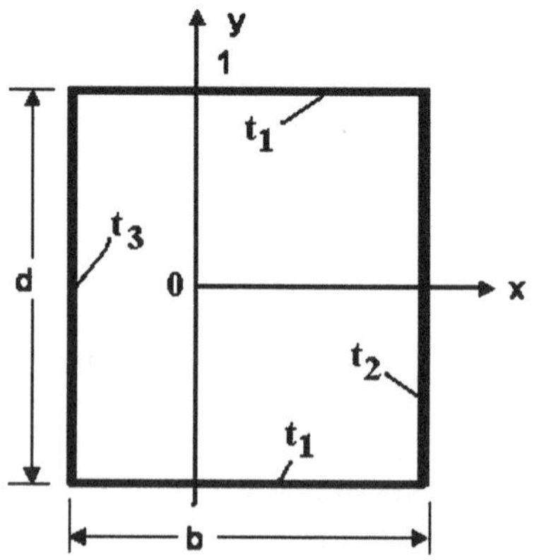

Fig. 7.18 Asymmetrical thin-walled box-girder

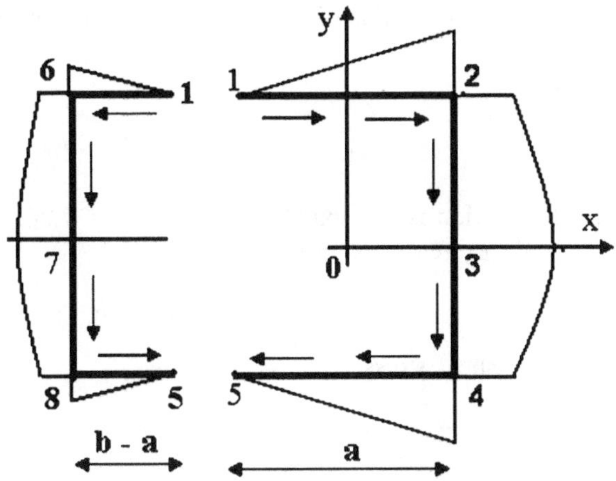

Fig. 7.19 Shear flow distribution over an asymmetrical thin-walled box-girder

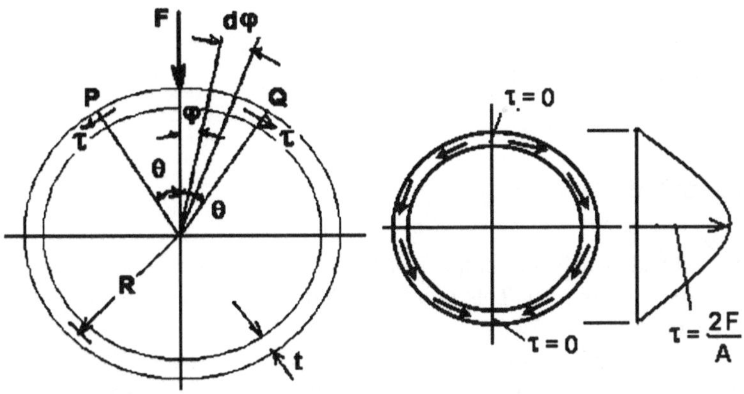

Fig. 7.20 Shear stress distribution over a thin-walled tube section

7.2.3.4 Shear Flow and Stresses in Thin-Walled Closed Tube Section

The shear stress distribution over the tube-section shown in Fig. 7.20 is calculated as follows.

Let P and Q be two symmetrically placed positions, where the shear stress is τ.

$$\tau = F/Ib \cdot A\bar{y}$$

$$A \cdot \bar{y} = \int_{-\theta}^{\theta} R \cdot t \cdot \mathrm{d}\varphi \cdot R \cdot \cos(\varphi) = 2 \cdot R^2 \cdot t.\sin(\theta)$$

$$\mathrm{I} = \int_{0}^{2\pi} R \cdot t \cdot \mathrm{d}\varphi \cdot R^2 \cdot \cos^2(\varphi) = R^3 \cdot t \cdot \int_{0}^{2\pi} \cos^2(\varphi) \cdot \mathrm{d}\varphi$$

$$= R^3 t[\varphi/2 + \sin(2\varphi)/4]_0^{2\pi} = \pi \cdot R^3 \cdot t$$

The maximum value of the shear stress occurs at the N.A. of the section where $\theta = \pi/2$.

Thus

$$\tau_{\max} = F/\pi Rt = 2F/A$$

The distribution is sinusoidal as shown in Fig. 7.20.

The shear area is given by

$$A_s = 2 \cdot \pi \cdot R \cdot t$$

The mean shear stress, t_{mean}, is given by

$$\tau_{\text{mean}} = F/2\pi Rt$$

Thus

$$\tau_{\max} = 2\tau_{\text{mean}}$$

Consider the following examples of application.

7.3 Shear Centre

It is the point at which the moment of the tangential forces for a section under transverse bending is *zero*. This point is called shear center, center of flexure or flexural center.

It should be realized that for sections having two axes of symmetry, the shear centre coincides with the centroid of the section. For asymmetrical sections, the position of the shear centre could be easily calculated using sectorial coordinates instead of Cartesian. Examples of these sections are shown in Fig. 7.21.

For sections having only one axis of symmetry, the shear centre lies on the axis of symmetry, but does not coincide with the centroid. Examples of these sections are shown in Fig. 7.22.

For sections having no axes of symmetry, the position of the shear center is as shown in Fig. 7.23.

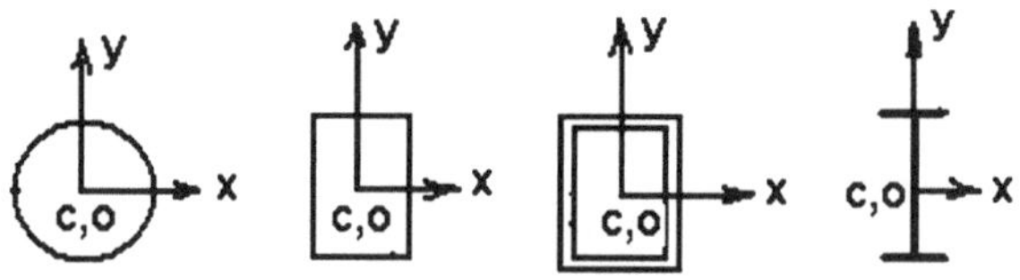

Fig. 7.21 Shear center for symmetrical thin-walled sections

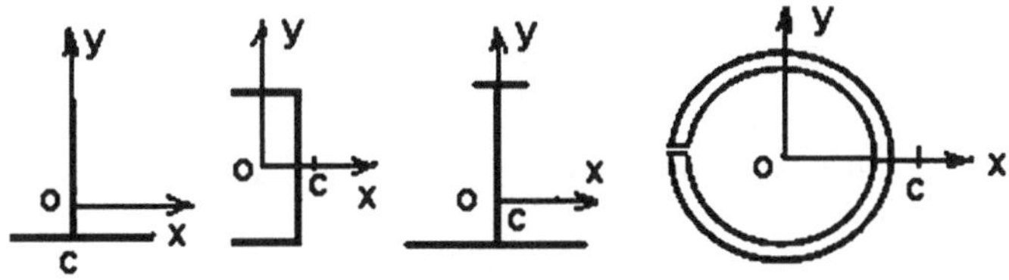

Fig. 7.22 Shear center for thin-walled sections having only one axis of symmetry

Fig. 7.23 Shear center for thin-walled sections having no axis of symmetry

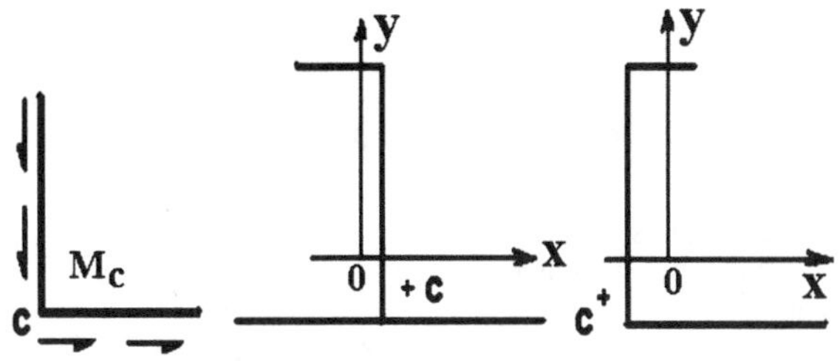

Example 1.2 Determine the position of the shear centre for the open circular section shown in Fig. 7.24.

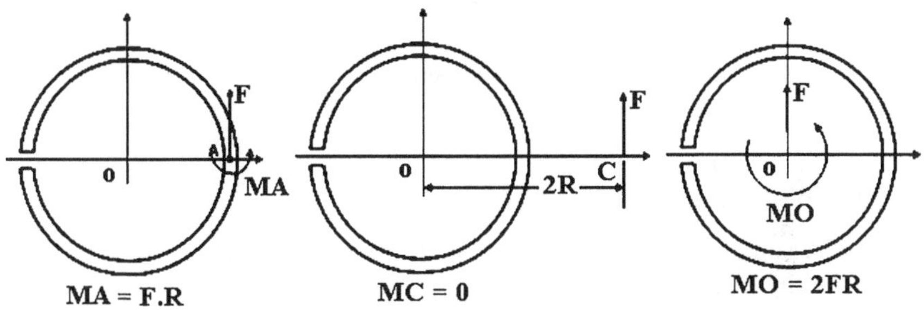

Fig. 7.24 Shear center for an open thin-walled tube section

Solution For the open circular section shown in Fig. 7.24, the shear stress is given by

$$\tau = F/\pi Rt \cdot [1 + \cos(\varphi)]$$

Hence moment of tangential forces about 0,

$$M_0 = \int_A \tau \cdot R \cdot dA = \int_{-\pi}^{+\pi} F/\pi Rt \cdot [1 + \cos(\varphi)] \cdot R \cdot Rd\varphi \cdot t$$

$$= FR/\pi \cdot \int_{-\pi}^{\pi} [1 + \cos(\varphi)]d\varphi = 2FR$$

If the moments are taken about point A

$$M_A = M_O - F \cdot R$$

The point at which the moment of the tangential forces is zero is determined from, see Fig. 7.25.

$$M_C = M_O - F \cdot x = 0.0$$

Hence

$$x = M_O/F = 2FR/F = 2R$$

i.e., C is at a distance $2R$ from 0.

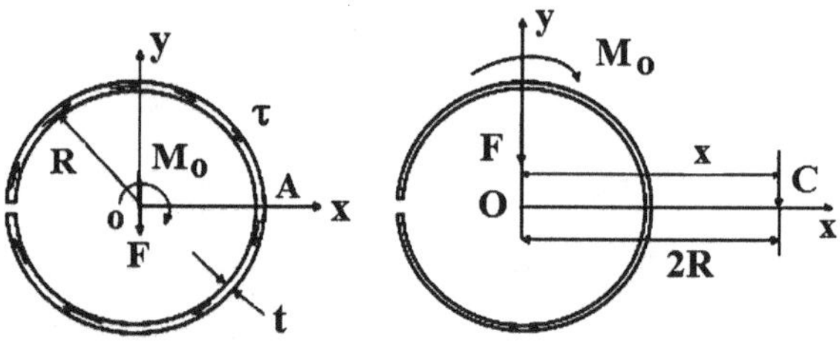

Fig. 7.25 Shear center for an open thin-walled circular section

The point C is the shear center of the section. It is sometimes called the center of flexure or flexural center.

Example 1.3 Determine the location of the shear centre for a channel section subjected to a vertical force, see Fig. 7.26.

Solution Assume that the shear center is at a distance x from the web of the section. Taking moments about point O

$$M_O = \int \tau.\mathrm{d}A.\mathrm{d}x$$

$$= F/I \cdot b^2 d^2 t/4$$

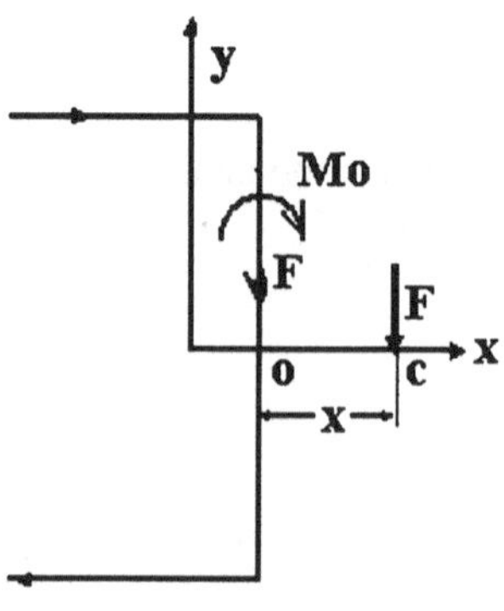

Fig. 7.26 Shear center for a thin-walled channel section

Hence, position of the shear center is given by

$$x = M_O/F$$

$$= \left(b^2d^2t/4\right)/\left(tbd^2/2 + td^3/12\right)$$

$$= 3b^2/(6b + d)$$

7.4 Shear Deflection

Shear deflection results from the sliding of adjacent cross-sections along each other, see Fig. 7.27. Consider the beam shown in Fig. 7.27 the work done by F over an element of length dx should equal the strain energy stored in the element, i.e.,

$$1/2 \cdot F \cdot \mathrm{d}y = \int_A 1/2 \cdot \tau \cdot \mathrm{d}A \cdot \gamma \mathrm{d}x$$

but

$$\gamma = \frac{\tau}{G}$$

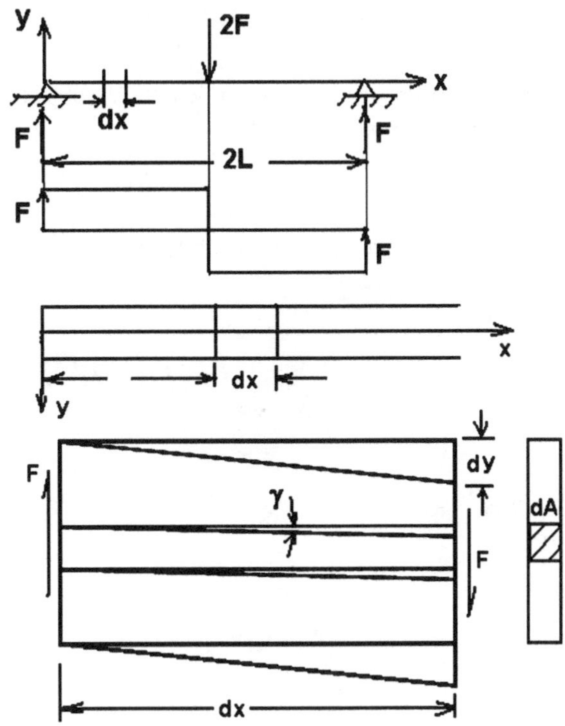

Fig. 7.27 Shear deflection of an element of a simply supported beam

Thus

$$1/2 \cdot F \cdot \mathrm{d}y = \int_{A} \tau^2/2G \cdot \mathrm{d}A \cdot \gamma \mathrm{d}x$$

Hence

$$\mathrm{d}y/\mathrm{d}x = 1/FG \cdot \int_{A} \tau^2 \mathrm{d}A = A/F^2 \cdot \int_{A} \tau^2 \mathrm{d}A$$

Let

$$A/F^2 \cdot \int_{A} \tau^2 \mathrm{d}A = k$$

We get

$$\mathrm{d}y/\mathrm{d}x = KF/GA$$

where k is the coefficient depending on the shape of the cross section.

The shear deflection is, therefore, given by

$$y_2 = \int_{0}^{x} KF/AG \cdot \mathrm{d}x$$

For a uniform section beam of length $2L$ subjected to a constant shear force F, the shear deflection is given by

$$y_2 = KFL/GA$$

Example 1.4 Obtain an expression for the beam section shown in Fig. 7.28.

Solution

$$y_2 = \int_{0}^{x} KF/AG \cdot \mathrm{d}x$$

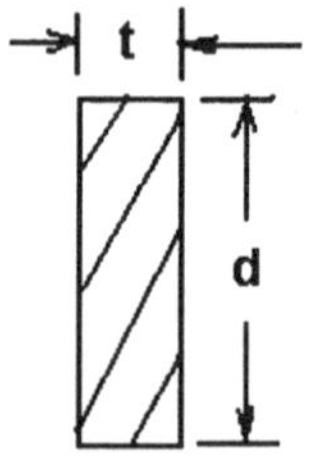

Fig. 7.28 A rectangular section

where

$$K = A/F^2 \cdot \int_A \tau^2 \mathrm{d}A$$

and

$$\begin{aligned} \tau &= F/It \cdot A\bar{y} \\ &= 6F/td^3 \cdot \left(d^2/4 - y^2\right) \end{aligned}$$

hence

$$\begin{aligned} k &= A/F^2 \cdot \int_A 36F^2/A^2d^4 \cdot \left(d^4/16 - d^2y^2/2 + y^4\right) \cdot t\mathrm{d}y \\ &= 36t/Ad^4 \cdot \left[d^4y/16 - d^2y^3/6 + y^5/5\right]_{-d/2}^{d/2} \\ &= 36t/Ad^4 \cdot \left[d^5/16 - 2d^5/(6 \times 8) + 2d^5/(5 \times 32)\right] \\ &= 36(1/16 - 1/24 + 1/80) \\ &= 1.2 \end{aligned}$$

thus

$$y_s = 1.2F\lambda/AG.$$

7.4.1 Shear Deformation

The shear deformation of a panel of plating is shown in Fig. 7.29.

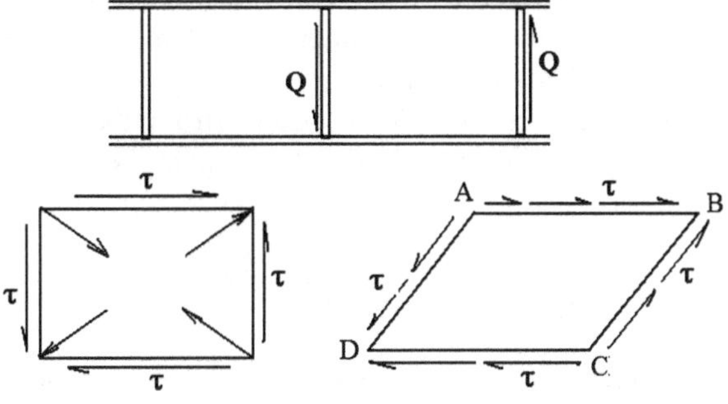

Fig. 7.29 Shear deformation of a rectangular plate subjected to pure shear loading

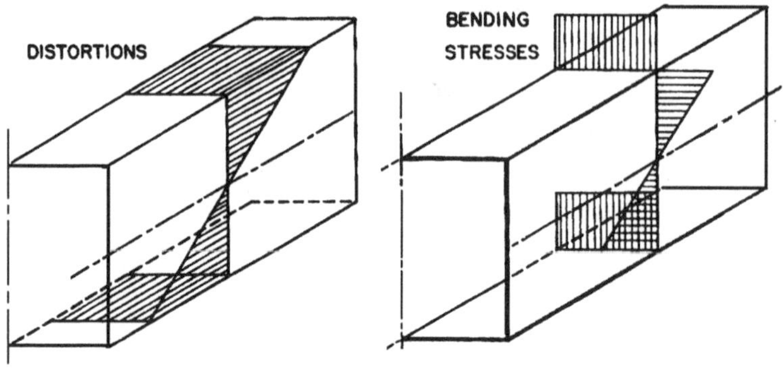

Fig. 7.30 Bending stress distribution and distortion of a plane section

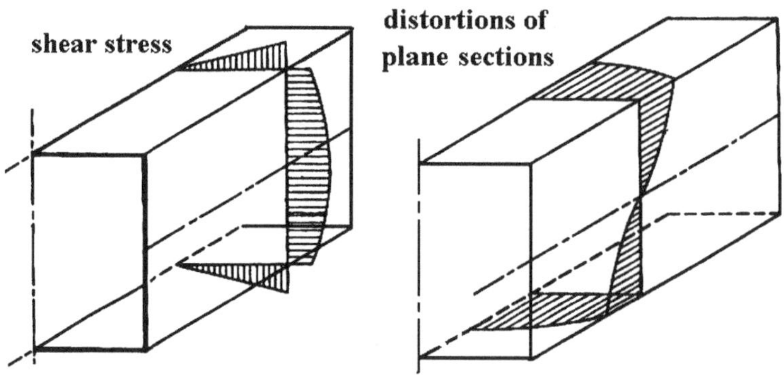

Fig. 7.31 Shear stress distribution and distortion of plane sections due to shear over a channel section

7.5 Shear Lag

The effect of shear stress, see Fig. 7.30, on bending stress, and also on the deflection of a beam, is a complex problem due to the shear lag effect, particularly for thin walled structures.

Shear lag, see Fig. 7.31, results from the influence of shear stress on the magnitude and distribution of the bending stress as calculated using the simple beam theory as given by

$$\sigma = My/I$$

In the presence of shear stresses, plane sections no longer remain plane.

Plane sections are distorted and the banding stresses calculated using the simple beam theory are no longer valid and need to be corrected, see Fig. 7.32.

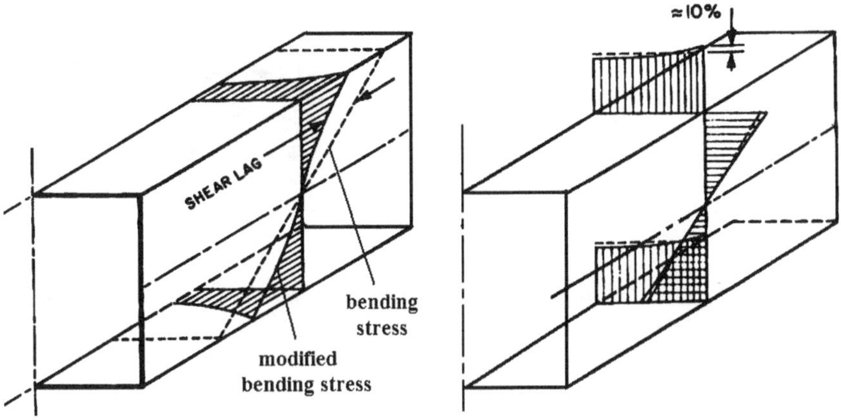

Fig. 7.32 Corrected bending stress over the channel section

Fig. 7.33 Effect of shear lag on hull girder bending stress

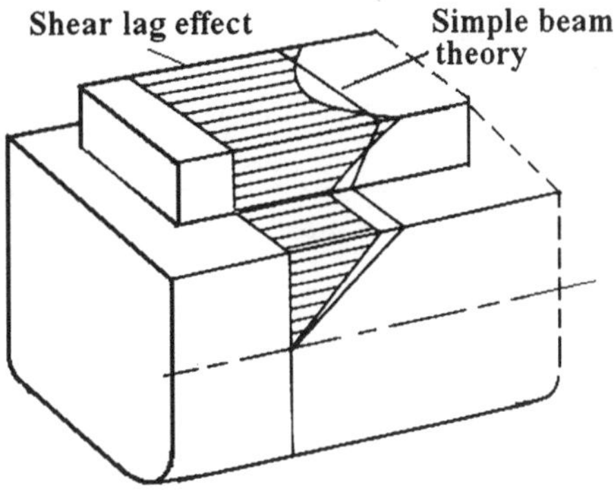

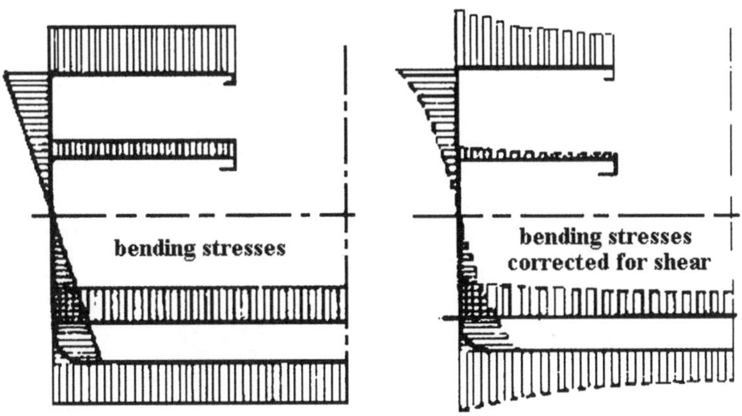

Fig. 7.34 Effect of shear lag on bending stress distribution over a ship section of a cargo ship

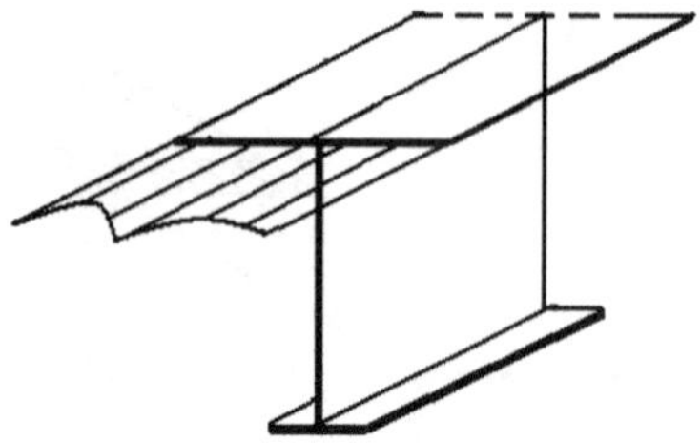

Fig. 7.35 Effect of shear lag on flange bending stress of a thin-walled girder

The modification to the bending stress arising from the presence of the shear stress is shown in Fig. 7.33.

The effect of shear lag on the hull girder bending stress distribution over a ship section is shown in Fig. 7.34.

The effect of shear lag on the stress distribution over the top flange of a T-section is shown in Fig. 7.35.

Chapter 8
Shear Flow and Stresses in Thin-Walled Box-Girders

8.1 Single Cell Box-Girder

Consider the single-cell tube shown in Fig. 8.1.

Assuming that

- Plane sections remain plane,
- The x and y axes are the principal centroidal axes of the section,
- The resultant shearing force passes through the shear center C of the section, i.e. these is no twisting.

The normal stress could then be determined from the simple beam theory i.e.

$$\sigma = M/I \cdot y$$

Let e be the origin of the coordinate S.

The equilibrium of the element gives

$$q\Delta z = q_e \Delta z - \partial P/\partial Z \cdot dz$$

where

$$q_e = \text{shear flow at } e$$

Hence

$$q = q_e - \partial P/\partial Z$$

If the section is open at e, then $q_e = 0.0$.

Thus

$$q = -\partial P/\partial Z$$

But

$$\delta p = \int_0^S \partial \sigma_Z \cdot t ds$$

M. Shama, *Torsion and Shear Stresses in Ships*,
DOI: 10.1007/978-3-642-14633-6_8, © Springer-Verlag Berlin Heidelberg 2010

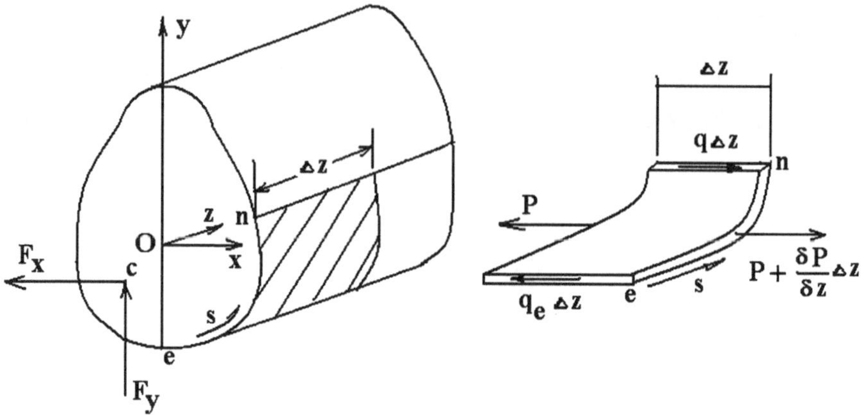

Fig. 8.1 Equilibrium condition of an element of a closed thin-walled box-girder

Hence

$$\partial P/\partial Z = \int_0^S \partial\sigma_Z/\partial Z \cdot t\mathrm{d}s = \int_0^S \mathrm{d}M/\mathrm{dZ} \cdot y/I \cdot t\mathrm{d}s = \int_0^S F \cdot y/I \cdot t\mathrm{d}s$$

$$q = F \cdot a \cdot \overline{y}/I$$

Therefore

$$q = \overline{q} + q_0$$

For a closed section under zero torque, the rate of twist should be *zero*. Hence

$$\theta = 1/2\mathrm{AG} \cdot \oint q \cdot \mathrm{d}s/t = 0$$

Substituting, we get

$$1/2\mathrm{AG} \cdot \oint (\overline{q} + q_0) \cdot \mathrm{d}s/t = 0$$

Thus, the equation of consistent deformation is given by

$$\delta_0 + \overline{q}.\delta_1 = 0$$

Hence

$$\overline{q} \cdot \oint \mathrm{d}s/t \cdot + \oint q_0 \cdot \mathrm{d}s/t = 0$$

where

$$\delta_0 = 1/\mathrm{G} \cdot \oint q_0 \cdot \mathrm{d}s/t$$

$$\delta_1 = 1/\mathrm{G} \cdot \oint \mathrm{d}s/t$$

Hence

$$\overline{q} = -\delta_0/\delta_1 = -\oint (q_0 \cdot \mathrm{d}s/t) \Big/ \oint \mathrm{d}s/t$$

where δ_0 and δ_1 are the warping flexibilities.

It is also possible to determine $\overline{q}$ from compatibility conditions as follows. The relative warping displacement is given by

$$\overline{\mu} = 1/\mathrm{G} \cdot \int \tau \cdot \mathrm{d}s - 2 \cdot \theta \cdot \omega$$

but

$$\theta = 0.0$$

Hence

$$\overline{\mu} = 1/\mathrm{G} \cdot \int \tau \cdot \mathrm{d}s$$

Substituting, we get

$$1/\mathrm{G} \cdot \oint q \cdot \mathrm{d}s/t = 1/\mathrm{G} \cdot \left[\oint q_0 \cdot \mathrm{d}s/t + \overline{q} \cdot \oint \mathrm{d}s/t \right] = 0$$

This gives

$$\overline{q} = -\oint (q_0 \cdot \mathrm{d}s/t) \Big/ \oint \mathrm{d}s/t.$$

8.2 Shear Flow in Asymmetrical Closed Box-Girders Subjected to a Vertical Shear Force F

Consider the thin-walled box-girder shown in Fig. 8.2.

The axes x–y are the principal centroidal axes. Assume the shear flow at A is zero, i.e.

$$q_{\mathrm{AB}} = q_{\mathrm{AH}} = 0$$

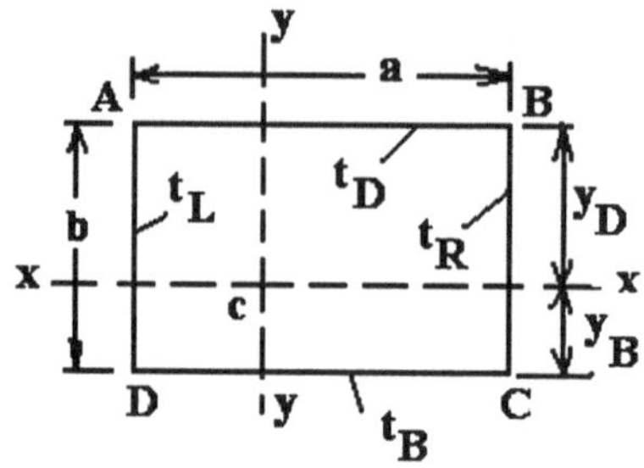

Fig. 8.2 An asymmetrical closed thin-walled box-girder

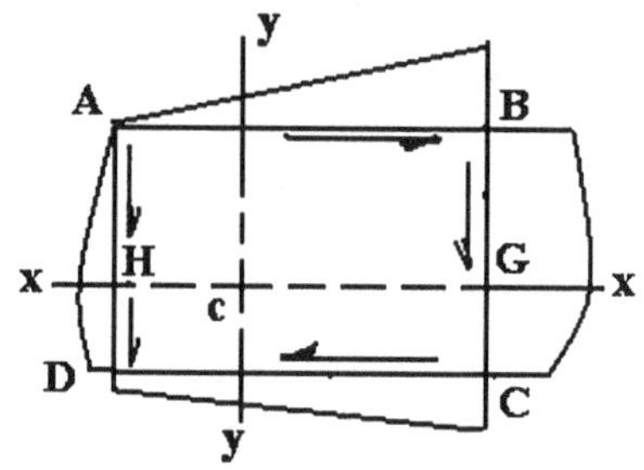

Fig. 8.3 Assumed shear flow distribution over the asymmetrical section

Then, the assumed shear flow around the closed section is as shown in Fig. 8.3. Thus

$$q_{BA} = F/I \cdot (a \cdot t_D \cdot y_D)$$

$$q_{BG} = q_{BA}$$

$$q_{GB} = q_{BG} + F/I \cdot (t_B \cdot y_D/2)$$

$$q_{GC} = q_{GB}$$

$$q_{CG} = q_{Gb} - F/2 \cdot \left(t_B \cdot y_B^2/2\right)$$

$$q_{CD} = q_{CG}$$

$$q_{DC} = q_{CD} - F/I \cdot (at_B \cdot y_B)$$

$$q_{DH} = q_{DC}$$

$$q_{HD} = q_{HA}$$

$$q_{HA} = F/I \cdot \left(t_L \cdot y_D^2/2\right)$$

The assumed shear flow will induce an angle of twist given by

$$\theta_a = 1/2\text{AG} \cdot \oint q\text{d}s/t$$

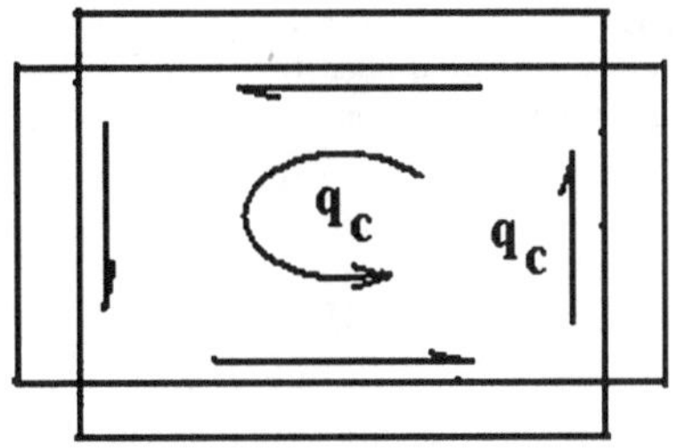

Fig. 8.4 Correcting shear flow distribution over the asymmetrical section

As the closed section is not subjected to an external torque *T*, a correcting shear flow q_c should be applied to cancel the angle of twist θ_a, see Fig. 8.4.

The correcting angle of twist θ_c is given by

$$\theta_c = q_C/2AG \cdot \oint ds/t$$

Since

$$\theta_C = -\theta_a$$

Then

$$q_c = -\oint q\frac{ds}{t} \Big/ \oint \frac{ds}{t}$$

$$q = q_a - q_c$$

$$q^C_{AB} = -q_C$$

$$q_{AH} = -q_c$$

$$q^C_{BA} = q_{BA} - q_C$$

$$q^C_{GB} = q_{GB} - q_C$$

$$q^C_{CB} = q_{CB} - q_C$$

$$q^C_{DC} = q_{DC} - q_C$$

$$q^C_{DH} = q_{DH} + q_C$$

$$q^C_{HD} = q_{HD} + q_C$$

Hence, the corrected shear flow distribution is given by, see Fig. 8.5.

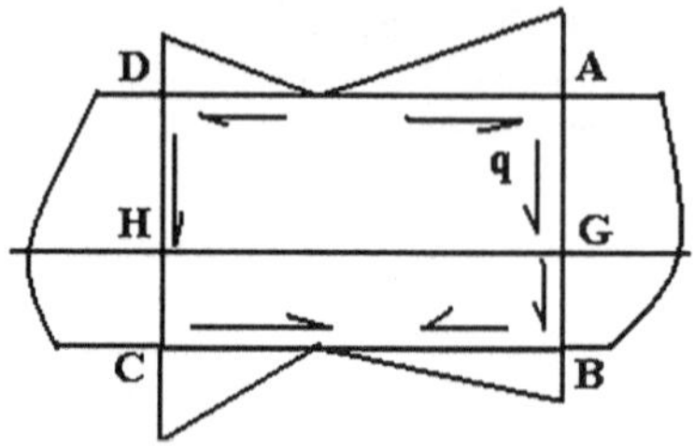

Fig. 8.5 Corrected shear flow distribution over the asymmetrical section

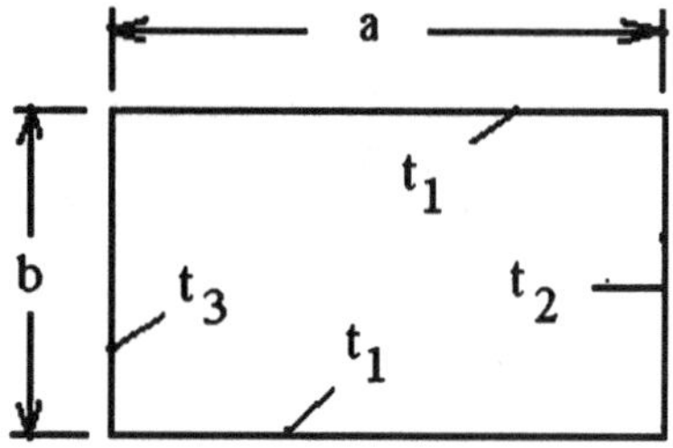

Fig. 8.6 Asymmetrical thin-walled box-girder

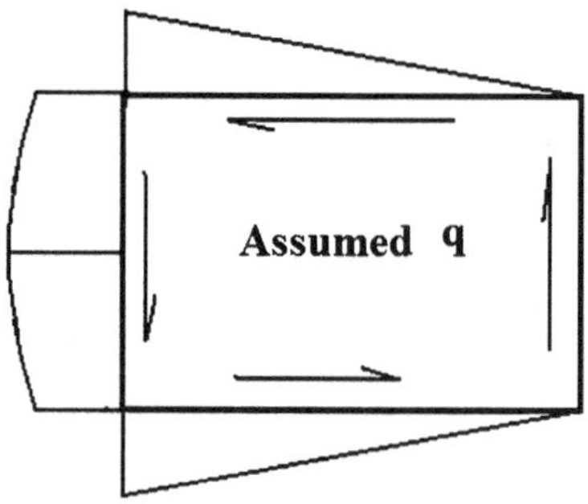

Fig. 8.7 Assumed shear flow distribution over the thin-walled box-girder

Example 8.1 Determine the shear flow distribution for the box-girder shown in the Fig. 8.6.

Solution Assume zero angle of twist.

$$\oint \Delta S/t = 2a/t_1 + b(1/t_2 + 1/t_3)$$

The assumed shear flow is shown in Fig. 8.7 and is given in Table 8.1. The correcting shear flow over the section of the box-girder is calculated as follows, see Fig. 8.8. Thus

$$\oint q \cdot \Delta S/t = a^2b/4 + ab^2t_1/4t_3 + b^3/24 + ab^2t_1/4t_3 + b^3/24$$
$$- b^3/24 - b^3/24 + a^2b/4 = a^2b/2 + ab^2t_1/2t_3$$

Hence

$$\bar{q} = -\left(a^2b/2 + ab^2t_1/2t_3\right)/\left[2a/t_1 + b(t_2 + t_3)/t_2t_3\right]$$

Table 8.1 Procedure of calculating shear flow

Member	q	q_m	$q_m \cdot \Delta s/t$
1–6	0.0	$abt_1/4$	$ab^2/4$
6–1	$abt_1/2$		
6–5	$abt_1/2$	$abt_1/2 + b^2t_3/12$	$ab^2t_1/4t_3 + b^3/24$
5–6	$abt_1/2 + b^2t_3/8$		
4–5	$abt_1/2$	$b^2t_3/12$	$ab^2t_1/4t_3 + b^3/24$
1–2	0.0	$b^2t_3/12$	$b^3/24$
2–1	$b^2\ t_2/8$		
3–2	0.0	$b^2t_3/12$	$b^3/24$
4–3	$abt_1/2$	$abt_1/4$	$ab^2/4$

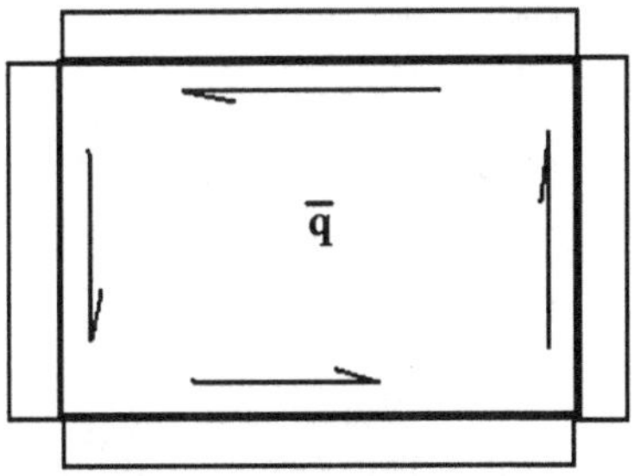

Fig. 8.8 Correcting shear flow distribution over the thin-walled box-girder

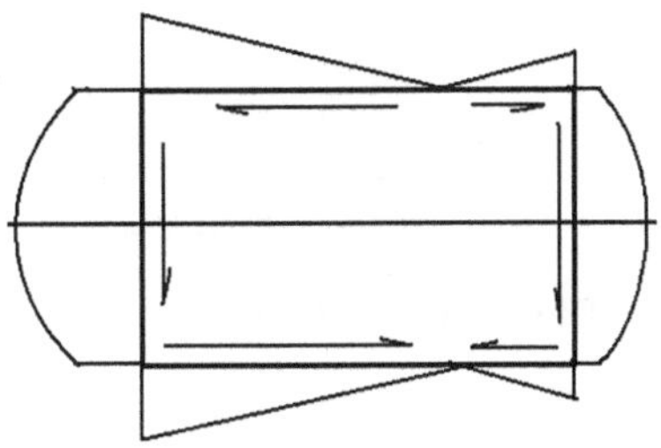

Fig. 8.9 Corrected shear flow distribution

where

$$t_1 = t_2 = t_3 = t$$

$$\begin{aligned}\overline{q} &= -\left(a^2b/2 + ab^2t_1/2t_3\right)/\left(2a/t + b2t/t^2\right)\\ &= -[ab(a+b)]/[4(a/t+b/t)] = -abt/4.\end{aligned}$$

The corrected shear flow distribution over the section of the box-girder is the summation of Figs. 8.7 and 8.8, see Fig. 8.9.

Example 8.2 Determine the shear flow distribution around the box girder shown in the Fig. 8.10 when it is subjected to shear loading without torsion.

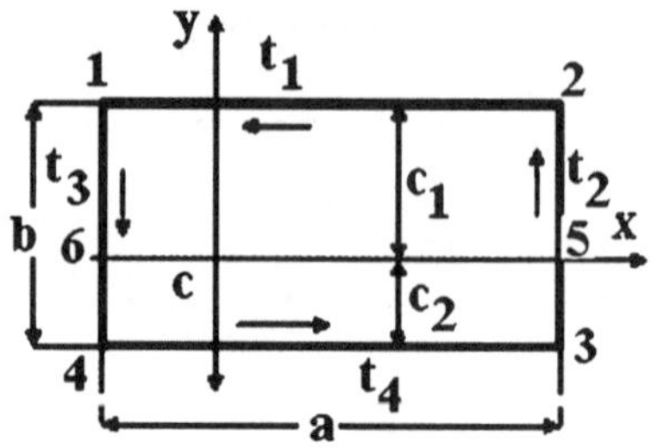

Fig. 8.10 Assumed shear flow distribution over an asymmetrical box-girder

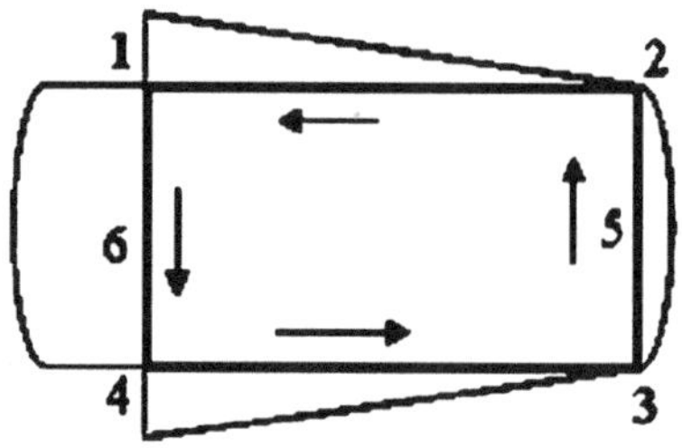

Fig. 8.11 Assumed shear flow distribution over an asymmetrical box-girder

Solution The axes x, y are principal centroidal axes and c is the shear centre of the section. The correcting shear flow q is given by, see Fig. 8.11.

$$\overline{q} = -\oint q\frac{\Delta s}{t} \Big/ \oint \frac{\Delta s}{t}$$

$$\oint \Delta s/t = b(1/t_2 + 1/t_3) + a(1/t_1 + 1/t_4)$$

The assumed shear flow is given in Table 8.2

Table 8.2 Calculation of the assumed shear flow

Member	q	q_m	$q_m \cdot \Delta s/t$
2–1	0	$A_{12} \cdot c_1/2$	$A_{12} \cdot c_1/2 \cdot a/t_1$
1–2	$A_{12} \cdot c_1$		
1–6	$A_{12} \cdot c_1$	$A_{12}c_1 + 2/3 \cdot A_{16} \cdot c_1/2$	$(A_{12}c_1 + 2/3 \cdot A_{16} \cdot c_1/2) \cdot c_1/t_3$
6–1	$q_{16} + A_{16}.C_1/2$		
4–6	$q_{61} - A_{46} \cdot c_2/2$	$q_{61} - 1/3 \cdot A_{46} \cdot c_2/2$	$(q_{61} - 1/3 \cdot A_{46} \cdot c_2/2) \cdot c_2/t_3$
4–3	q_{46}	$q_{46}/2$	$q_{46}/2 \cdot a/t_4$
3–4	0		
2–5	0	$2/23 \cdot A_{25} \cdot c_1/2$	$2/3 \cdot A_{25} \cdot c_1/2 \cdot c_1/t_2$
5–2	$A_{25}.c_1/2$		
3–5	0	$A_{35} \cdot c_2/3$	$A_{35} \cdot c_2/3t_2$

$$\oint q \cdot \Delta S/t = \mathrm{A}_{12} \cdot c_1/2 \cdot a/t_1 + (\mathrm{A}_{12}c_1 + 2/3 \cdot \mathrm{A}_{16} \cdot c_1/2) \cdot c_1/t_3$$
$$+ (q_{61} - 1/3 \cdot \mathrm{A}_{46} \cdot c_2/2) \cdot c_2/t_3 + q_{46}/2 \cdot a/t_4$$
$$+ 2/3 \cdot \mathrm{A}_{25} \cdot c_1/2 \cdot c_1/t_2 + 2/3 \cdot \mathrm{A}_{35} \cdot c_2/2 \cdot c_2/t_2$$

Hence the correcting shear flow is calculated by, see Fig. 8.12

$$\overline{q} = -\oint q\frac{\Delta s}{t} \Big/ \oint \frac{\Delta s}{t}$$

Fig. 8.12 Correcting shear flow over the asymmetrical box-girder

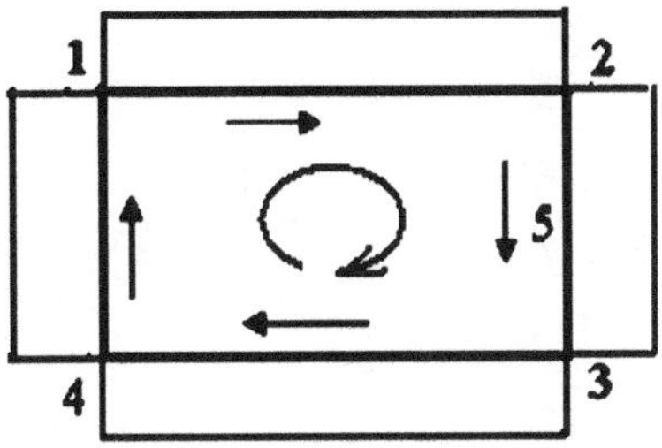

Example 8.3 Determine the shear flow distribution over the combined closed and open section shown in Fig. 8.13.

Solution Due to symmetry, only half the structural section need be considered, see Fig. 8.14.

Fig. 8.13 Combined closed and open thin- walled section

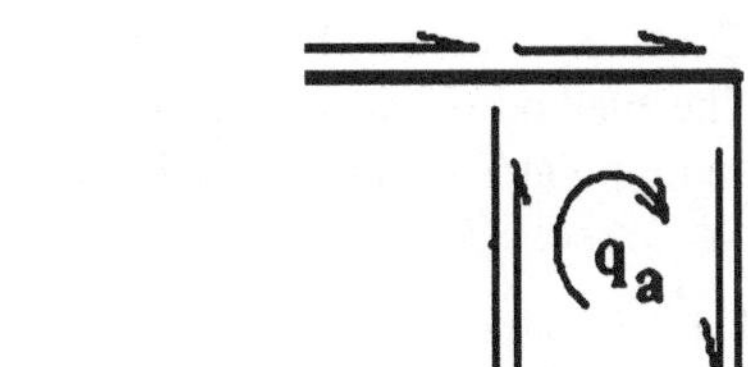

Fig. 8.14 The assumed shear flow over half the thin-walled section

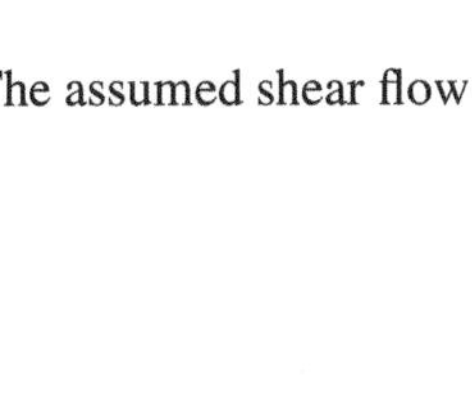

Assume a cut at the top left corner of the box structure. The assumed shear flow qa will cause a twist angle θ_a, given by

$$\theta_a = \frac{1}{2\mathrm{AG}} \oint q\frac{\mathrm{d}s}{at}$$

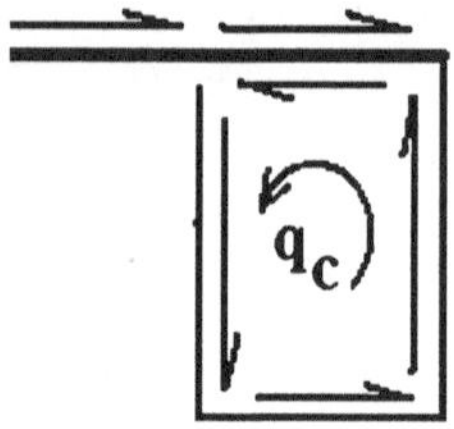

Fig. 8.15 The correcting shear flow over half the thin-walled section

A correcting shear flow qc, see Fig. 8.15, is introduced to create an opposite angle of twist θ_c so as to eliminate the induced angle of twist θ_a.

$$\theta_c = \frac{q_c}{2AG} \oint q\frac{ds}{t}$$

Hence

$$\theta_a + \theta_c = 0.0$$

$$\frac{1}{2AG} \oint q_a ds/t = -\frac{qc}{2AG} \oint ds/t$$

The correcting shear flow is therefore given by

$$q_c = -\oint q\frac{ds}{t} \bigg/ \oint \frac{ds}{t}$$

The shear flow at any point i on the closed section is given by

$$q_i = q_{ai} - q_c.$$

8.3 Shear Stresses in Thin-Walled Two-Cell Box-Girders

Consider a two cell box-girder, see Fig. 8.16, subjected to a vertical shear force F. The method of calculation is carried out by assuming the shear flow $q = 0.0$ at some point.The shear flow is calculated using the following equation

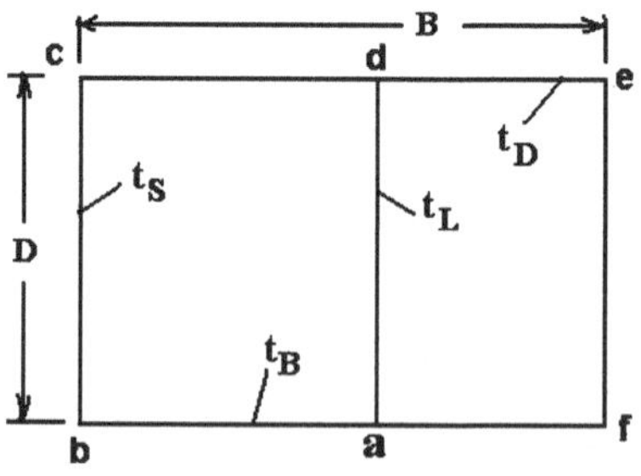

Fig. 8.16 Idealized section of a 2-cell thin-walled structure

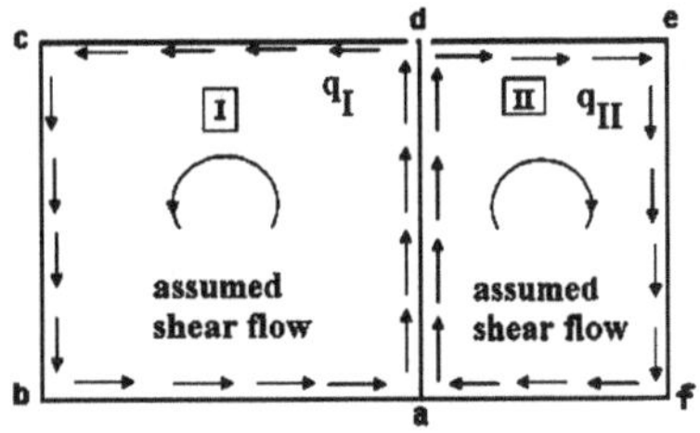

Fig. 8.17 Assumed shear flow distribution over the two-cell structure

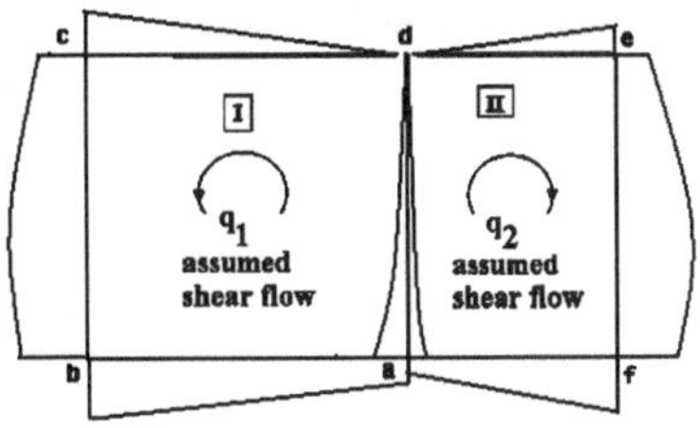

Fig. 8.18 Assumed shear flow distribution over the two cell structure

$$q = \frac{F}{It}\sum A\bar{y}$$

where

$$F = \text{shear force}$$

In order to solve this indeterminate structure, a cut is assumed at point dm see Fig. 8.17, to transform the two closed cells into open cells. The shear flow at the cut point is zero.

Assume shear flow in cell I = q_1
Assume shear flow in cell II = q_2

The assumed shear flow distribution is shown in Fig. 8.18. Since the box-girder is not subjected to any torsional moments, the angle of twist induced by the assumed shear flow distribution must be zero. This could be achieved by introducing a correcting torque that will produce a correcting shear flow that will induce an equal and opposite angle of twist.

Assume correcting shear flow in cell I = q_{c1}, see Fig. 8.19
Assume correcting shear flow in cell II = q_{c2}

The angle of twist resulting from the assumed and correcting shear flows must be equal.

$$\text{i.e}\quad \theta_1 - \theta_I = 0$$

$$\theta_2 - \theta_{II} = 0$$

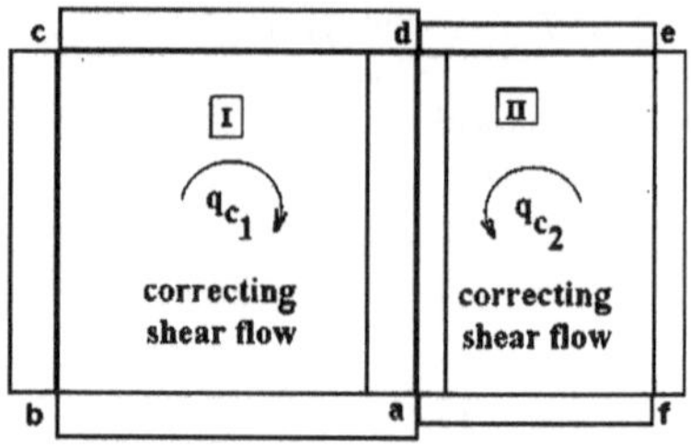

Fig. 8.19 Correcting shear flow distribution over the two cell structure

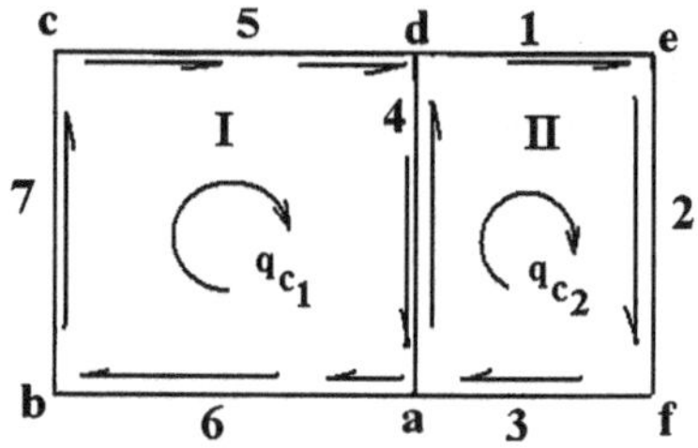

Fig. 8.20 Correcting shear flow distribution over the two cell structure

where

θ_1 = angle of twist induced by the assumed shear flow distribution in cell I
θ_{I} = angle of twist induced by the correcting shear flow distribution in cell I
θ_2 = angle of twist induced by the assumed shear flow distribution in cell II
θ_{II} = angle of twist induced by the correcting shear flow distribution in cell II.

Hence for cell I we have

$$q_{c1}\sum\cdot\Delta s/t\cdot\left\{\begin{matrix} a\rightarrow b\\ b\rightarrow c\\ c\rightarrow d\end{matrix}\right\}+(q_{c1}-q_{c2})\times\sum\cdot\Delta s/t\cdot\{d\rightarrow a\}-\sum q_{\mathrm{I}}\cdot\Delta s/t=0 \tag{8.1}$$

And for cell II we have

$$q_{c2}\sum\cdot\Delta s/t\cdot\left\{\begin{matrix} d\rightarrow e\\ e\rightarrow f\\ f\rightarrow a\end{matrix}\right\}+(q_{c2}-q_{c1})\times\sum\cdot\Delta s/t\cdot\{a\rightarrow d\}-\sum q_{\mathrm{II}}\cdot\Delta s/t=0 \tag{8.2}$$

where q_{c1} and q_{c2} are the correcting shear flows in cells I and II. Solving Eqs. (8.1) and (8.2) q_{c1} and q_{c2} can be determined.

Solving Eqs. (8.1) and (8.2) q_{c1} and q_{c2} can be determined.

The correcting shear flows, see Fig. 8.20, are given by

$$q_{\mathrm{C}}=-\oint q\frac{\Delta s}{t}\Big/\oint\frac{\Delta s}{t}=-\Delta/\delta$$

Thus

$$q_c \delta = -\Delta$$

where

$$q_c = (q_{c1} \rightarrow q_{c2})$$

$$\Delta = (\Delta_1 \rightarrow \Delta_2)$$

and

$$\delta = \begin{bmatrix} \delta_{11} & \delta_{12} \\ \delta_{21} & \delta_{22} \end{bmatrix}$$

where

$$\Delta_1 = \sum_{i=1}^{4} (q\lambda/t)_\mathrm{i} \quad \text{for cell I}$$

$$\Delta_2 = \sum_{i=4}^{7} (q\lambda/t)_\mathrm{i} \quad \text{for cell II}$$

$$\delta_{11} = \sum_{i=1}^{4} (\lambda/t)_\mathrm{I} \quad \text{for cell I}$$

$$\delta_{12} = \delta_{21} = -(\lambda/t) \quad \text{for member between cell I and II}$$

$$\delta_{22} = \sum_{i=1}^{4} (\lambda/t)_\mathrm{II} \quad \text{for cell II}$$

where q = assumed shear flow distribution; l_i = length of member "I"; t_i = thickness of member "I"; $\delta = \oint \Delta s/t$; $\Delta = \oint q\Delta s/t$.

The corrected shear flow at any point j in cell 1 is given by

$$q_{1j} = \left(q_{aj} - q_{c1}\right).$$

And the corrected shear flow at any point j in cell 2 is given by

$$q_{2j} = \left(q_{aj} - q_{c2}\right).$$

The distribution of the corrected shear flow over the thin-walled section is shown in Fig. 8.21. The method of calculation is summarized in Tables 8.3 and 8.4

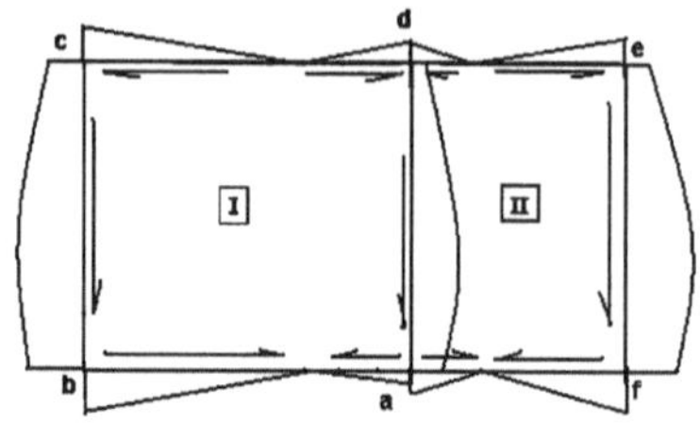

Fig. 8.21 Corrected shear flow distribution over the two cell structure

Table 8.3 Calculation of the assumed shear flow distribution

Member	Dimensions	lever $\bar{y}$,m	Area A (cm)^2	$A\bar{y}$ (m m^2)	$\sum A\bar{y}$ (m cm^2)	cm	$\sum \frac{A\bar{y}}{t}$ (m cm^2)	$\frac{F}{I} \cdot \sum \frac{A\bar{y}}{t}$ (m/cm^2)

Table 8.4 Calculation of the correcting shear flow

Member	Dimensions	Lever $\bar{y}$ (m)	Area A (cm^2)	$A\bar{y}$ (m cm^2)	Assumed $\sum A\bar{y}$	Aver. q'	$\frac{\Delta S}{t}$	$q'\frac{\Delta S}{t}$	q' Final	q

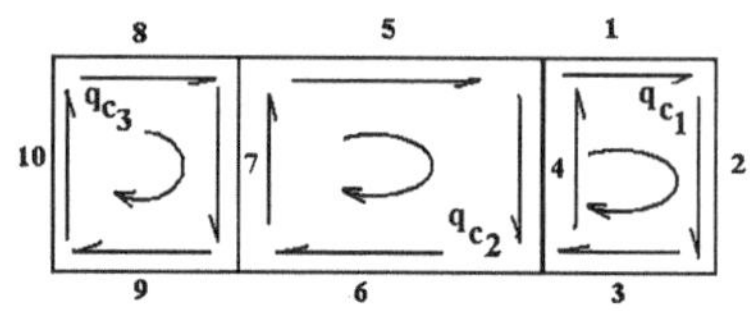

Fig. 8.22 Correcting shear flow distribution over a 3-cell structure

8.4 Calculation of the Correcting Shear Flow for 3-Cell Box-Girders Subjected to Shear Load

The correcting shear flows are given by, see Fig. 8.22

$$q_C = -\oint q\frac{\Delta s}{t} \Big/ \oint \frac{\Delta s}{t} = -\Delta/\delta$$

Thus

$$q_c\delta = -\Delta$$

where

$$q_c = (q_{c1} \cdots q_{c2} \rightarrow q_{c3})$$

$$\Delta = (\Delta_1 \ldots \Delta_2 \rightarrow \Delta_3)$$

$$\Delta_1 = \sum_{i=1}^{4} (q\lambda/t)_{\mathrm{i}}$$

$$\Delta_2 = \sum_{i=4}^{7} (q\lambda/t)_{\mathrm{i}}$$

$$\Delta_3 = \sum_{7}^{10} (q\lambda/t)_{\mathrm{i}}$$

$$\delta = \begin{bmatrix} \delta_{11} & \delta_{12} & 0 \\ \delta_{21} & \delta_{22} & \delta_{23} \\ 0 & \delta_{32} & \delta_{33} \end{bmatrix}$$

where

$$\delta_{11} = \sum_{i=1}^{4} (\lambda/t)_{\mathrm{i}}$$

$$\delta_{22} = \sum_{i=4}^{i=7} (\lambda/t)_{\mathrm{i}}$$

$$\delta_{12} = \delta_{21} = (\lambda/t)_4$$

and

$$\delta_{23} = \delta_{32} = (\lambda/t)_7$$

$$\delta_{33} = \sum_{c=7}^{i=10} (\lambda/t)_{\mathrm{i}}.$$

Chapter 9
Shear Flow and Stresses in Ships

9.1 Introduction

The importance of calculating shear stress distribution for different ship types could be demonstrated by calculating the equivalent stresses at certain locations over the ship section. For certain types of ships such as container ships and bulk carriers, the shear stress could reach high values at certain sections along the ship length and also at certain locations over the ship depth.

It should be emphasized here that measurements taken on actual ships have shown good agreement with the results obtained from using the presented s methods of calculation of shear flow distribution over ship sections, see Fig. 9.1.

9.2 Procedure of Calculation of Shear Flow Distribution

The procedure of calculation of shear flow distribution over a ship section is based on the following main steps.

9.2.1 Ship Section Idealization

9.2.1.1 General Cargo Ships

Figure 9.2 shows a longitudinal profile of a general cargo ship having four cargo holds and aft accommodation and machinery space. Figure 9.3 shows a ship section of a twin-deck general cargo ship. The bottom and decks structure are longitudinally framed.

Figure 9.4 shows a simplified idealized section of a transversely framed general cargo ship. The idealization, in this case uses combined open and closed sections.

M. Shama, *Torsion and Shear Stresses in Ships*,
DOI: 10.1007/978-3-642-14633-6_9,

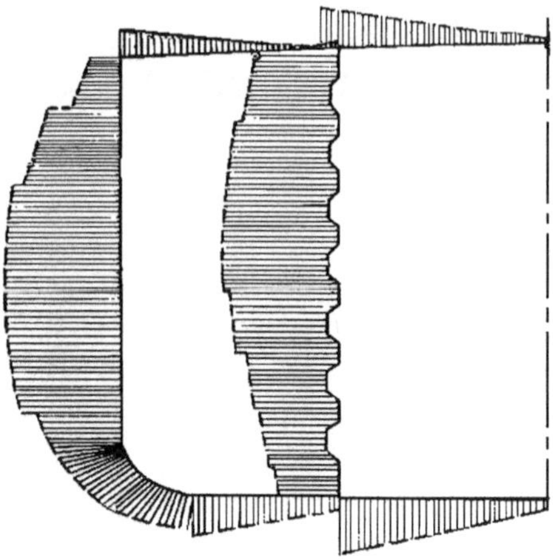

Fig. 9.1 Calculated and measured shear stresses over a ship section of a conventional oil tanker

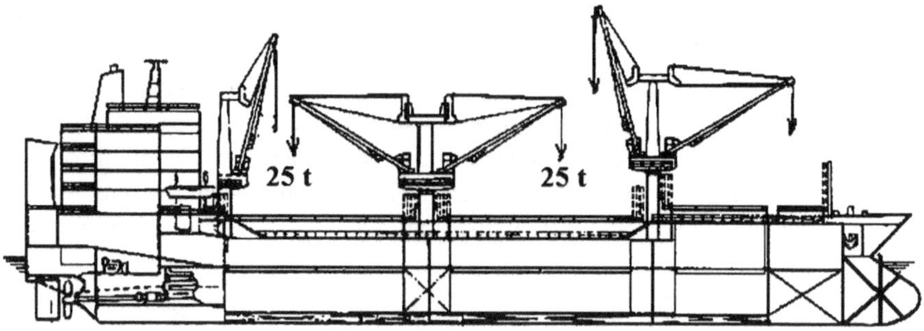

Fig. 9.2 A longitudinal section of a general cargo ship

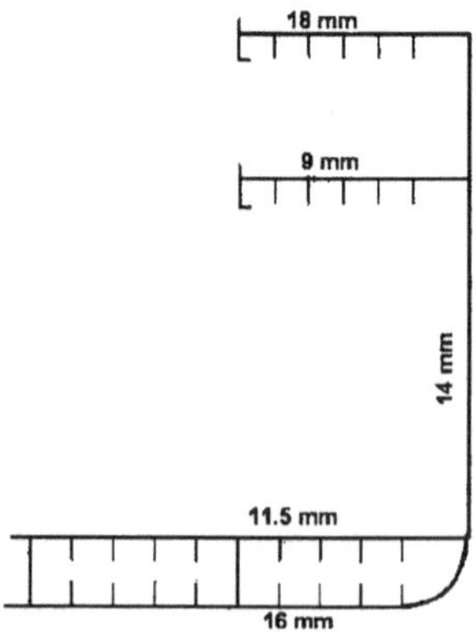

Fig. 9.3 A ship section of a general cargo ship

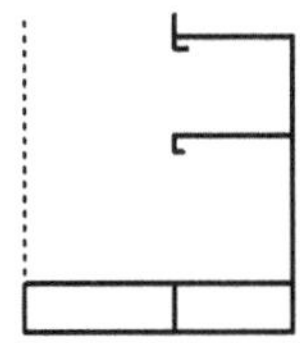

Fig. 9.4 Idealized section of a general cargo ship using open and closed sections

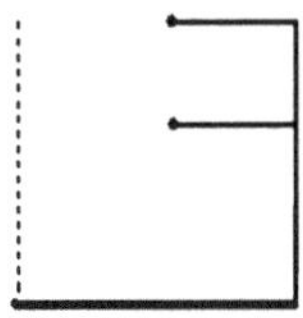

Fig. 9.5 A simplified idealized section of a general cargo ship using only open sections

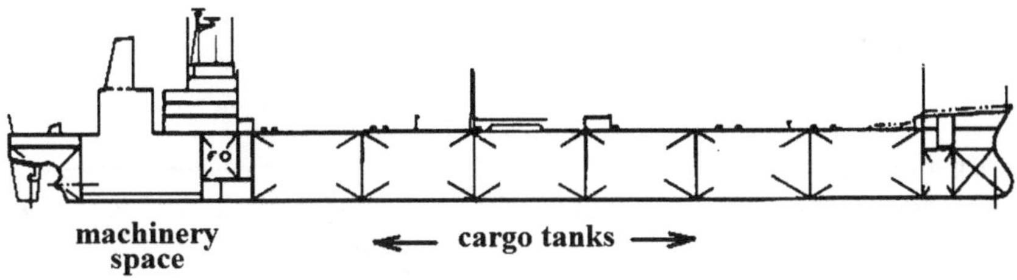

Fig. 9.6 A longitudinal section of a conventional oil tanker

However due to symmetry, the calculation of shear flow is greatly simplified by using only half ship section. The ship section of a general cargo having the bottom and decks structure longitudinally framed could be idealized by a simplified configuration as shown in Fig. 9.5.

9.2.1.2 Oil Tankers

Conventional Oil Tanker with Two Longitudinal Bulkheads

A longitudinal section of a crude oil tanker having an aft engine room is shown in Fig. 9.6. A typical ship section of a conventional longitudinally framed crude oil tanker is shown in Fig. 9.7.

Figure 9.8 shows a simplified configuration of a conventional twin longitudinal bulkhead crude oil tanker. A simplified idealization of the conventional twin longitudinal bulkheads oil tanker is shown in Fig. 9.9. This is the case of multi-web box girder composed of a combination of open and closed sections, see Fig. 9.9.

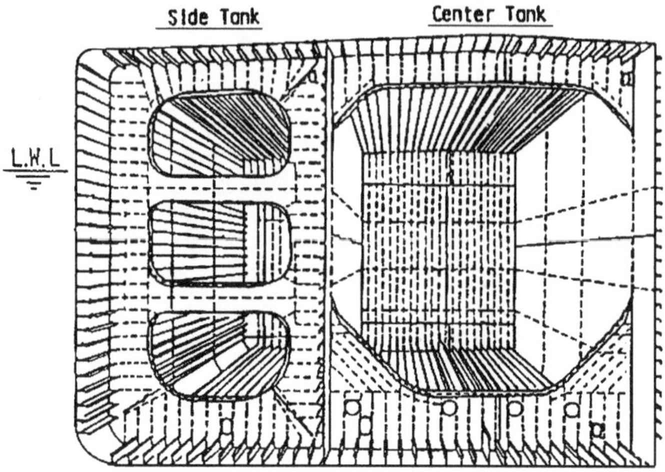

Fig. 9.7 3-D presentation of a conventional oil tanker

Fig. 9.8 Configuration of a twin bulkhead oil tanker

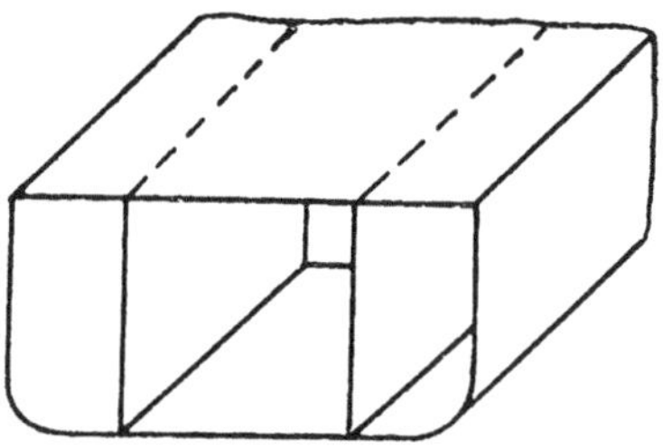

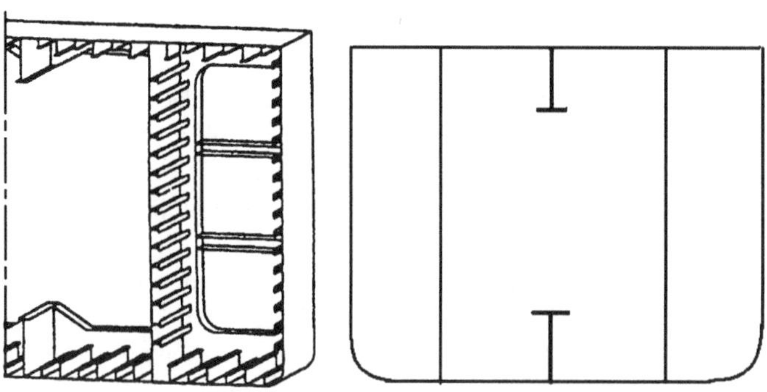

Fig. 9.9 A ship section of a conventional twin bulkhead oil tanker

Conventional Oil Tanker with Three Longitudinal Bulkheads

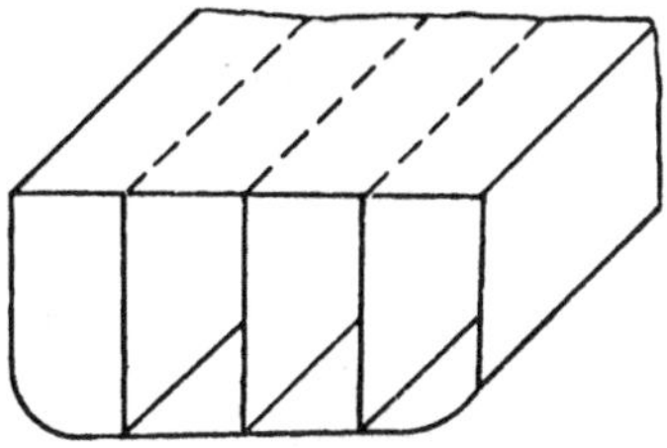

Fig. 9.10 Conventional oil tanker with 3 longitudinal bulkheads

Conventional Coastal Oil Tanker with One Longitudinal Bulkhead

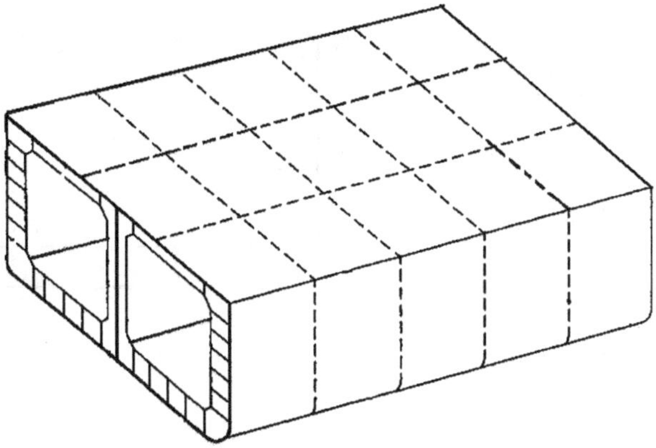

Fig. 9.11 Coastal oil tanker with one centerline longitudinal bulkhead

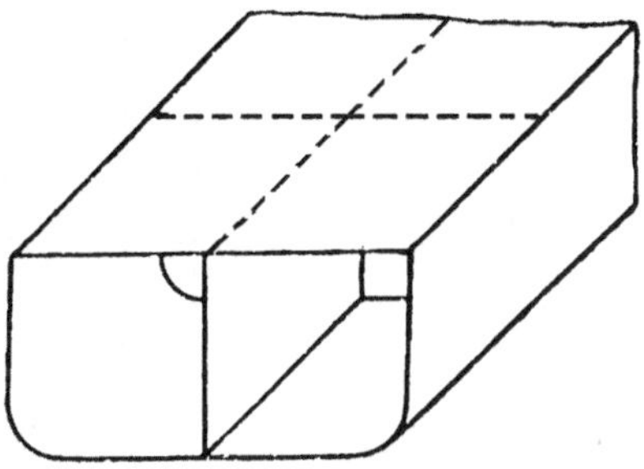

Fig. 9.12 A simplified idealization of a ship section of a coastal oil tanker

An Oil Tanker Fitted with Double Side

Fig. 9.13 Idealized section of a double side oil tanker

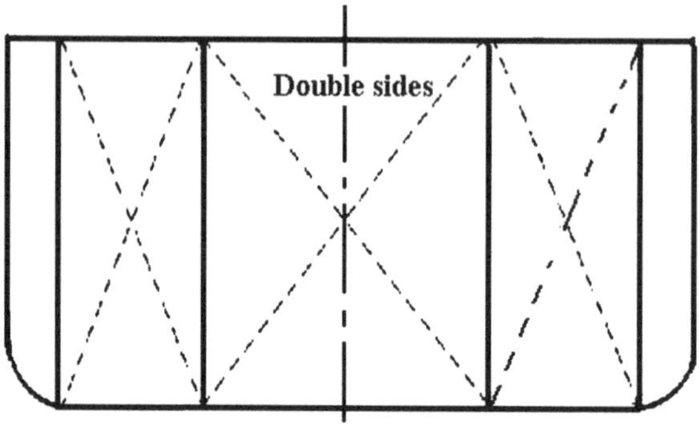

An Oil Tanker Fitted with Double Bottom

Fig. 9.14 Idealized section of an oil tanker fitted with a double bottom

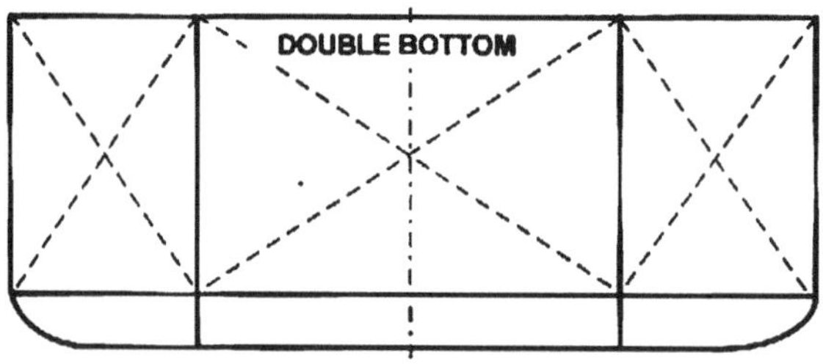

An Oil Tanker Fitted with Double Skin

Fig. 9.15 An idealized section of a double skin oil tanker

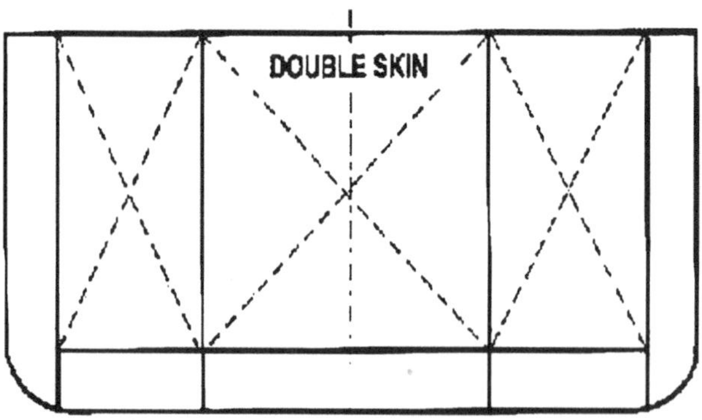

9.2.1.3 RO-RO ships

The ship section of a RO-RO ship is shown in Fig. 9.16. The idealized ship section combines open and closed multi-box thin-walled structure, see Fig. 9.17.

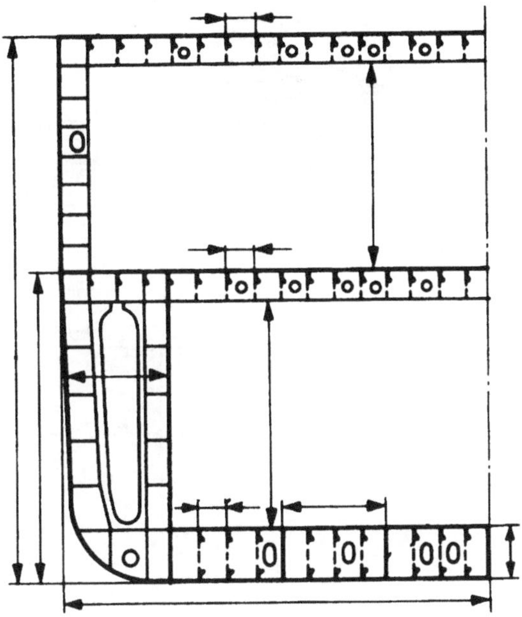

Fig. 9.16 A ship section of a RO-RO ship

Fig. 9.17 An idealized ship section of a RO-RO ship

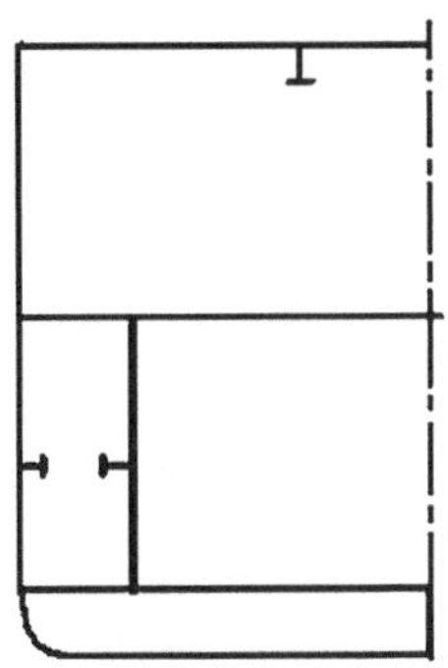

9.2.1.4 Container Ships

A longitudinal section of a loaded feeder container ship is shown in Fig. 9.18. A typical ship section of a container ship is composed of a thin-walled multi-box structure, see Fig. 9.19. A simplified multi-box idealization of a half ship section of container ship is shown in Fig. 9.20.

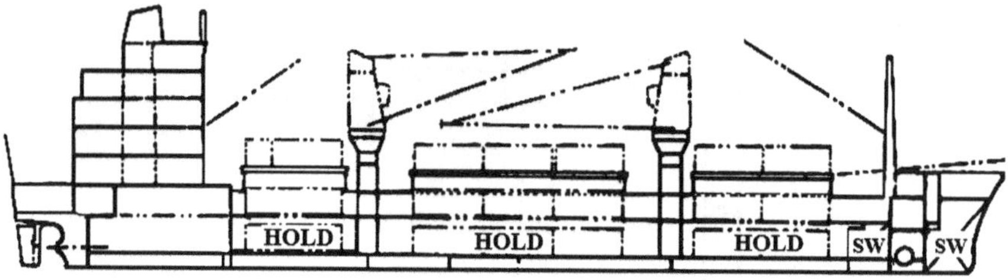

Fig. 9.18 A ship section of a container ship with full load

Fig. 9.19 A ship section of a container ship

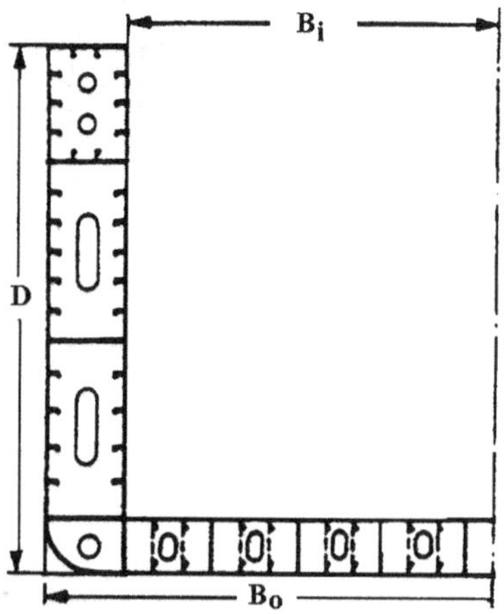

Fig. 9.20 Idealized section of a container ship

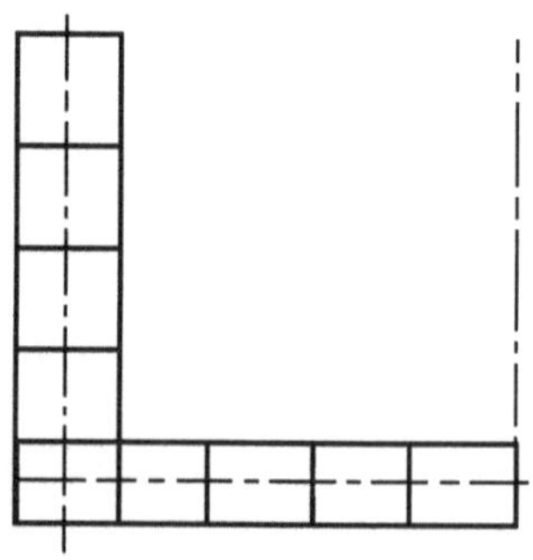

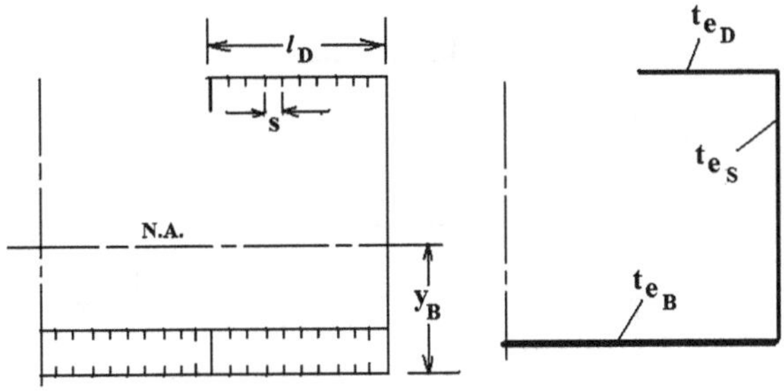

Fig. 9.21 Idealized ship section of a cargo ship

9.3 Determination of the Effective Thickness

The idealized ship section is based on using effective thickness concept for the decks, sides, and bottom structure, see Fig. 9.21. The effective thickness takes account of longitudinal stiffeners and girders and is calculated as follows

$$t_e = t + \sum a_{\lambda j}/\lambda$$
$$t_{De_1} = t_{D_1} + \sum a_{\lambda_1}/b_1$$
$$t_{De_2} = t_{D_2} + \sum a_{\lambda_2}/b_1$$
$$t_{Se_1} = t_{S_1} + \sum a_{\lambda S_1}/d_1$$
$$t_{Se_2} = t_{S_2} + \sum a_{\lambda S_2}/d_2$$
$$t_{Be} = t_{Bie} + t_{Boe}$$
$$t_{B_{ie}} = t_{B_i} + \sum a_{\lambda i}/B/2$$
$$t_{B_{oe}} = t_{B_o} + \sum a_{\lambda o}/B/2$$
$$t_{eD} = t_D + \sum_D a_{\lambda j}/\lambda_D$$
$$\sum_D a_{\lambda j} = n_\lambda \times a\lambda_D$$

a_{lD} is the sectional area of deck longitudinals; n_l is the number of Deck long; $nl = \lambda_D/S - 1$; S is the spacing of deck long.

9.4 Shear Flow Calculation

The calculation of shear flow around an idealized ship section is based on the following equation

$$q = FA\bar{y}/I$$

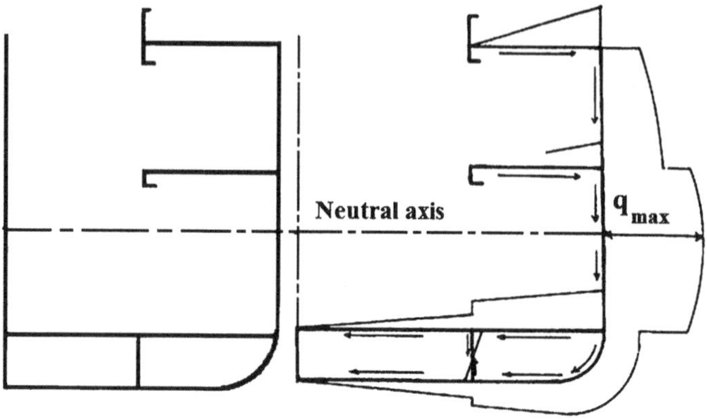

Fig. 9.22 Shear stress distribution over a ship section of a cargo ship

Fig. 9.23 Equilibrium conditions of an element in the deck panel of a ship

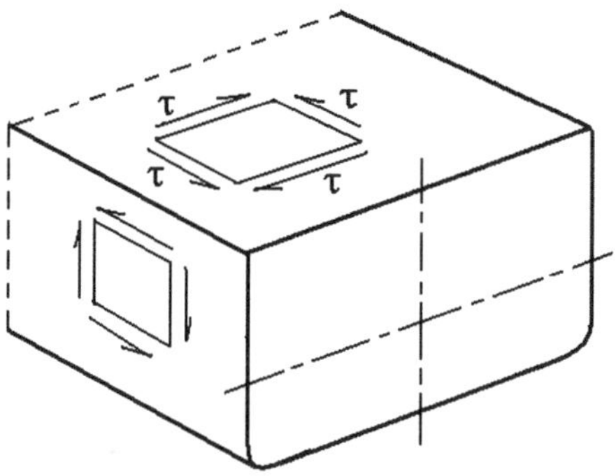

The shear flow over an idealized ship section of a general cargo ship is shown in Fig. 9.22.

For indeterminate thin-walled structures, the above equation cannot not be used directly to calculate the distribution of shear stresses over a section of the structure. The equilibrium conditions of an element in the deck panel of a ship is shown in Fig. 9.23 where the direct and complementary shear stress are presented.

9.4.1 Procedure of Calculation of Shear Flow Distribution

In order to clarify the calculation procedure, the following steps are used

– Assume a distribution of shear flow q′ given by

$$q' = \tau \cdot t = F/I \cdot \sum A\bar{y}$$

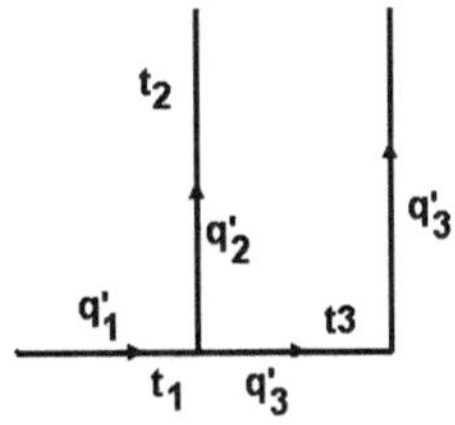

Fig. 9.24 Basic concept of shear flow over a plate structural connection

- This shear flow distribution must satisfy the force condition at each joint, see Fig. 9.24.

It should be noted that the sum of the shear flows going into the joint = sum of the shear flows coming out from the joint see Fig. 9.24.

Hence

$$q_1' = q_2' + q_3'$$

The procedure of calculating the shear flow distribution is based on the following steps

- Assume a fictitious value for the shear stress at some pints so as to proceed with the calculations.
- Calculate the angle of twist induced by the unbalance of forces in each box.
- Calculate the resulting torque in each box of the girder.
- If the ship is undergoing a bending moment without torsion then there will be no internal torque and the calculated torque must be removed by the application of an equal and apposite torque round the section.
- The final shear flow distribution = Assumed shear flow + q_c; where q_c is the shear flow due to correcting torque.

Note:

(a) steps (1) and (2) are generally simplified by the symmetry of the section
(b) Angle of twist is given by

$$\theta = 1/2AG \cdot \oint q'\, ds/t$$

(c) The angle of twist required to remove θ could be obtained by applying a correcting uniform shear flow q_c
The correcting angle of twist is given by

$$\theta_c = \theta = q_c/2AG \cdot \oint \Delta s/t$$

Hence, the correcting shear flow is given by

$$q_c = -\oint q\frac{\Delta s}{t} \Big/ \oint \frac{\Delta s}{t}$$

(d) The corrected shear flow distribution is given by

$$q_{final} = q' - q_c$$

(e) The shear stress is given by

$$\tau = q/t.$$

9.4.2 Shear Flow Distribution over a Ship Section of a Two-deck Cargo Ship

The shear flow is given by

$$q = F/I \cdot \sum a \cdot \bar{y}$$

Since points 1 and 3 are free ends, see Fig. 9.25, we have

$$q_1 = q_2 = 0$$

Since points 5 and 7 are located on the axis of symmetry, we have

$$q_5 = q_7 = 0$$

The shear flow at point 4 is obtained from

$$q_{4-9} = q_{4-2} + q_{4-3}$$

The maximum shear flow occurs at point 9 at the neutral axis of the ship section.

Hence

$$q_9 = q_{max}.$$

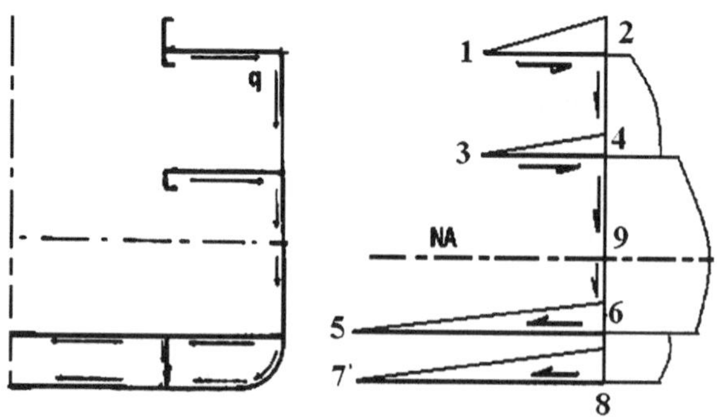

Fig. 9.25 Shear flow distribution over a ship section of a cargo ship

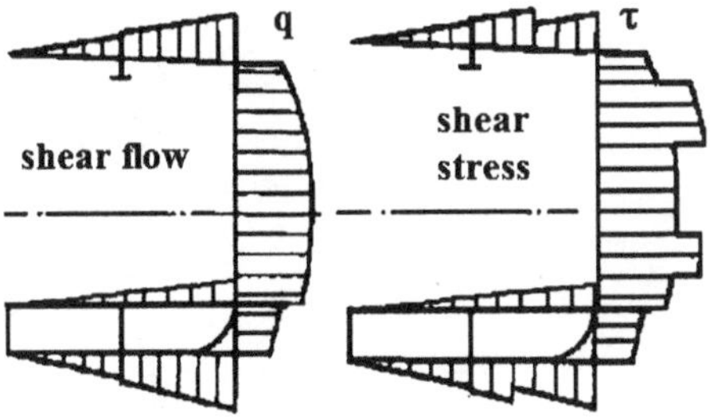

Fig. 9.26 Distribution of shear flow and shear stress over the ship section

9.5 Calculation of Shear Stress Distribution

The calculation of shear stresses should be based on actual thickness and not on equivalent thicknesses. The shear stress at any point on the section could be calculated using the following equation

$$\tau = \mathrm{FA\overline{y}/It}$$

The distribution of shear stress over the ship section could be significantly different from the distribution of the shear flow over the ship section. This is due to the variation of shell plating thicknesses over the ship section. Figure 9.26 shows the distribution of shear flow and shear stress over a ship section of a general cargo ship.

9.5.1 Equivalent Stress

The importance of calculating the magnitude and distribution over the ship section stems from the need to calculate the magnitude of the equivalent stresses at certain locations over the ship section.

The equivalent stress at any location over a ship section subjected to hull girder vertical bending stress and shear stress is given by

$$\sigma_{\mathrm{e}} = \sqrt{\sigma^2 + 3\tau^2}$$

where σ is the hull girder bending stress, τ is the shear stress

The equivalent stresses at points (1) and (2) over the ship section shown in Fig. 9.27 are given by

$$\sigma_{\mathrm{e}_1} = \sqrt{\sigma_1^2 + 3\tau_1^2}$$

and

$$\sigma_{\mathrm{e}_2} = \sqrt{\sigma_2^2 + 3\tau_2^2}$$

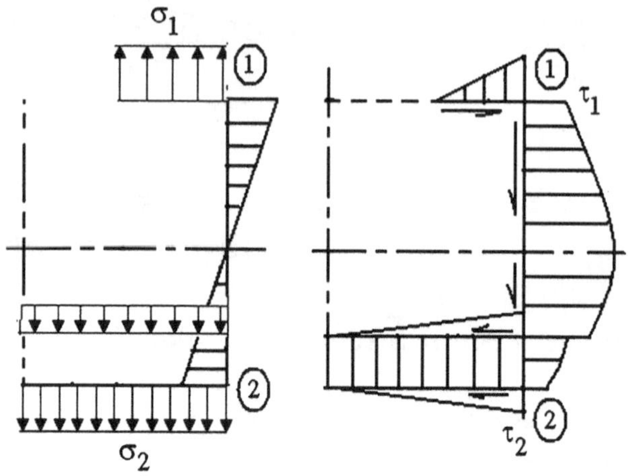

Fig. 9.27 Hull girder bending and shear stress distribution over the ship section

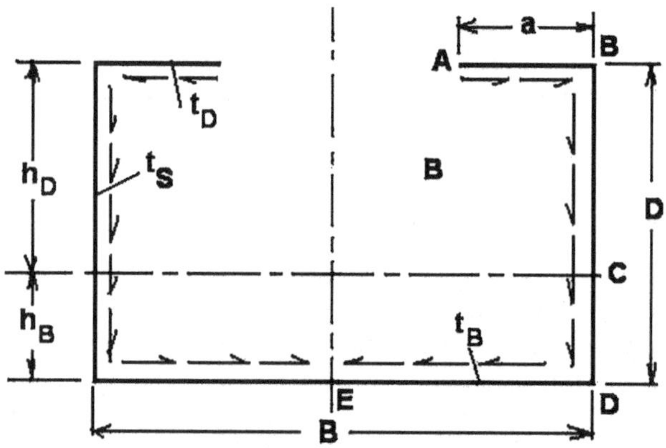

Fig. 9.28 Shear flow around an idealized ship section of a single deck ship

9.6 Calculation of Shear Stress Distribution over a Ship Section

The shear flow around the ship section is shown in Fig. 9.28.

The values of h_D and h_B are calculated as follows

$$t_D \cdot a \cdot D + D^2/2 \cdot t_S = h_B \left[a t_D + D t_S + B/2 \cdot t_B\right]$$

and

$$h_B = \left(D \cdot a t_D + D^2 \cdot t_S/2\right)/\left(a t_D + D t_s + B/2 \cdot t_B\right)$$

Table 9.1 Procedure of calculation of shear flow

Member or Point	Area a	ay	q′	τ
A	–	–	–	–
AB	at_D	ah_Dt_D		
$(B)_{BA}$	–	–	ah_Dt_D	ah_D
$(B)_{BC}$	–	–	ah_Dt_D	ah_Dt_D/t_S
BC	h_Dt_S	$h_D^2t_S/2$	$ah_Dt_D + h_D^2t_S/2$	–
$(C)_{CB}$	–	–	$ah_Dt_D + h_D^2t_S/2$	$ah_Dt_D/t_S + h_D^2/2$
E	–	–	–	–
ED	$Bt_B/2$		$Bh_Bt_B/2$	–
$(D)_{DE}$	–	–	$Bh_Bt_B/2$	$Bh_B/2$
	–	–	–	$Bh_B/2 + t_B/t_S$
$(D)_{DC}$	h_Bt_S	$h_B^2t_S/2$	$Bh_Bt_B/2 + h_B^2t_S/2$	$Bh_Bt_B/2t_S + h_B^2/2$

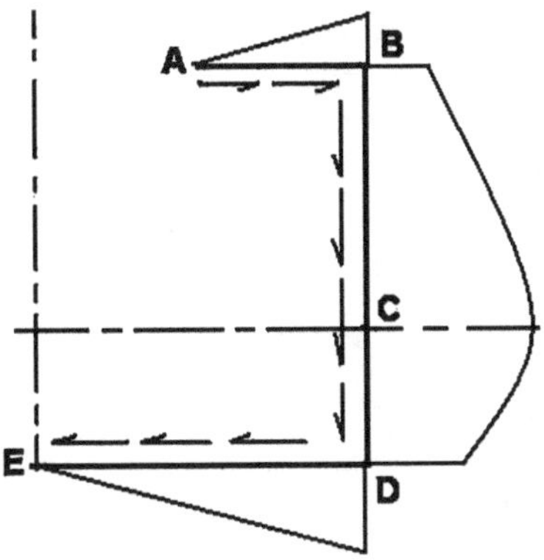

Fig. 9.29 Shear flow distribution over an idealized half section of a single deck ship

Due to symmetry, only half the ship section needs to be considered. The calculation of shear flow for the single deck cargo ship is given in Table 9.1.

The shear flow distribution over the section is shown in Fig. 9.29.

9.6.1 *Calculation of Shear Flow Distribution over a Twin Deck Cargo Ship*

The dimensions, idealized ship section and effective thicknesses of the main and second deck, sid and bottom structure are shown in Fig. 9.30. The position of the neutral axis is given by

$$h_B = \frac{a(Dt_D + H_Dt_T) + H_Tt_T(D - H_T/2) + H_D^2 \cdot t_S/2}{a(t_D + t_T) + H_T \cdot t_T + H_D \cdot t_S + Bt_B/2}$$

The shear flow distribution is shown in Figs. 9.31 and Fig. 9.32.

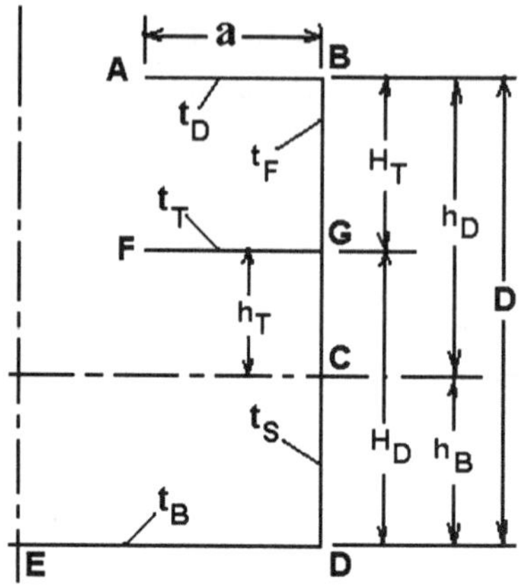

Fig. 9.30 The idealized ship section of the twin deck ship

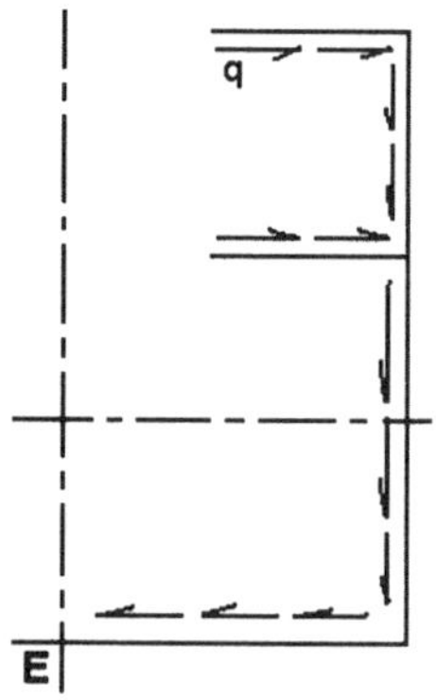

Fig. 9.31 Shear flow over the idealized ship section of the twin deck ship

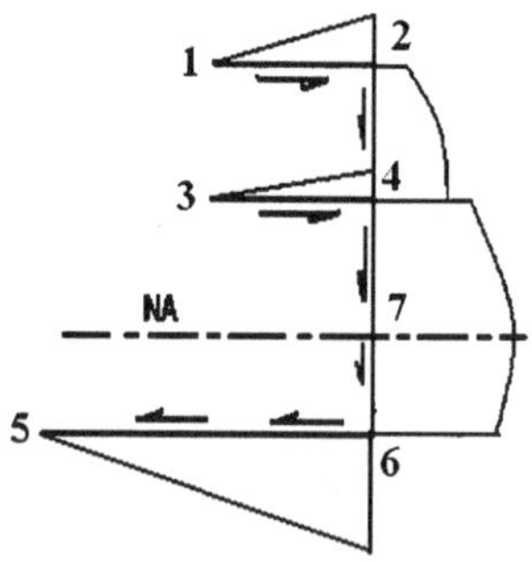

Fig. 9.32 Shear flow distribution over the idealized ship section

9.7 Shear Flow Distribution over a Catamaran Section

The calculation of shear flow distribution over a ship section of a catamaran vessel follows the same procedure previously outlined. Because the ship section is

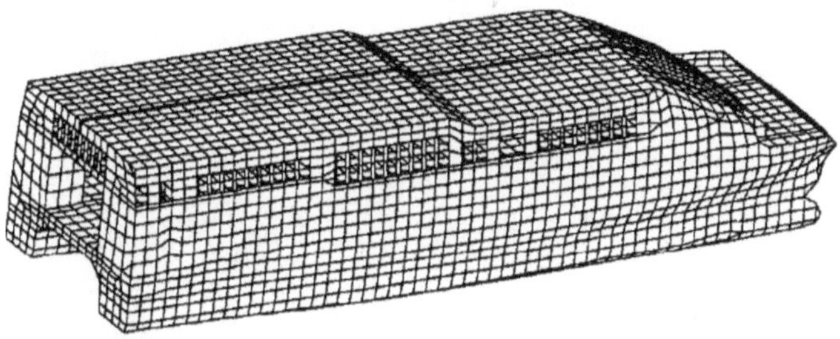

Fig. 9.33 A picture of a catamaran vessel

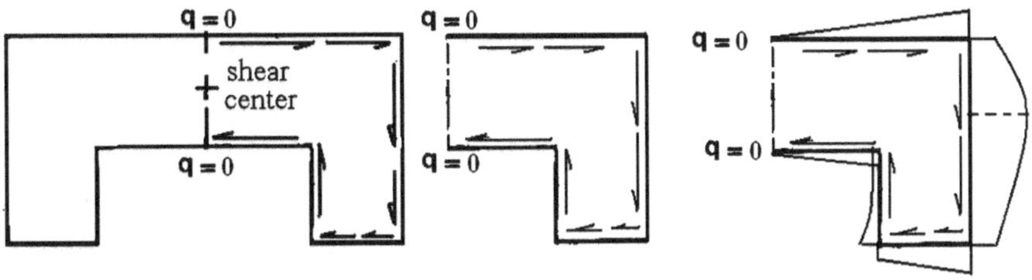

Fig. 9.34 An idealized section of a catamaran vessel

Fig. 9.35 Shear flow distribution over half ship section

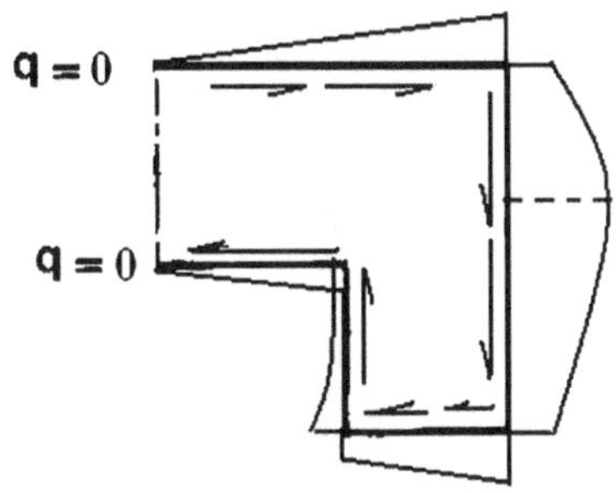

symmetrical about the vertical centerline axis, the shear flow is zero at the axis of symmetry, see Fig. 9.33.

The distribution of shear flow over the idealized section of the catamaran shown in Fig. 9.34 is shown in Fig. 9.35. Due to symmetry, the shear flow is zero at points A and B.

Chapter 10
Calculation of Shear Stresses in Tankers Subjected to Longitudinal Vertical Shear Forces

10.1 Coastal Tankers Having One Longitudinal Bulkhead

A typical configuration of a conventional coastal tanker with one centerline longitudinal bulkhead is shown in Fig. 10.1.

A simplified idealization of the one centerline bulkhead tanker is shown in Fig. 10.2. The thicknesses of the shell plating and longitudinal bulkheads are the effective thicknesses.

The procedure adopted for calculating the shear flow distribution over the idealized ship section is summarized as follows:

- Assume correcting shear flow in cell $I = q_{c1}$.
- Assume correcting shear flow in cell $II = q_{c2}$.
- A zero shear flow is assumed at point C, see Fig. 10.3 $q_{CA} = 0$.
- The shear flow distribution is calculated for cell ABBE, see Fig. 10.2.
- The angle of twist resulting from the assumed shear flow is calculated from Eq. (10.3) for cell ABDE.
- A correcting shear flow q_c is assumed in each cell and is calculated from the following two equations.

$$\theta_1 - \theta_c = 0 \tag{10.1}$$

$$\theta_2 - \theta_c = 0 \tag{10.2}$$

where θ_1 and θ_2 are the angles of twist for cells ABDE and GAEH, respectively.

Due to symmetry

$$\theta_1 = \theta_2 \tag{10.3}$$

Substituting for θ_1 from Eq. (10.3) and θ_c from

$$\theta_C = q_c/2AG \cdot \oint ds/t$$

M. Shama, *Torsion and Shear Stresses in Ships*,
DOI: 10.1007/978-3-642-14633-6_10, © Springer-Verlag Berlin Heidelberg 2010

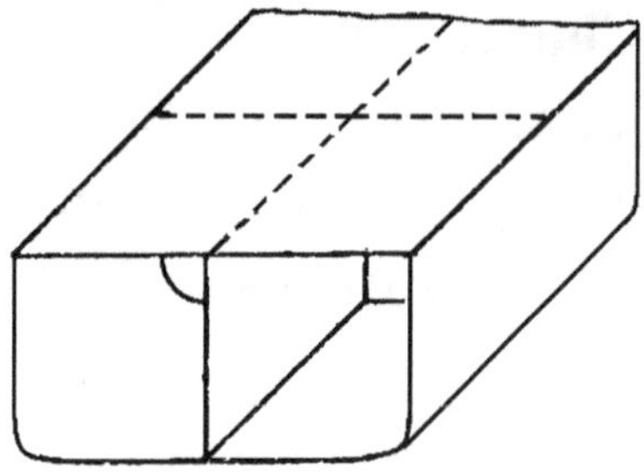

Fig. 10.1 Ship section configuration of a coastal tanker with one centerline longitudinal bulkhead

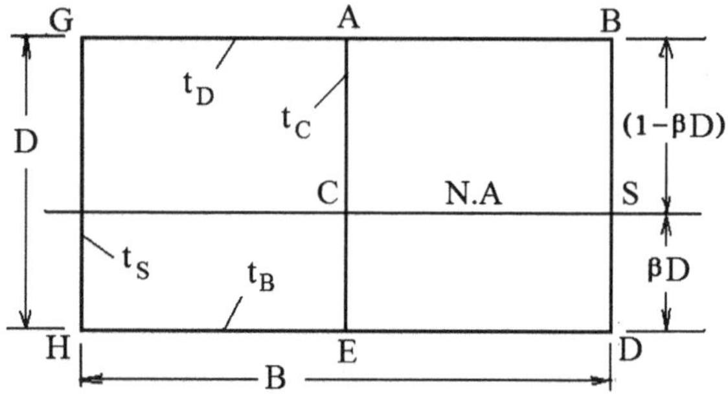

Fig. 10.2 Idealized section of a coastal tanker with one centerline longitudinal bulkhead

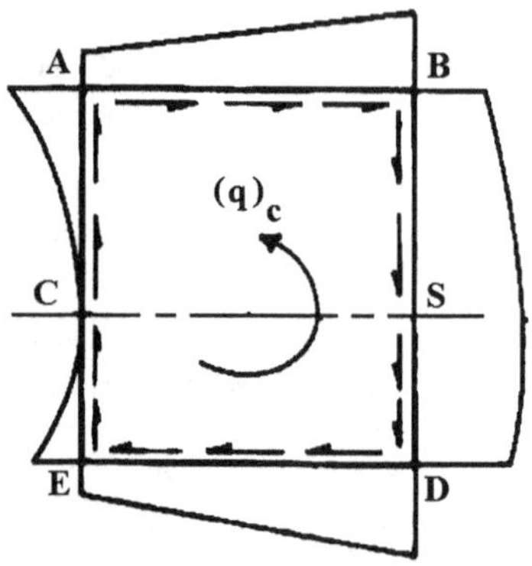

Fig. 10.3 Assumed shear flow distribution over the idealized half section of the coastal tanker

Equations (10.1) and (10.2) are reduced to

$$q_c\left\{\sum \Delta s/t\cdot\begin{bmatrix}E\rightarrow D\\ D\rightarrow B\\ B\rightarrow A\end{bmatrix}+2\sum \Delta s/t\cdot[A\rightarrow E]\right\}-\oint q\Delta s/t=0$$

Hence

$$q_C=\oint q\frac{\Delta s}{t}\Bigg/\left(\int_A^E \Delta s/t+2\int_E^A \Delta s/t\right)$$

The distribution of the correcting shear flow is shown in Fig. 10.4.

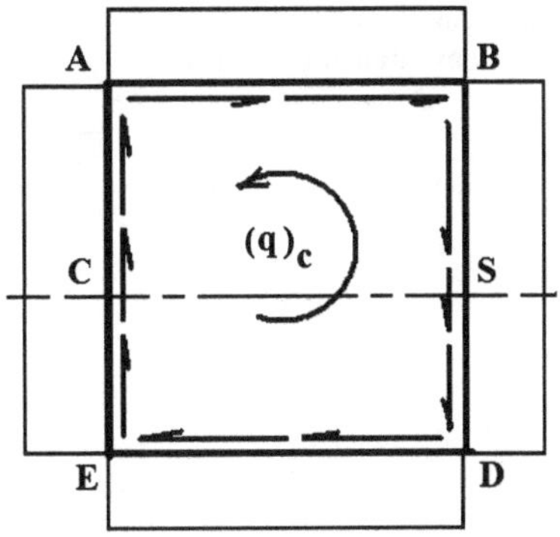

Fig. 10.4 Correcting shear flow distribution over half the idealized section of the coastal tanker

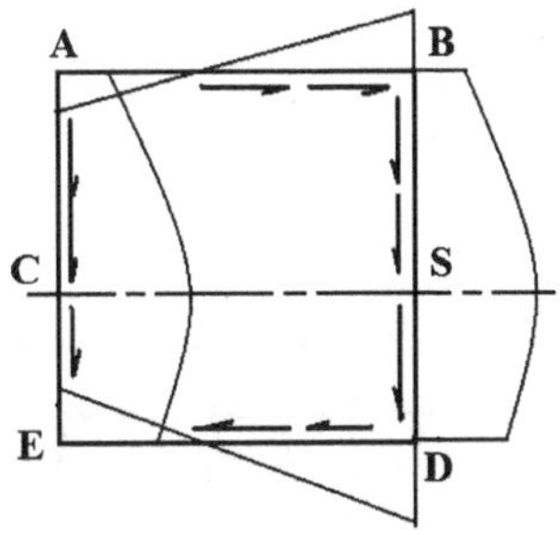

Fig. 10.5 Resultant shear flow distribution over half the idealized section of the coastal tanker

The resultant shear flow at any point i in each cell is given by, see Fig. 10.5.

$$(q_i)_r = q_i - q_c.$$

10.2 Calculation of Shear Flow Distribution for Twin Longitudinal Bulkhead Tankers

The actual and idealized ship section of an oil tanker are shown in Fig. 10.6 and 10.7, respectively.

Because of symmetry and absence of a longitudinal center-line bulkhead, the shear flows at points A and E on the ship center-line are *zero*, i.e., see Fig. 10.8. Hence

$$q_{AH} = q_{EI} = 0$$

The method of calculation is summarized as follows:

- A zero shear flow is assumed at point L, see Fig. 10.9. $q_{LH} = 0$.
- Shear flow at HB, see Fig. 10.9, is given by $q_{HB} = q_{HA} + q_{HL}$.
- Shear flow distribution is calculated for cell ABDE.

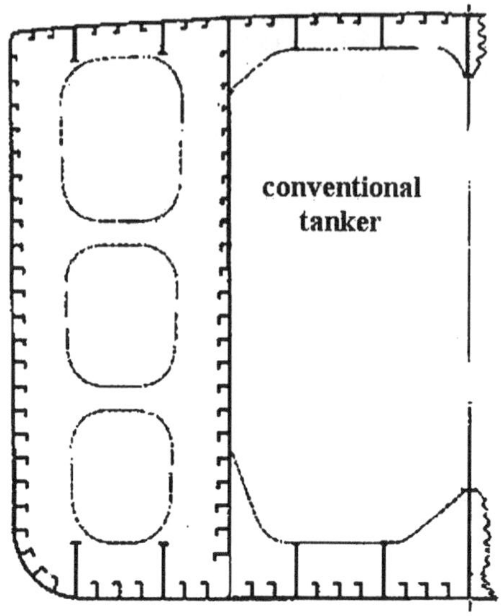

Fig. 10.6 Half ship section of a conventional oil tanker

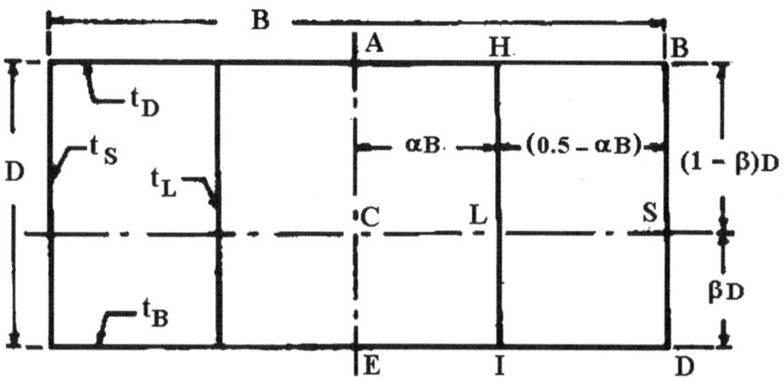

Fig. 10.7 A simplified idealization of a ship section with two longitudinal bulkheads of a conventional oil tanker

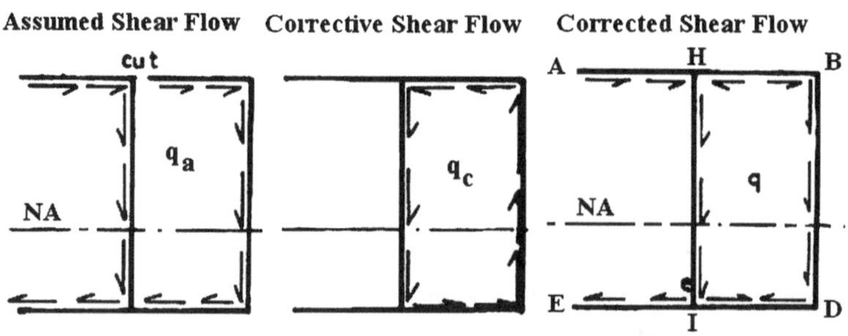

Fig. 10.8 Selection of the location of cut

Fig. 10.9 Assumed shear flow distribution

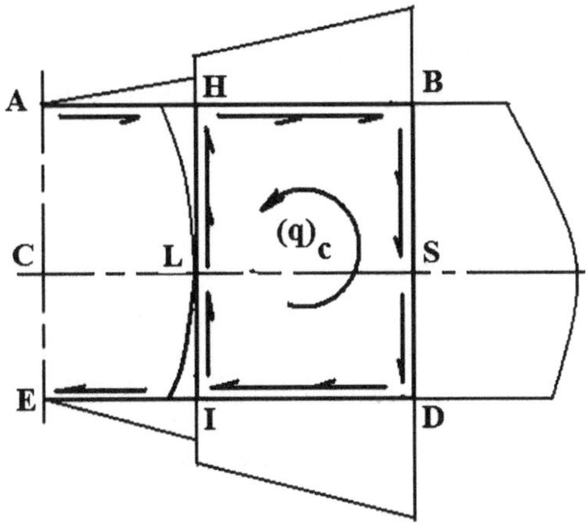

Fig. 10.10 Correcting shear flow

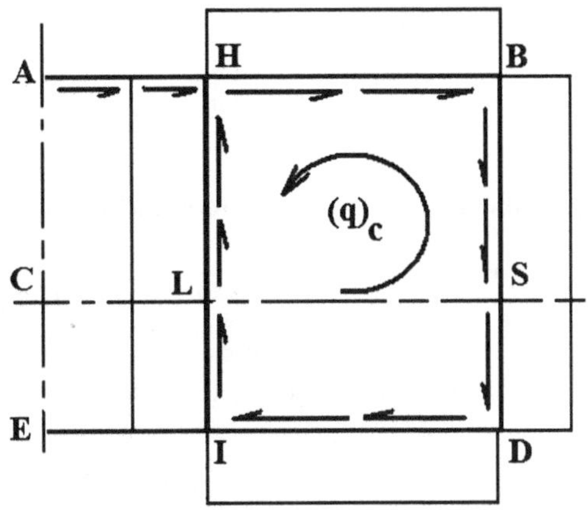

- The angle of twist θ in cell HBDI, due to the assumed shear flow is calculated from Eq. (10.3).
- A correcting shear flow q_c is assumed in cell HBDI. See Fig. 10.10. The corresponding angle of twist θ_c is given by $\theta_C = q_c/2AG \cdot \oint \Delta s/t$.

The value of q_c should satisfy the equilibrium condition for cell HBDI, i.e., $\theta = \theta_c$.

Hence q_c is given by

$$q_c = \oint q\frac{\Delta s}{t} \Big/ \oint \frac{\Delta s}{t}$$

The resultant shear flow at any point i in cell HBDI given by

$$(q)_{ir} = q_i - q_c$$

Since it is assumed that the section is not subjected to any torsional moments and is subjected only to longitudinal vertical shear force, the angle of twist for each cell and for the total section must be zero. Thus

$$\theta_{I} = \theta_{II} = 0.0$$

where

$$\theta_{I} = \theta_{1} - \theta_{c1} \tag{10.4}$$

$$\theta_{II} = \theta_{2} - \theta_{c2} \tag{10.5}$$

θ_1 is the angle of twist due to shear flow q_1, i.e.,

$$\theta_1 = 1/2A_1G \cdot \oint q_1 ds/t \tag{10.6}$$

θ_{c1} is the angle of twist due to correcting shear flow q_{c1}, i.e.,

$$\theta_{c1} = q_{c1}/2A_1G \cdot \oint ds/t \tag{10.7}$$

θ_2 is the angle of twist due to shear flow q_2, i.e.,

$$\theta_2 = 1/2A_2G \cdot \oint ds/t$$

θ_{c2} is the angle of twist due to correcting shear flow q_{c2}, i.e.,

$$\theta_{c2} = q_{c2}/2A_2G \cdot \oint ds/t$$

From Eqs. (10.1)–(10.7) we have

$$q_{c1} \sum \Delta S/t \cdot \begin{Bmatrix} a \rightarrow b \\ b \rightarrow c \\ c \rightarrow d \end{Bmatrix} + (q_{c1} - q_{c2}) \times \sum \Delta S/t \cdot \{d \rightarrow a\} - \oint q_1 \Delta S/t = 0 \tag{10.8}$$

$$q_{c2} \sum \Delta S/t \cdot \begin{Bmatrix} d \rightarrow e \\ e \rightarrow f \\ f \rightarrow a \end{Bmatrix} + (q_{c1} - q_{c2}) \times \sum \Delta S/t \cdot \{a \rightarrow d\} - \oint q_2 dS/t = 0 \tag{10.9}$$

By solving Eqs. (10.8) and (10.9), we obtain q_{c1} and q_{c2}.

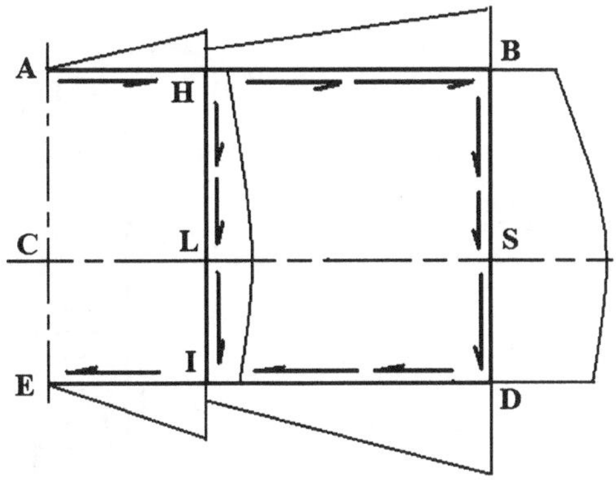

Fig. 10.11 Corrected shear flow

The resultant shear flow at any point in cell I and cell II are given by

$$q_{r1} - q_1 - q_{c2}$$

$$q_{r2} - q_2 - q_{c2}$$

The selection of the location of the assumed cut on the periphery of the closed cell is arbitrary. The procedures of calculation is the same irrespective of the position of the point of cut, see Fig. 10.11.

10.3 Shear Load Carried by Longitudinal Bulkheads and Side Shell Plating

10.3.1 Sea-Going Tankers with Two Longitudinal Bulkheads

The shear forces carried by the two longitudinal bulkheads and the side shell plating must satisfy the following condition, see Fig. 10.12

$$2F_L + 2F_S = F$$

where F is the shear force

$$F_L = K_L \cdot F$$

$$F_S = K_S \cdot F$$

F_L is the shear force carried by longitudinal bulkhead plating as shown in Fig. 10.13.

$$F_L = \int_D q_y \cdot d_y = (q_L)_m \cdot D$$

F_S is the shear force carried by side shell plating as shown in Fig. 10.14.

$$F_S = \int_D q_y \cdot dy = (q_S)_m \cdot D$$

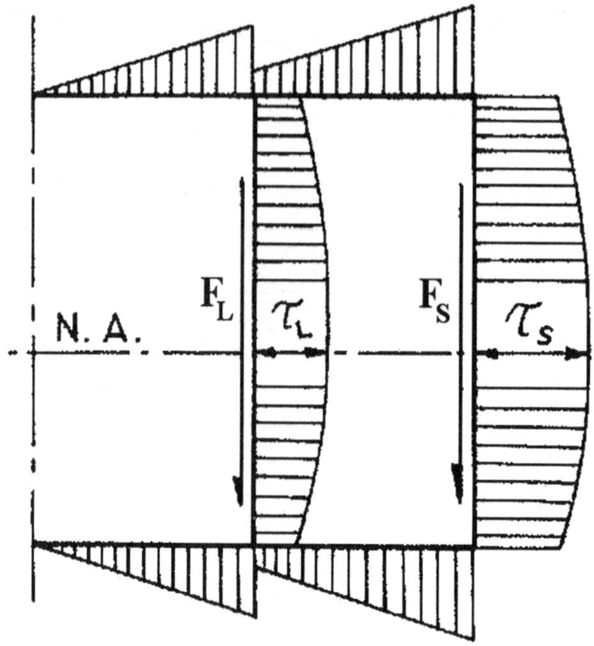

Fig. 10.12 Typical shear flow distribution

Fig. 10.13 Shear force carried by the longitudinal bulkhead plating

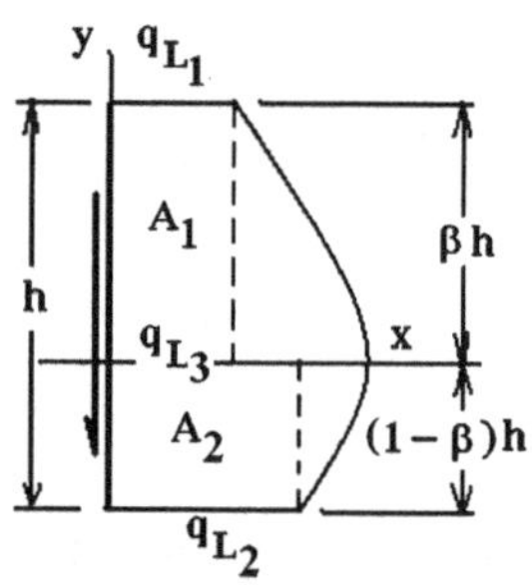

Fig. 10.14 Shear force carried by side shell plating

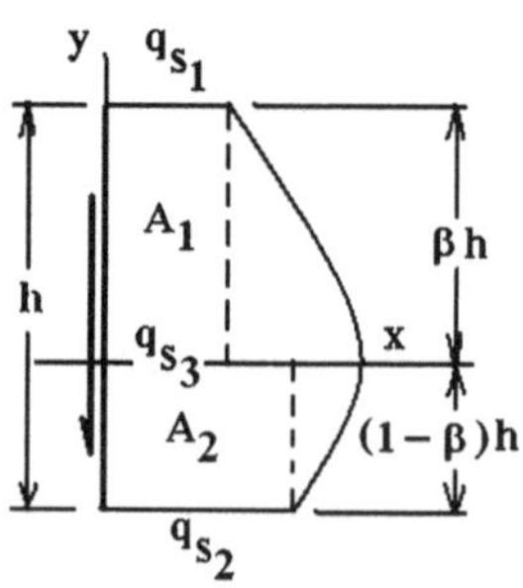

$$(q_L)m = 1/3 \cdot [\beta q_{IL} + (1-\beta)q_{HL} + 2q_L]$$

$$(q_S)_m = 1/3 \cdot [\beta q_{DS} + (1-\beta)q_{BS} + 2q_S]$$

see Fig. 10.12.

10.3.2 Coastal Tankers with One Longitudinal Bulkhead

The total shear force is given by

$$F = F_L + 2F_S$$

The shear load carried by one side shell plating is given by

$$F_S = \int_o^h q_S dy$$

The shear force carried by the longitudinal bulkhead plating is given by

$$F_L = \int_o^h q_L dy$$

Consider the shear load carried by one side shell plating as shown in Fig. 10.15.

$$F_S = A_1 + A_2$$

$$\begin{aligned} A_1 &= q_{S1} \times \alpha h + 2/3 \cdot (q_{S3} - q_{S1}) \cdot \alpha h \\ &= 1/3 \cdot q_{S1} \alpha h + 2/3 \cdot q_{S3} \alpha h \\ &= 1/3 \cdot \alpha h \cdot (q_{S1} + 2q_{S3}) \end{aligned}$$

$$\begin{aligned} A_2 &= q_{S2}(1-\alpha)h + 2/3 \cdot (q_{S3} - q_{S2})(1-\alpha)h \\ &= 1/3 \cdot (1-\alpha)h \cdot [q_{S2} + 3q_{S3}] \end{aligned}$$

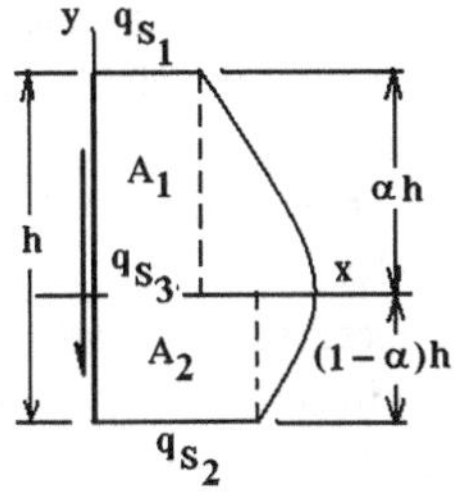

Fig. 10.15 Shear load carried by the side shell

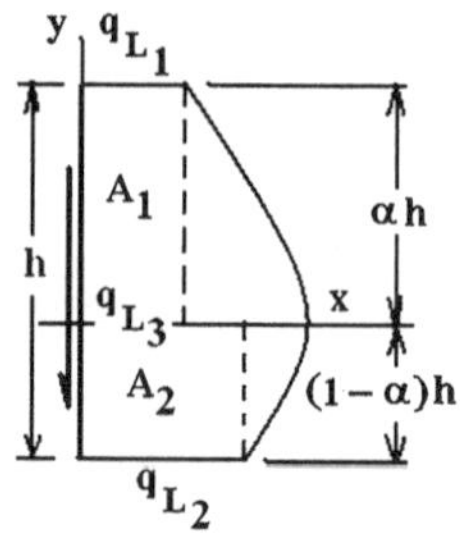

Fig. 10.16 Shear load carried by the longitudinal bulkhead

$$F_S = \left[1/3 \cdot \alpha h \cdot q_{S1} + 1/3 \cdot (1-\alpha)h \cdot q_{S2} + 2/3 \cdot \alpha h \cdot q_{S3} + 2/3 \cdot (1-\alpha)h \cdot q_{S3} \right]$$
$$= h/3 \cdot [\alpha q_{s1} + (1-\alpha)q_{s2} + 2q_{s3}]$$

Similarly, the shear load F_L carried by the longitudinal bulkhead plating is given by, see Fig. 10.16.

$$F_L = h/3 \cdot [\alpha q_{s1} + (1-\alpha)q_{s2} + 2q_{s3}].$$

10.4 Shear Flow Distribution Over a Ship Section of an Oil Tanker Experiencing a Local Damage in the Shell Plating or Longitudinal Bulkhead

10.4.1 Introduction

Structural failures in oil tankers are generally attributed to failures of structural details in areas of high stress concentration, particularly at bracket toes and longitudinal connections to transverse web frames.

The main types of structural defects affecting directly the strength of local structural details and indirectly the strength of ship hull girder are: sources of crack initiation, weld failures, inadequate local buckling strength, excessive deformations, corrosion, etc. The most probable consequences of the presence of these structural defects are: increased size of the defect, crack propagation, buckling, reduction in thickness, deficient load carrying capacity, etc. The initiation and propagation of these local failures may subsequently cause major structural failures.

Small failures of oil tankers that may not immediately threaten ship structural safety may subsequently cause serious economical and pollution problems. Therefore, in order to reduce/prevent the risk of a pollution hazard to the marine environment, oil tankers should be designed and maintained to a level of structural safety compatible with economic operations and environmental protection.

The presence of a local damage to the shell plating or the longitudinal bulkheads of an oil tanker will cause a redistribution of the shear and bending stresses over the ship section. Some structural members will be over stressed and others may experience stress reduction. Therefore, some panels of the shell plating may experience high values of the equivalent stresses or severe buckling. It is necessary therefore to study the redistribution of stresses over the ship section after a set of assumed local damages occurring at some selected locations over the ship section.

10.4.2 Shear Stress Distribution Over a Tanker Section Experiencing a Local Damage

The bending and shear stress distribution over ship sections experiencing local damages should be examined to improve the strength of the ship section so as to avoid any overloading of any of its structural members. Several investigations have been conducted to examine the strength and consequences of the presence of a local structure damage of the ship hull girder. The effect on the magnitude and distribution of bending and shear stresses over the ship section of a general cargo ship subjected to assumed local damage occurring at specified locations should be examined. The redistribution of the bending and shear stresses over the ship section could have significant on the magnitude of the hull girder bending and shear stresses.

10.4.3 Scenarios of Assumed Damage Locations on the Tanker Section

The effect on shear stress distribution and maximum shear stresses for coastal and sea-going tankers experiencing assumed damages at certain locations over the ship section are examined. The method of calculation is based on the procedures given in Chap. 3. The presence of a small local damage at any location over the tanker section will transform the ship section structure from a closed thin-walled box to an open and closed thin-walled structure.

In order to simplify the analysis and calculations, it is assumed that the presence of a small damage will not induce any significant values of torsional loading.

10.4.3.1 Coastal Oil Tankers

The main dimensions of the coastal tanker considered for this study are, see Fig. 10.17, Length = 65 m, Breadth = 12.85 m, and Depth = 5.8 m. The idealized ship section is shown in Fig. 10.18.

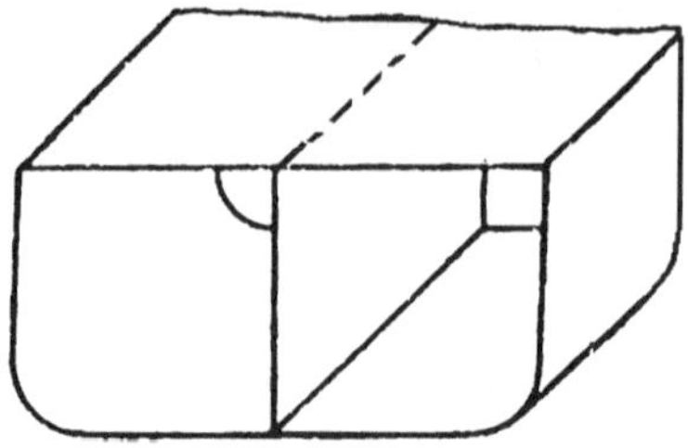

Fig. 10.17 A simplified configuration of a conventional coastal oil tanker

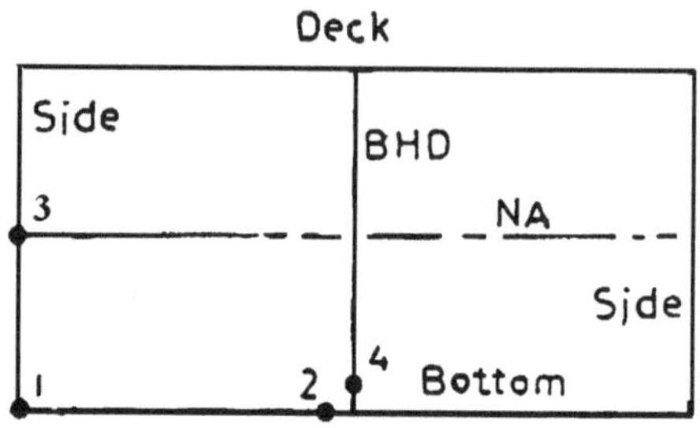

Fig. 10.18 Idealized ship section of the oil tanker and the assumed local damage locations

The various assumed damage locations are, see Fig. 10.18.

- Local damage at the bilge strake in the Port side (*Point 1*),
- Local damage at the keel plate close to the centerline bulkhead (*Point 2*),
- Local damage at mid-depth of the side shell (*Point 3*),
- Local damage to the lower edge of the centerline longitudinal bulkhead (*Point 4*),

The shear stress distribution over the idealized tanker section in the intact condition is shown in Fig. 10.19. The shear stress distribution over the ship section of the tanker in various local damage scenarios are shown in Figs. 10.20, 10.21, 10.22 and 10.23.

A comparison between the shear stress values in the intact and the various assumed local damage scenarios is given in Table 10.1. It is shown that the redistribution of the shear stress over of a ship section of a coastal oil tanker subjected to an assumed local damage in the shell plating gives very high values of shear stresses in the side shell, longitudinal bulkhead, deck and bottom plating, see Figs. 10.20, 10.21 and 10.23.

These high values of shear stresses when combined with hull girder bending and local stresses may induce unacceptable high values incompatible with *Von Misses* equivalent stresses.

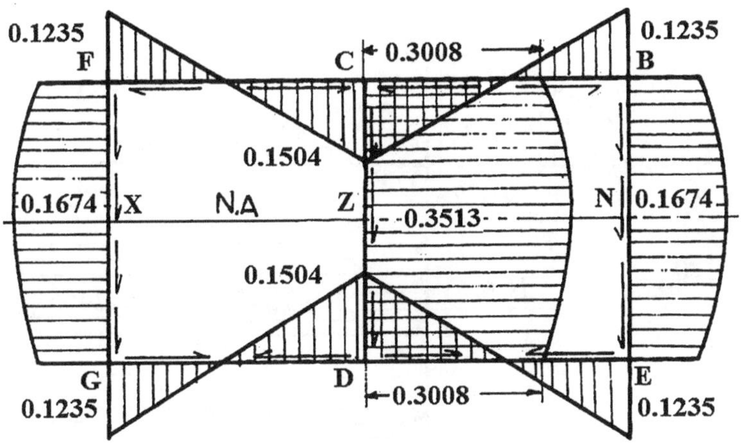

Fig. 10.19 Shear stress distribution over the ship section of a coastal oil tanker in the intact condition

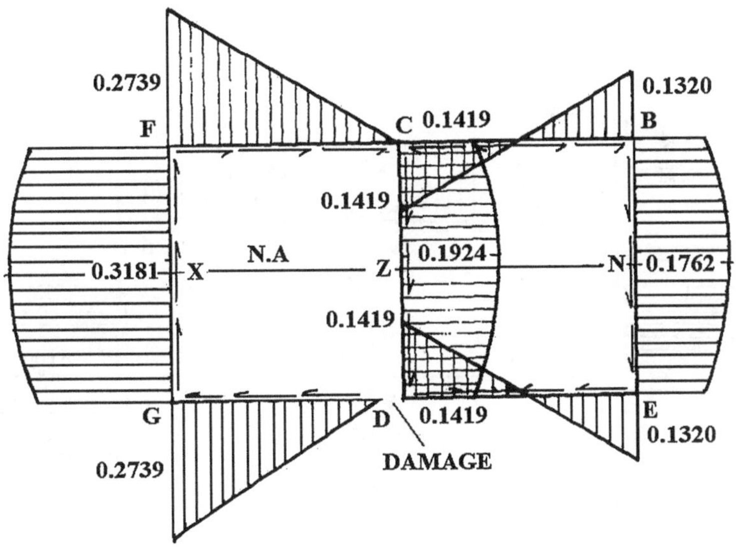

Fig. 10.20 Shear stress distribution over the ship section after damage has occurred to the port bilge plate

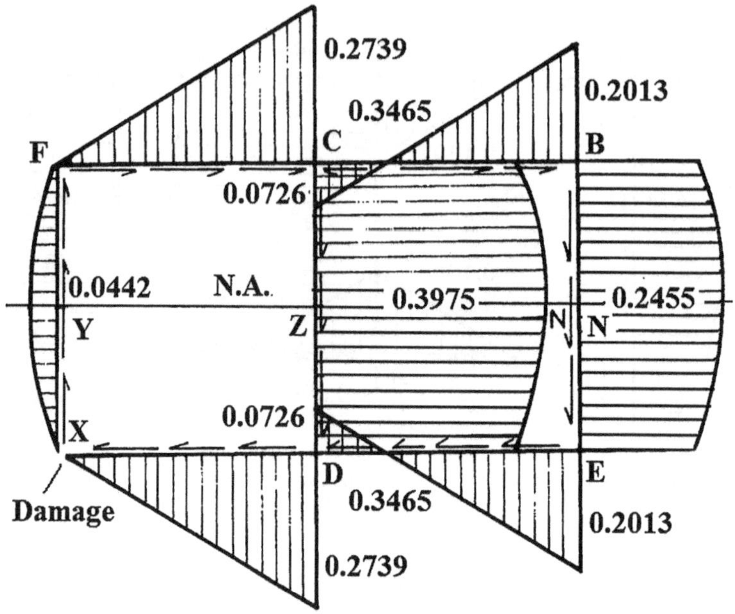

Fig. 10.21 Shear stress distribution over the ship section after damage has occurred at the bottom plating close to the centerline longitudinal bulkhead

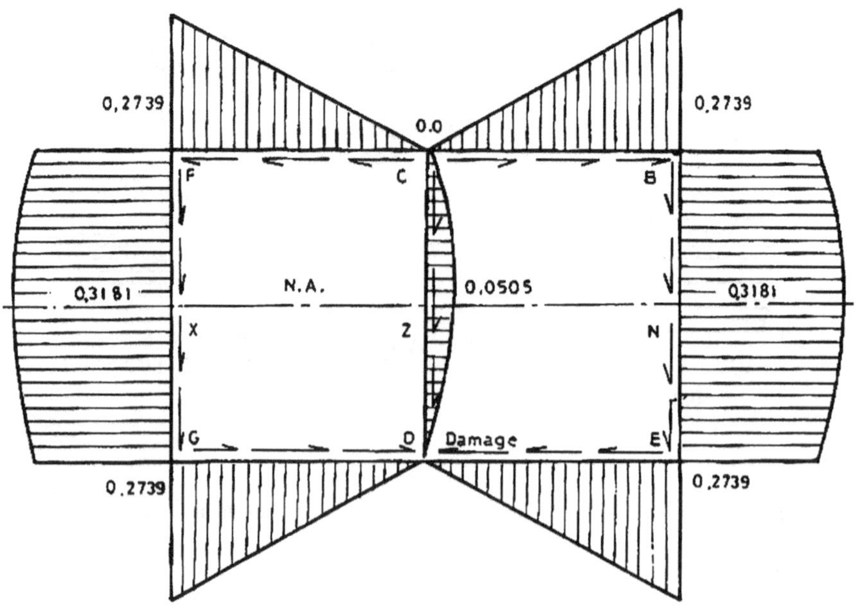

Fig. 10.22 Shear stress distribution over the ship section after damage has occurred at mid depth of the side shell structure

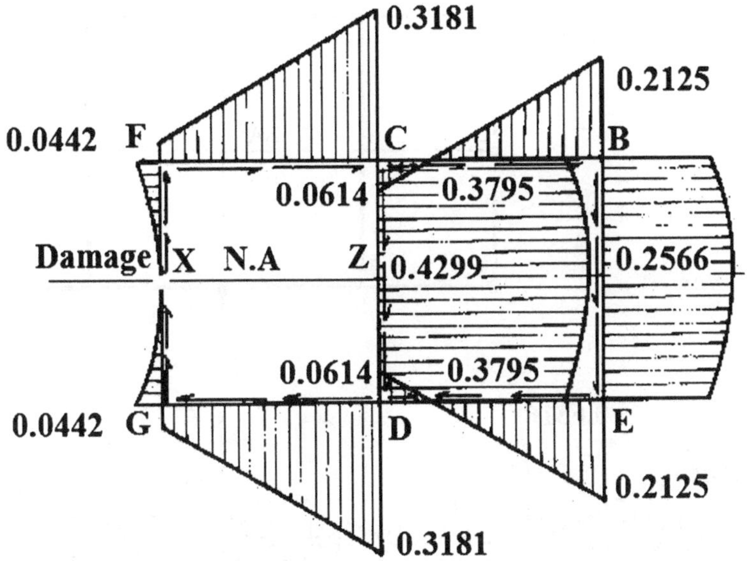

Fig. 10.23 Shear stress distribution over the ship section after damage has occurred at the bottom edge of the centerline longitudinal bulkhead

Table 10.1 Comparison between the shear stresses in the intact and the various assumed local damage locations over the tanker section

Case	τ_j/τ_o						
	1	2	4	5	6	7	8
1	0.358	*2.11*	1.26	*1.72*	*1.53*	1.22	–
2	–	*1.82*	1.15	*1.63*	*1.46*	1.13	0.264
3	*2.218*	–	–	2.22	*1.90*	0.14	*1.900*
4	*2.220*	–	0.47	1.07	1.05	0.55	*1.900*

10.4.3.2 Seagoing Oil Tankers

The main dimensions of the conventional sea-going tanker considered in this study are, see Fig. 10.24, Length = 260 m, Breadth = 40 m, and Depth = 20 m.

Fig. 10.24 A simplified idealization of a conventional sea-going tanker

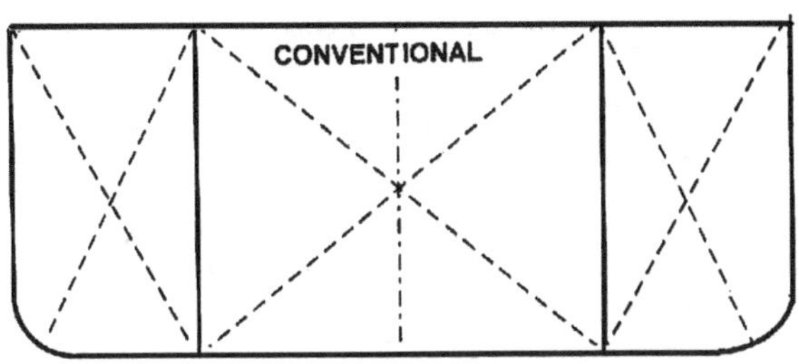

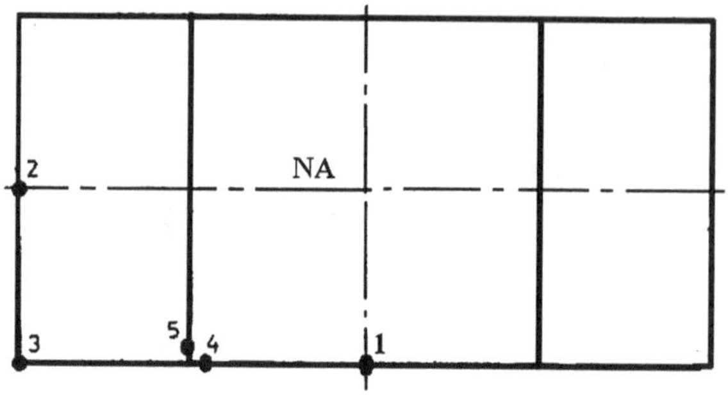

Fig. 10.25 Idealized ship section of the sea-going oil tanker

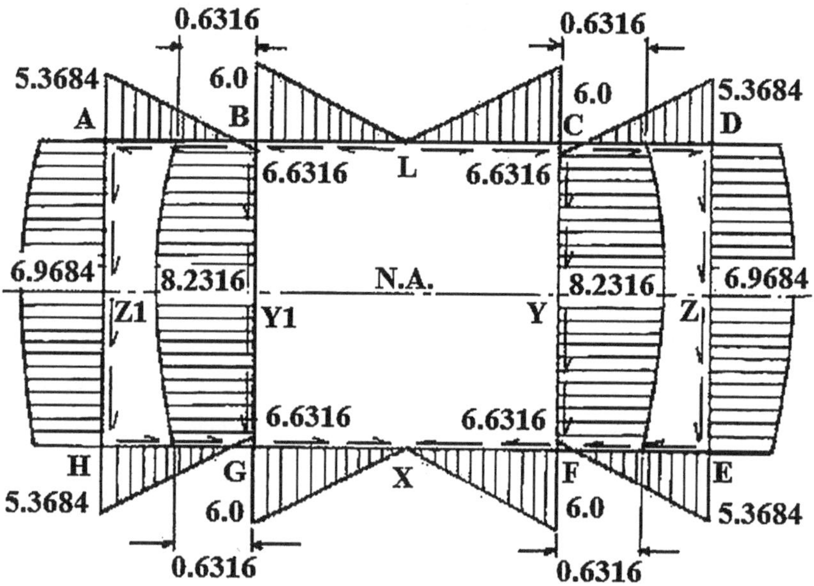

Fig. 10.26 Shear stress distribution over the ship section of a sea going oil tanker in the intact condition

Table 10.2 Comparison between the shear stress in the intact and the assumed various local damage locations over the tanker section

Case	τ_j/τ_o									
	1	2	4	5	6	8	9	10	11	12
1	0.298	*12.03*	*1.853*	1.22	*5.680*	*1.78*	–	*1.680*	0.646	*1.60*
2	–	*9.500*	*1.703*	1.12	*6.050*	*1.83*	0.226	*1.570*	0.544	*1.64*
3	1.300	*1.510*	–	*1.84*	*11.65*	*2.49*	1.230	0.194	0.644	*2.15*
4	0.559	*4.750*	0.452	*2.00*	*2.750*	*1.43*	0.660	0.559	*1.440*	1.34

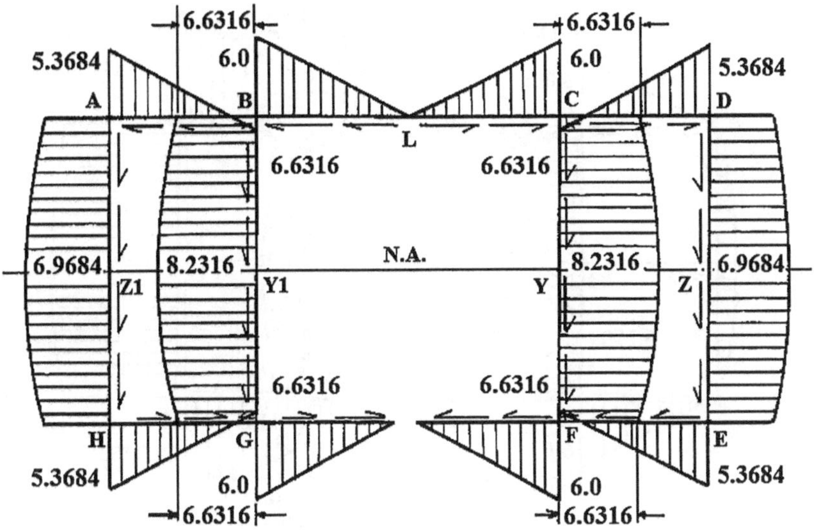

Fig. 10.27 Shear stress distribution over the ship section of a sea going oil tanker after an assumed damage has occurred at the keel plate

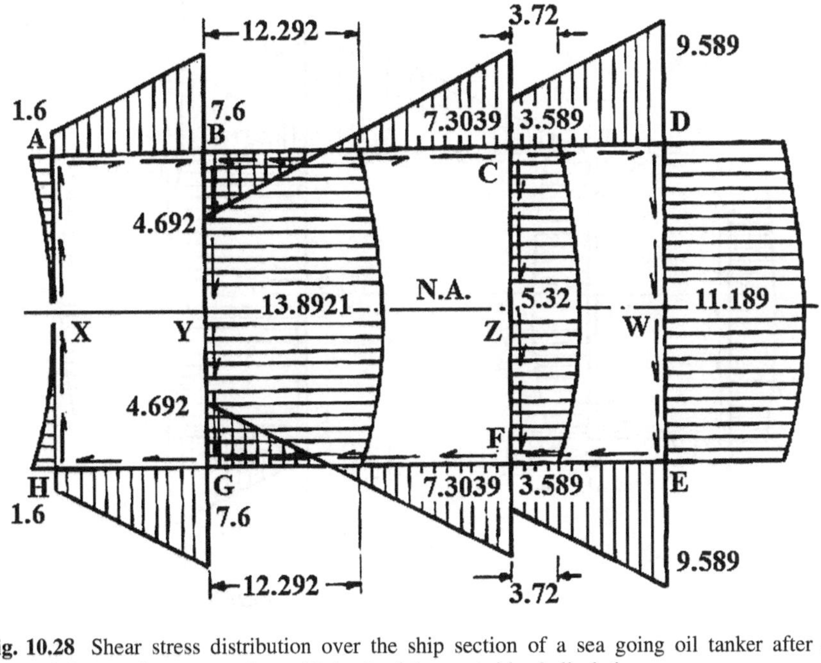

Fig. 10.28 Shear stress distribution over the ship section of a sea going oil tanker after an assumed damage has occurred at mid depth of the port side shell plating

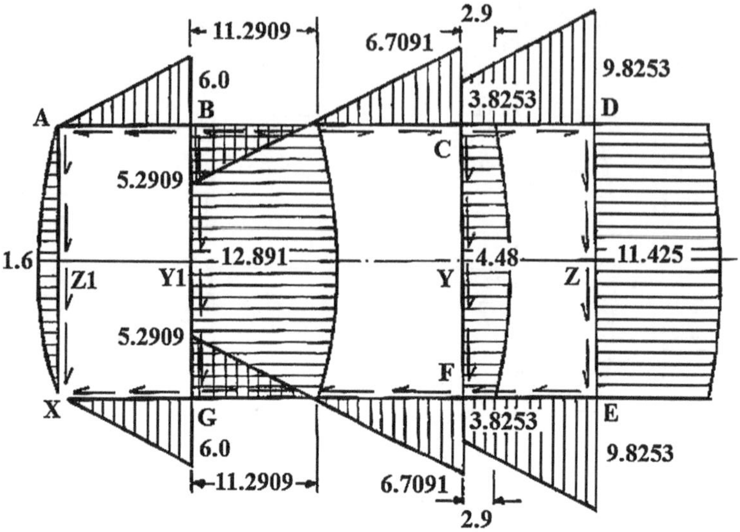

Fig. 10.29 Shear stress distribution over the ship section of a sea going oil tanker after an assumed damage has occurred at the port side of the bilge plating

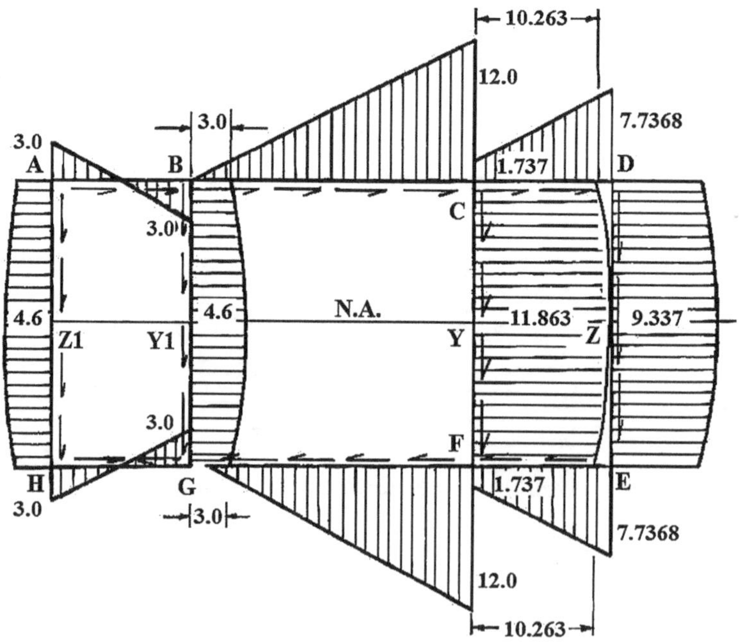

Fig. 10.30 Shear stress distribution over the ship section of a sea going oil tanker after an assumed damage has occurred at the bottom plating close to the port longitudinal bulkhead

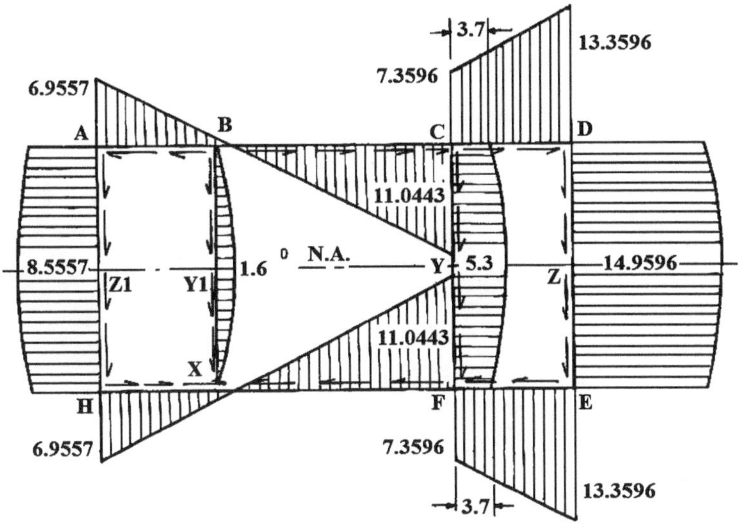

Fig. 10.31 Shear stress distribution over the ship section of a sea going oil tanker after an assumed damage has occurred at the bottom plating close to the port longitudinal bulkhead

The idealized ship section of the sea-going oil tanker is shown in Fig. 10.25. The various assumed damage locations are, see Fig. 10.25.

- Local damage at the keel plate (*Point 1*),
- Local damage at mid-depth of the side shell (*Point 2*),
- Local damage at the bilge strake at the Port side (*Point 3*),
- Local damage at the bottom plating close to the port side longitudinal bulkhead (*Point 4*),
- Local damage to the lower edge of the Port side longitudinal bulkhead (*Point 5*).

The shear flow distribution over the idealized ship section for the intact condition of the sea-going oil tanker is shown in Fig. 10.26. Table 10.2 gives a comparison between the shear stress values in the intact and the various assumed damage scenarios. It is shown that the redistribution of the shear stress over of a ship section of the sea-going tanker subjected to an assumed local damage in the shell plating or in the bottom end of the side longitudinal bulkhead give very high values of shear stresses in the side shell, longitudinal bulkhead, deck and bottom plating, see Figs. 10.27, 10.28, 10.29, 10.30 and 10.31.

Chapter 11
Shear Loading and Stresses in Bulk Carriers

11.1 Introduction

Bulk carriers are single decker vessels having normally an odd number of holds to carry cargo in bulk, such as grain. Cargo loading could be transported either in all the holds or in alternative holds, depending on the cargo type and density. The side structure could be designed and constructed either with single or double skin. For high density cargo, the double bottom is designed to have heavy structure to sustain the heavy cargo load.

A typical longitudinal section of a conventional single side bulk carrier is shown in Fig. 11.1.

11.2 Structural Configuration

Bulk carriers could be designed with either single sided structure or double sided structure. Conventional bulk carriers are single sided. The ship section of single sided structure configuration is composed of combined open and multi-box thin-walled structures, see Fig. 11.2.

Figure 11.2 shows a typical Structural configuration of conventional single side bulk carrier. The structural weaknesses of single side configuration results from the presence of the high rigidity of the top wing tank and hopper tank at the top and bottom ends of the side structure. Double sided structural arrangements are used to overcome the structural deficiency and weakness of single side configuration. Figure 11.3 shows a typical section of a bulk carrier having a double skin side structure. The ship section of double sided configuration is composed mainly of multi-box arrangement.

M. Shama, *Torsion and Shear Stresses in Ships*,
DOI: 10.1007/978-3-642-14633-6_11,

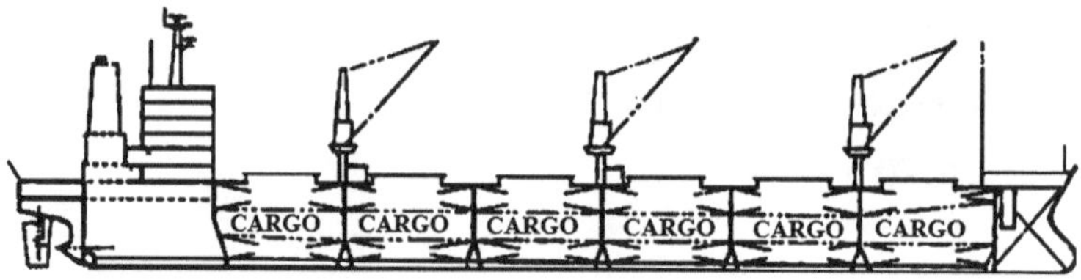

Fig. 11.1 A longitudinal section of a bulk carrier

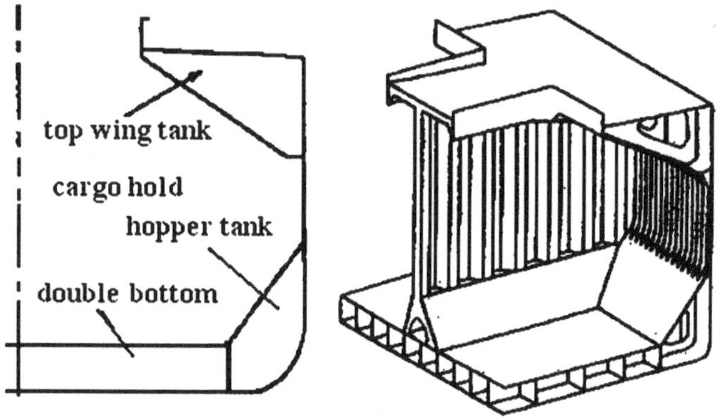

Fig. 11.2 Structural configuration of conventional bulk carriers

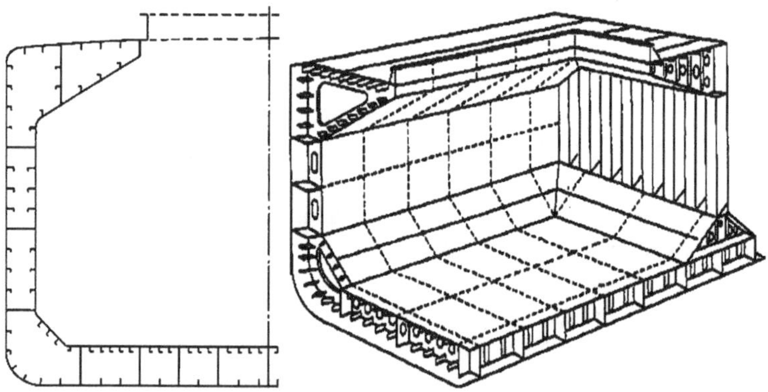

Fig. 11.3 Ship section of a double skin side structure

11.2.1 Upper and Lower Stools of Transverse Bulkheads

The bulk cargo requires that the transverse bulkheads have lower and upper stools to improve loading and unloading of the bulk cargo and also to improve structural strength of transverse bulkheads. Figure 11.4 shows the structural configuration of the upper and lower stools of the transverse bulkhead.

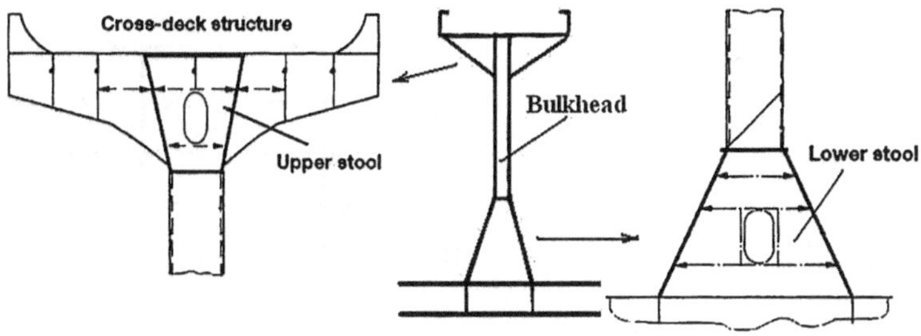

Fig. 11.4 Transverse bulkhead and stool

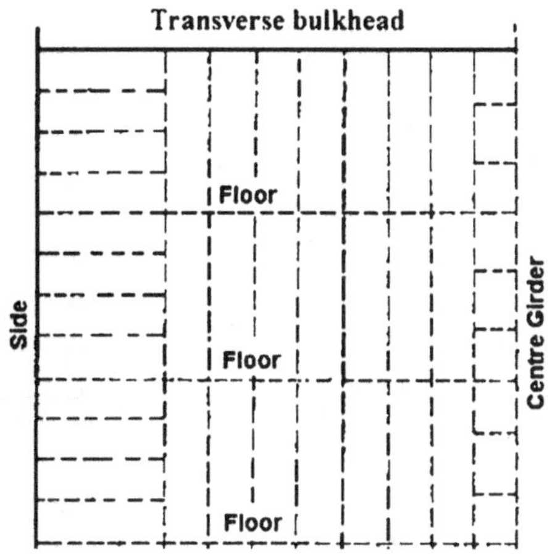

Fig. 11.5 Transverse system of framing of the double bottom structure

11.2.2 Double Bottom Structure

The double bottom structure of a general bulk carrier is normally stiffened by a combined system of framing comprising longitudinal girders and solid floors, see Figs. 11.5 and 11.6. The stiffness and rigidity of these members may be increased when the ship is designed for carrying ore. The attachments of the double bottom to both transverse bulkheads and side shell are normally designed in the form of a closed box-girder (a stool for the transverse forces bulkheads and a hopper tank for the side shell).

11.3 Hull Girder Loading

The hull girder of a ship is subjected to a very complex system of loading comprising shear forces, bending moments and torsion moments induced by cargo loads, wave induced forces, moments and torques and dynamic loading induced by ship motions among waves see Fig. 11.7. Figure 11.8 shows hull girder bending and shear forces acting on a middle part of the ship.

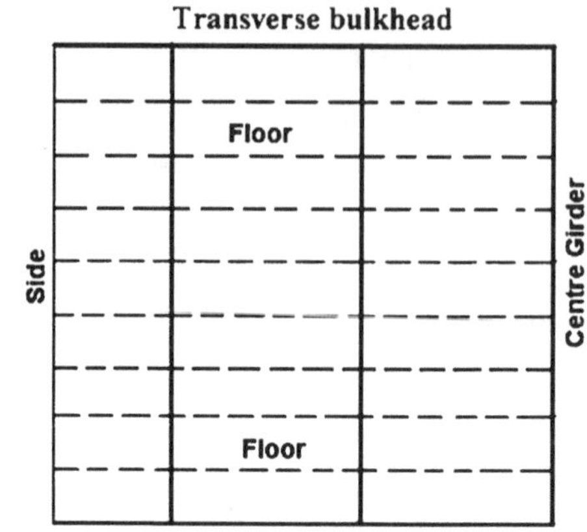

Fig. 11.6 Combined system of framing of the double bottom structure

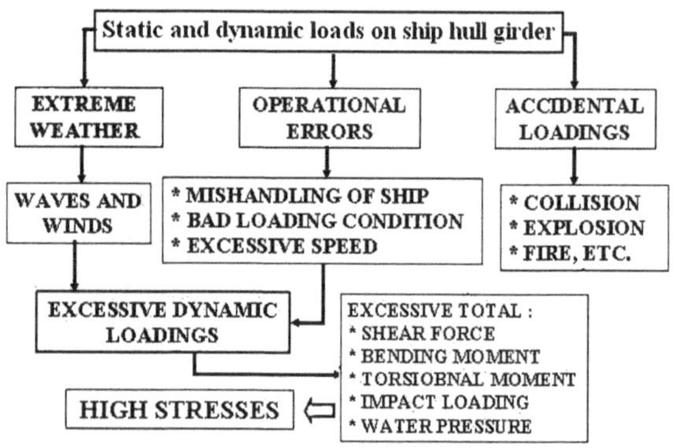

Fig. 11.7 Static and dynamic loads on ship hull girder

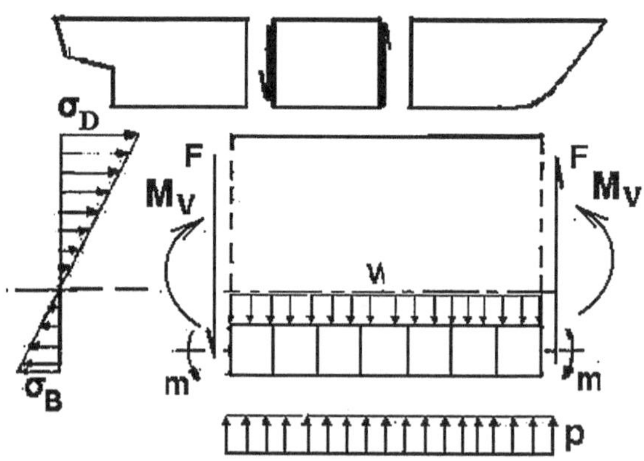

Fig. 11.8 Hull girder bending and shear forces

11.4 Longitudinal Vertical Shearing Force

Bulk carriers experience unique problems which result mainly from the particular structural configuration and the commonly used alternate hold loading arrangement, see Fig. 11.9.

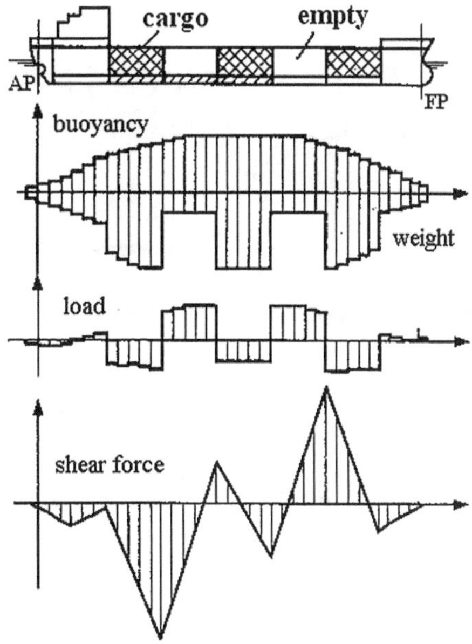

Fig. 11.9 Stillwater shear force distribution along ship length for alternate

The conventional calculation of longitudinal vertical shearing force and bending moment, along the ship length, gives results pertinent only to the general behavior of the idealized ship girder, without due regard to the effect of local cargo loading. Figure 11.9 shows the shear force distribution along ship length for the general condition of cargo loading in alternate holds.

11.4.1 Stillwater Component (FS)

The effect on the magnitude and distribution of the Stillwater shear force and bending moments along ship length of loading cargo in all holds is shown in Fig. 11.10 and loading in alternate holds is shown in Fig. 11.11. It is shown that loading of cargo in alternate holds could induce high values of Stillwater shear forces and bending moments in comparison with uniform cargo loading in all holds.

11.4.1.1 Effect of Cargo Hold Length on the Magnitude and Distribution of the Vertical Shear Force

The effect on the magnitudes of the Stillwater bending moments and shear forces for alternate loading system of the length of loaded cargo holds relative to the length of unloaded holds is shown in Fig. 11.12. In configuration (a), all cargo holds have the same length. In configuration (b), cargo holds have shorter lengths than empty holds. The magnitude of the maximum shear force and bending moment in configuration (b) are larger than those for configuration (a).

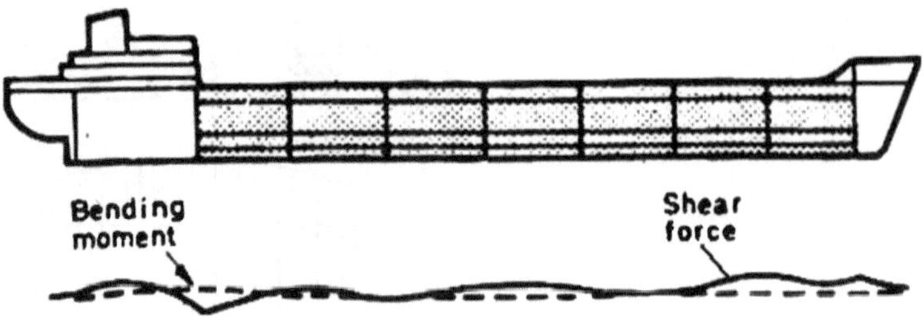

Fig. 11.10 Shear force distribution for uniform loading of cargo holds

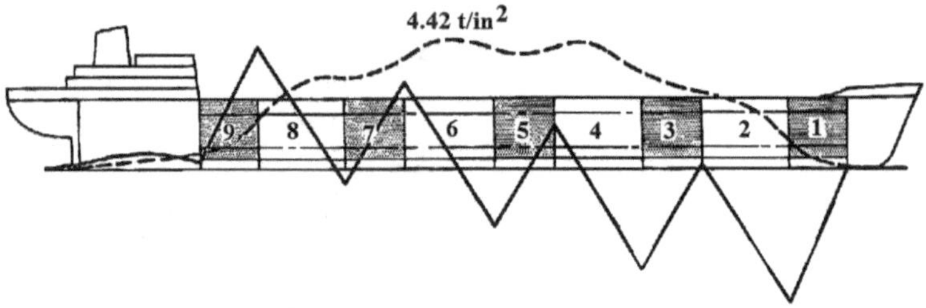

Fig. 11.11 Shear force distribution for alternate cargo hold loading

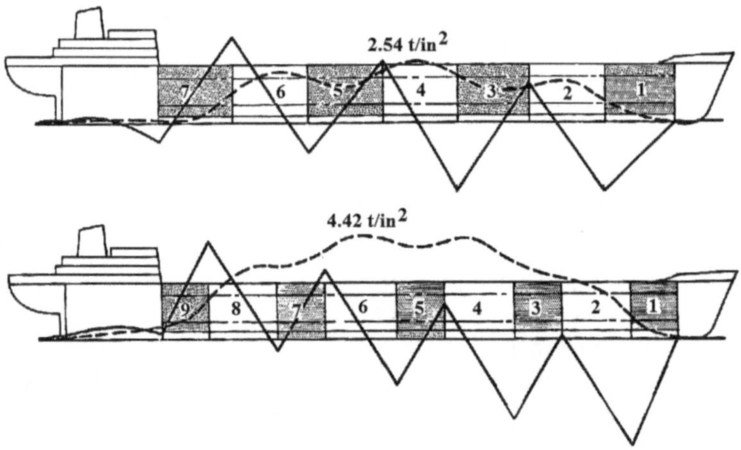

Fig. 11.12 Effect of cargo holds length on the bending moments and shear forces for alternate loading

11.4.1.2 Mechanism of Load Transmission

The net local loading on the double bottom is the difference between the cargo loading (grain, ore, ... or ballast) acting on the tank top and the upward water pressure acting on the outer skin. For ships having a single bottom, such as oil tankers, the effect of cargo loading in the holds may not be very significant, whereas for ships having a double bottom, such as bulk carriers and *OBO* ships

(*Ore Bulk Oil*), the effect of cargo loading may become very significant, especially when the alternate hold loading arrangement is used and particularly when the transverse bulkheads structure are of double skin.

In bulk carriers, the longitudinal vertical shearing force is carried by the side shell plating, top wing tanks, and hopper tanks. The side shell, therefore, may carry a high proportion of the longitudinal vertical shearing force. Consequently, shear stresses in the side shell plating may reach unfavorable values. Shear buckling, or high values of combined or equivalent stresses, may also occur in some panels in the side shell plating. Adequate measures should, therefore, be taken to prevent instability and high stresses in the side shell plating.

Because of the high rigidity of the double bottom and transverse bulkheads, the shearing force carried by the side shell plating should be corrected for local cargo loading, particularly when the cargo loading is carried in alternate holds, see Figs. 11.13 and 11.14 shows the shear forces carried by the longitudinal girders in the double bottom structure and the shear force carried by the transverse bulkhead structure.

Any section along the ship length is subjected to a shear free *F* and a bending moment *M*. The shear force is carried by the side shell and the double bottom structure. The proportion of the shear force carried by the side shell depends upon:

- The stiffness of the double bottom structure and particularly bottom girders.
- Aspect section, i.e., length breadth of hold.
- Loading condition, i.e., alternate hold loading ballast, …, etc.

The bending moment, on the other hand, is carried by the whole section of the ship i.e., no correction to the bending moment diagram is necessary.

The double bottom structure will transmit this local loading to the side shell and transverse bulkheads in the form of shear forces and bending moments, see Fig. 11.15. The mechanism of load transmission depends mainly upon the construction of the double bottom, that is, the stiffness and rigidity in the longitudinal

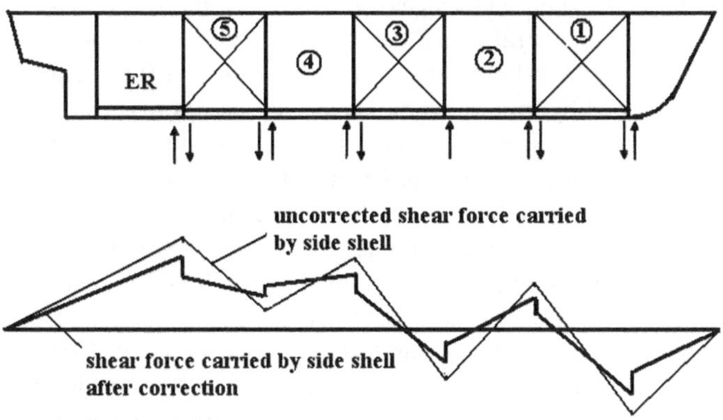

Fig. 11.13 Shear force distribution over a defined length for a combined system of framing in the double bottom structure

Fig. 11.14 Shear force correction

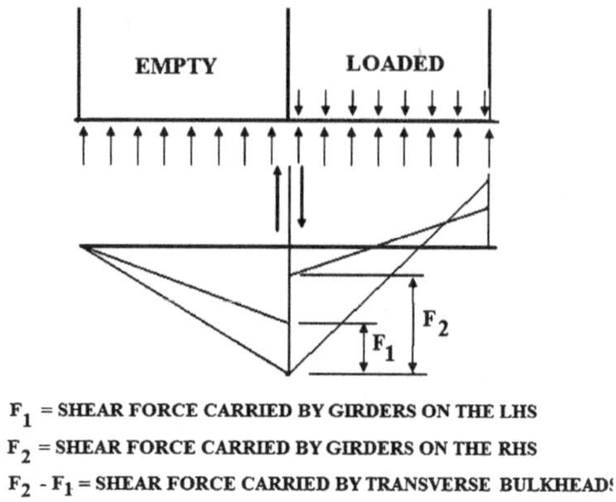

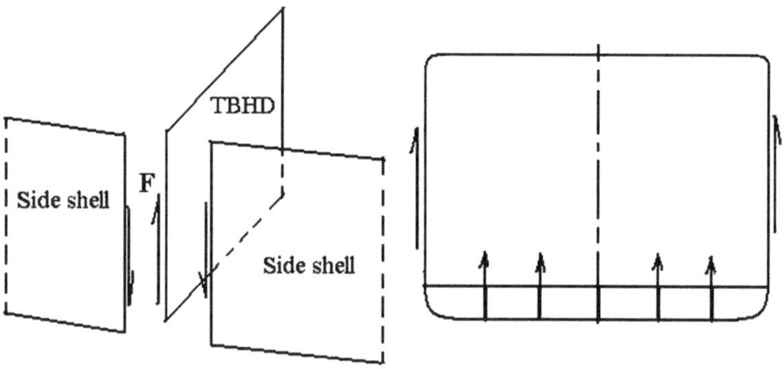

Fig. 11.15 Interaction of forces between transverse bulkheads, side shell and double bottom

and transverse directions, the construction of the attachments between the double bottom and both side shell and transverse bulkheads, the shear area of transverse bulkheads, the hold aspect ratio, and the loading condition (uniform or in alternate holds). For relatively short holds (that is, of *aspect ratio* = 0.5), nearly 80% of the net loading is carried by transverse bulkheads as shear forces.

The shear forces transmitted to the side shell as concentrated shear forces represent the local corrections to the general distribution of the shear force diagram obtained from longitudinal strength calculation.

The mechanism of load transmission is best explained by Fig. 11.15. It is shown that, for a transversely framed double bottom, the shear loading is entirely carried by the side shell, and the effect of local loading is almost nonexistent. Because of the stiffness of double bottom, the shear force diagram obtained from longitudinal strength should be corrected to give the actual shear force carried by the side shell.

When longitudinal girders are used in the construction of the double bottom, the shear forces carried by the side shell will be reduced by the contribution of these girders to the local net loading, see Fig. 11.16. The proportion of the local shear

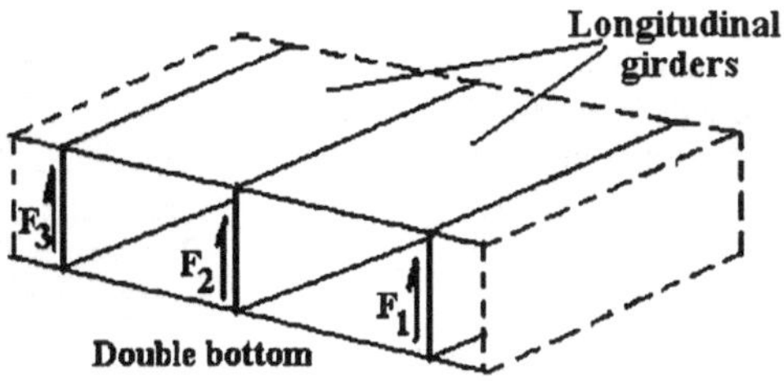

Fig. 11.16 Local loading on double bottom hull girders

forces carried by the transverse bulkheads could be estimated by solving the double bottom as a grillage structure.

11.4.1.3 Ship Section Structural Capability for Sustaining Stillwater Shear Force

The procedure for ensuring adequate capability of the ship section of a bulk carrier to sustain the induced still water shear force for both uniform loading or for alternate loading is shown in Fig. 11.17. The maximum values of the Stillwater shear force and bending moment should not exceed the allowable values given in the Rules of Classification Societies.

11.4.1.4 Shear Loading on Local Structural Elements

The hydrostatic loading on the bottom structure of oil tankers is transmitted to the transverse webs via the bottom longitudinals in the form of shear forces. Figure 11.18 shows the mechanism of load transmission with and without a lug and a vertical stiffener.

11.5 Wave-Induced Component (*FW*)

The wave induced shear force component results from the distribution of the forces of support throughout the length of a ship during her passage among waves, see Fig. 11.19. The variability of this component results from the variability of the sea state and ship parameters.

An approximate estimate of the maximum value of the wave induced bending moment M_W could be obtained from the maximum values of the wave bending moments of box-shaped and diamond-shaped vessels as follows:

Let

$M_b = M_W$ for box shaped vessel,
$M_b = \gamma BhL^2/4\pi^2$,
$M_d = M_W$ for diamond shaped,

Fig. 11.17 Procedure for determining an acceptable still-water shear force for uniform and alternate loading

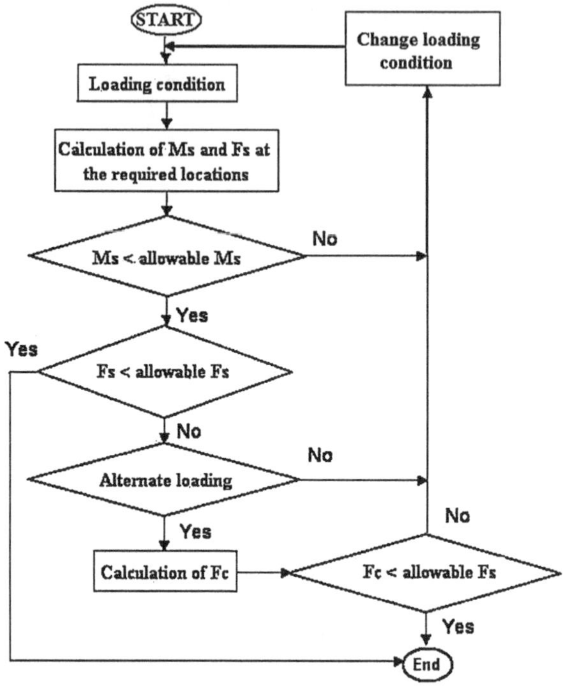

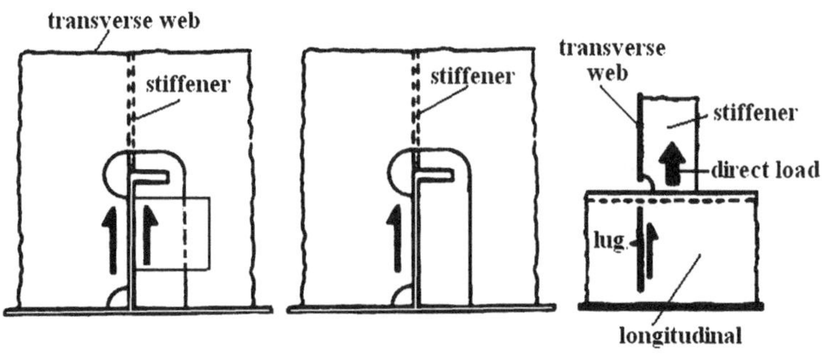

Fig. 11.18 Local shear forces in a longitudinal-transverse connection

Fig. 11.19 Wave induced bending moment and shear force

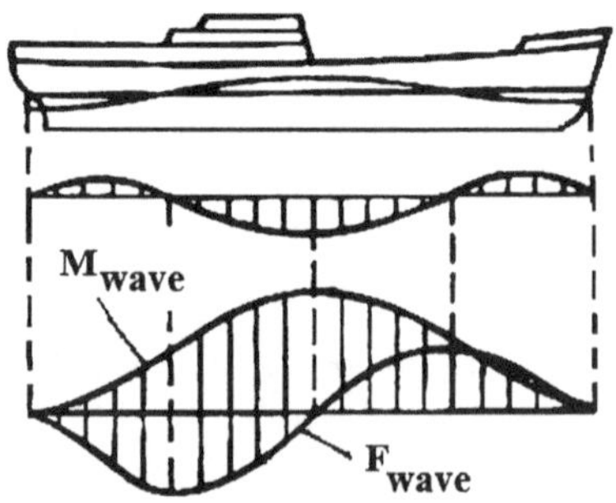

$M_d = \gamma BhL^2/12\pi^2 = 1/3 \cdot M_b$,
$M_{sh} = M_W$ for ship shaped vessel.

Hence the wave induced bending moment for a ship is given by

$$M_{sh} = \alpha M_b + (1 - \alpha)M_d \quad \text{For } 0 \leq \alpha \leq 1.0$$

where α is a factor that could be determined as follows, $\alpha = 2(C_b - 0.5)$, C_b is the block coefficient of the ship.

Hence

$$\begin{aligned} M_{sh} &= 2M_b \cdot (C_b - 0.5) + [1 - 2(C_b - 0.5)] \cdot M_d \\ &= 2[M_b \cdot (C_b - 0.5) + (1 - C_b) \cdot M_d] \end{aligned}$$

Substituting for M_b and M_d we get

$$\begin{aligned} M_{sh} &= 2[M_b \cdot (C_b - 0.5) + (1 - C_b) \cdot M_b/3] \\ &= M_b/3 \cdot (4C_b - 1) \\ &= \gamma BhL^2/12\pi^2 \cdot (4C_b - 1) \\ M_{sh} &= 0.61/(3 \times 40) \cdot B \cdot L^{2.5} \cdot (4C_b - 1) \qquad \text{For } h = 0.607\sqrt{L} \\ &\cong 5 \times 10^{-3} \cdot B \cdot L^{2.5} \cdot (4C_b - 1) \end{aligned}$$

11.5.1 The Distribution of the Largest Expected Vertical Wave-Induced Shearing Force

Using wave statistics, it is possible to obtain the long-term distribution function of *FW*. This technique has been used successfully for the prediction of bending moments-among several other responses experienced by a ship over her service life.

Wave induced shear force is given in the Rules of all Classification societies for both sagging and hogging conditions in the following formulation, see Fig. 11.20

$$\begin{aligned} F_{hog} &= +0.3 \cdot F_1 \cdot C \cdot L \cdot B \cdot (C_b + 0.7) \quad \text{kN} \\ F_{sag} &= -0.3 \cdot F_2 \cdot C \cdot L \cdot B \cdot (C_b + 0.7) \quad \text{kN} \end{aligned}$$

where L and B are in meters

$$\begin{aligned} C &= 10.75 \times [(L - 300/)150]^{1.5} && \text{For} \quad 90 \leq L \leq 300 \\ C &= 10.75\, C = 10.75 && \text{For} \quad 300 \leq L \leq 350 \\ C &= 10.75 - [(L - 350)/150]^{1.5} && \text{For} \quad 350 \leq L \end{aligned}$$

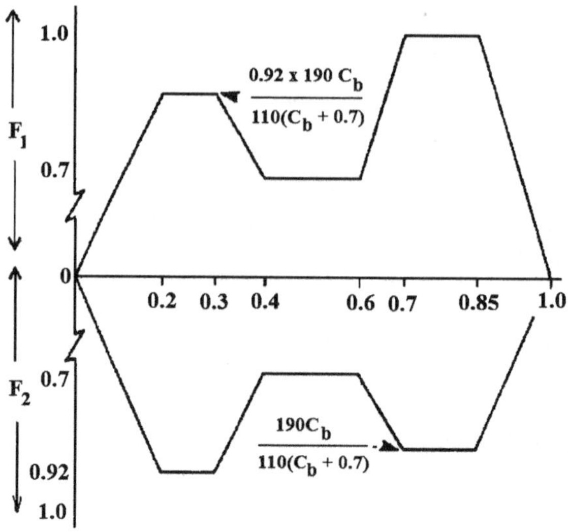

Fig. 11.20 Shear load coefficients

The variability of *FW* results also from the variability of the main ship parameters affecting her responses. These parameters are

- Main ship dimensions (particularly ship length),
- Form coefficients,
- Ship speed,
- Angle between ship course and direction of wave advance,
- Weight distribution along ship length,
- Sea state.

The sea state is influenced by various factors, among them the fetch, wind velocity, direction and duration. It is possible, however, to represent any particular sea state by its energy spectrum.

The mathematical formulations of the wave spectra are normally given in terms of some observable properties of the sea, such as the average wave height and average wave period.

Figure 11.21 shows the distribution of the largest expected vertical wave-induced shearing force for a probability of exceedence of 10^{-8}.

11.6 Dynamic Component (*FD*)

This component could be either impulsive, resulting mainly from ship motions among waves, see Fig. 11.22, slamming, or non-impulsive, resulting from the vibration of the hull girder, see Fig. 11.23. Its magnitude is influenced by ship's

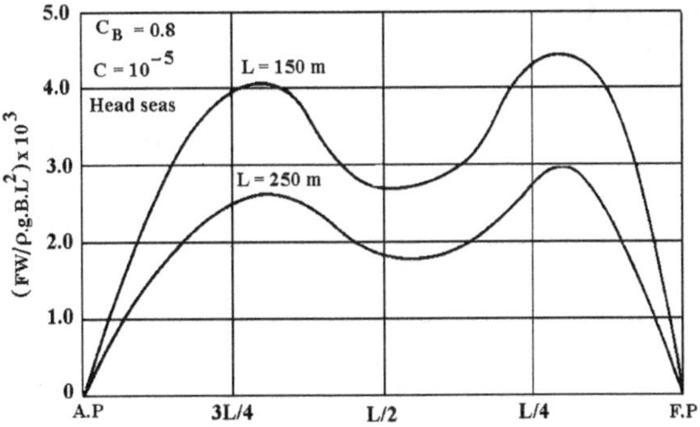

Fig. 11.21 Distribution of largest expected vertical shear force

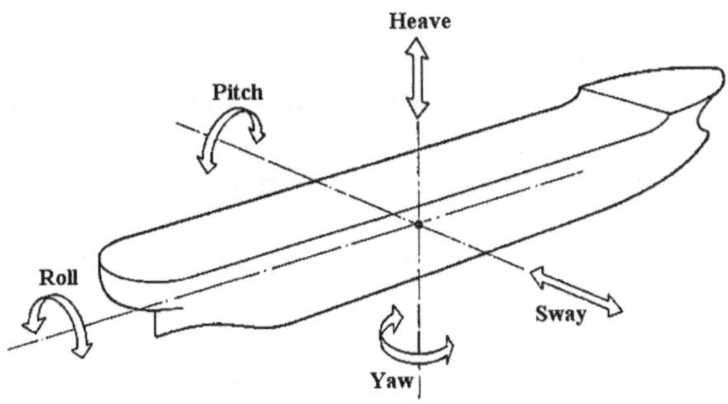

Fig. 11.22 Types of ship motions

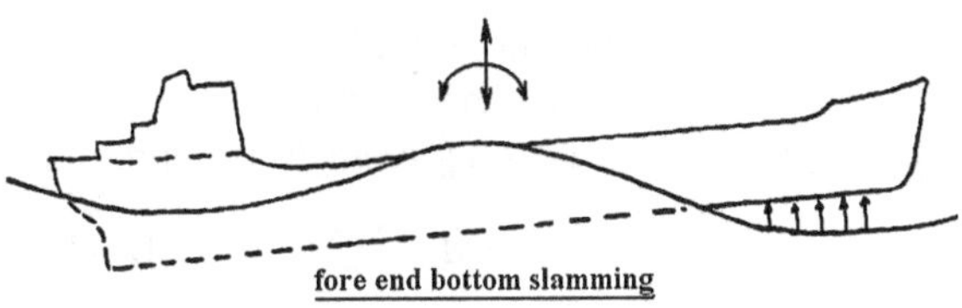

Fig. 11.23 Dynamic loads on the fore-end of the bottom structure of a ship

speed, heading, length, draft, trim, sea condition, and weight distribution. These factors affect the amplitudes of pitching and heaving, which affect the magnitude of the dynamic component.

The impulsive component increases directly as the area under the bow that appears above the water surface increases, see Fig. 11.24. Therefore, for ships having a full form in the bow and running at a shallow draft in a severe sea condition, the impulsive component may be rather significant. This is the case of a bulk carrier, or an oil tanker, in the ballast condition.

An approximate estimate of the maximum impulsive dynamic shear force component could be obtained from the dynamic bending moment component. It is shown that the maximum impulsive dynamic shearing force component could be estimated by

$$\left(M_{Dyn}\right)_{max} \cong 3\left(M_{dyn}\right)_{max}/L$$

where L is the ship length.

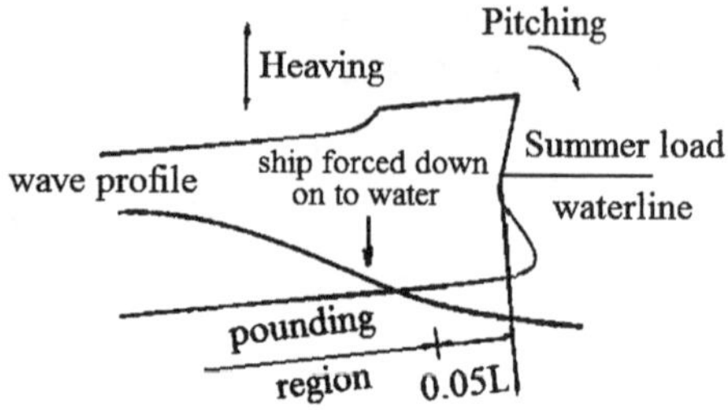

Fig. 11.24 Dynamic loads on the fore-end of the bottom structure of a ship

An approximate estimate of the impulsive dynamic shearing force component could be obtained from the available data on the dynamic bending moment component.

The maxim dynamic bending moment is given by

$$\left(M_{Dyn}\right)_{max} = K \cdot B^2 \cdot L \cdot h/\lambda$$
$$K = 0.2 \times [a + b \cdot (C_b - 0.6) +]C \cdot T/0.08$$

where a, b and c are coefficients depending on the d/L ratio and *Froude* No.; L, B and d are length of load line, breadth and draft of the ship; h and λ are the height and length of the resonant wave (wave having a period equal to the natural pitching period of the ship in Still-water; T is the trim; C_b is the block coefficient

Therefore, for bulk carriers having *zero* trim and a *Froude* No. 0.15, the maximum impulsive bending moment is approximately given by:

$$\left(M_{Dyn}\right)_{max} \cong 0.1 \cdot B^2 \cdot L \cdot h/\lambda$$

The maximum impulsive dynamic shearing force is approximately given by

$$\left(F_{Dyn}\right)_{max} = 0.3 \cdot B^2 \cdot h/\lambda$$

11.7 Total Vertical Shearing Force F

The total shear force acting on any section along the length of ship hull girder of a bulk carrier is composed of the following main components:

- Still water
- Wave induced
- Slamming
- Whipping
- Springing

However, for design purposes, the total longitudinal vertical shearing force, at any section, along a ship steaming in waves is the vectorial sum of the still-water component, wave-induced component, and the dynamic component.

The total vertical shear force is the vectorial summation of the various components as given by

$$F = F_s + F_w + F_h + F_i$$

where F_s is the vertical still water component; F_w is the vertical wave induced component; F_h is the springing component (high frequency component); F_i is the impulsive dynamic component.

Using this total vertical shearing force, the shear stress distribution around a typical ship section of a bulk carrier could be calculated.

11.8 Approximate Value to the Maximum Vertical Shear Force

The method is based on the assumption that the distribution of the wave induced bending moment could be given by

$$M = M_x/2 \cdot (1 - \cos(2\pi X/L)).$$

Since

$$M = dM/dx = F = 0 \qquad \text{at } x = 0 \text{ and } x = L.$$

Then

$$\begin{aligned} F = dM/dx &= M_x/2 \cdot [2\pi/L \cdot \sin(2\pi x/L)] \\ &= \pi/L \cdot M_x \cdot \sin(2\pi x/L). \end{aligned}$$

F_{max} occurs at approximately L/4 and 3L/4.

Then

$$F_{max} = \pi M/L \cong CM/L$$

where

$$\begin{aligned} C &= \text{factor depending on ship type} \\ &= 3.50 \text{ for Passenger Ships} \\ &= 3.75 \text{ for Cargo Ships} \\ &= 4.30 \text{ for Oil Tankers.} \end{aligned}$$

11.9 Variation of Various Shear Stress Components with Time

The variation of the various shear stresses (still water, wave induced and high frequency shear stresses) with time is shown in Fig. 11.25.

11.10 Shear Flow Distribution in Bulk Carriers

11.10.1 Structure Idealization

In order to calculate the shear stress distribution over a typical ship section of a bulk carrier, the ship section should be idealized by a simplified configuration so as to reduce the laborious calculations associated with shear flow distribution. The complex structure configuration of a conventional bulk carrier is shown in Fig. 11.26.

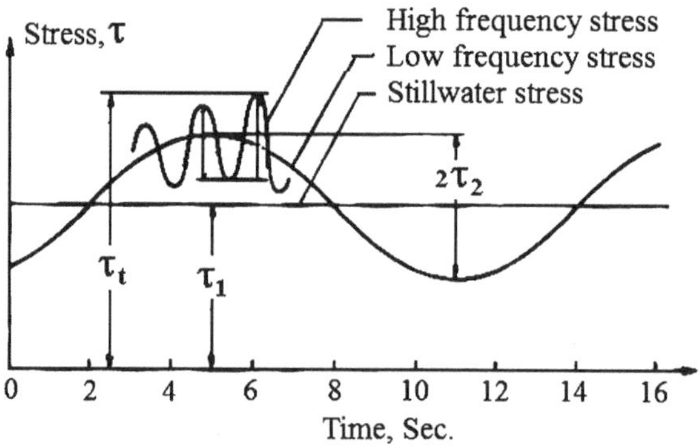

Fig. 11.25 Variation of various shear stress components with time

Fig. 11.26 Complex structure configuration of a conventional bulk Carrier

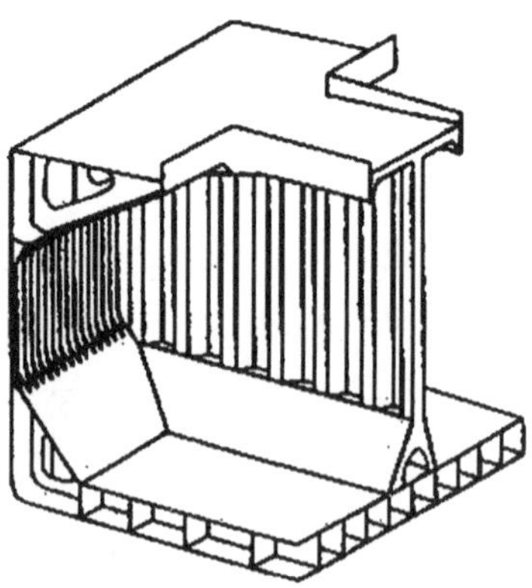

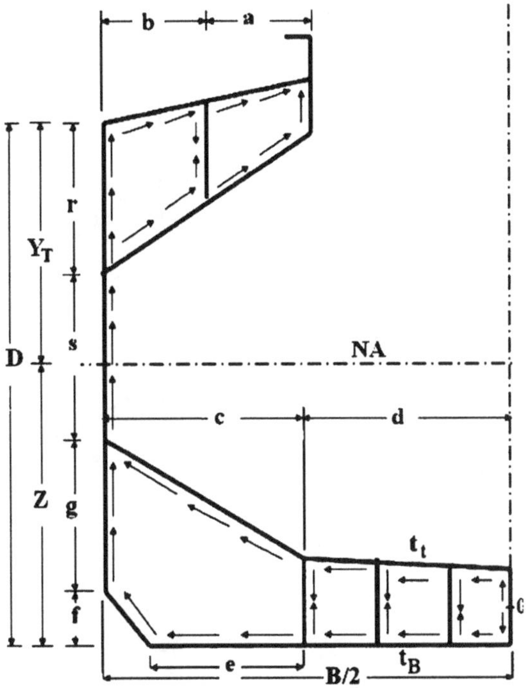

Fig. 11.27 Idealized ship section

The idealized structure should neither affect the magnitude nor the distribution of shear flow around the top wing tanks, hopper tanks, nor side shell.

Since the main objective is to present a simplified procedure for calculating shear flow distribution around a typical section of a bulk carrier, see Fig. 11.27, then the local shear flow distribution within the double bottom structure may be ignored for the sake of simplicity. This could be achieved by replacing the complex structure of the double bottom by an equivalent plate structure; see Fig. 11.28. The idealized ship section should retain the same geometrical properties; namely, total sectional area, shear area, position of neutral axis, and second moment of area of ship section about the neutral axis of the ship section.

The longitudinal stiffening members may not included in the calculation of shear flow distribution, as this would have a negligible effect on the results of shear flow calculations. These stiffening members, however, should be taken into account by using effective thicknesses instead of actual thicknesses.

11.10.2 Effective Thickness

The effective plating thickness of an element j is given by

$$t_{je} = t_j + \sum a_j / l_{ij}$$

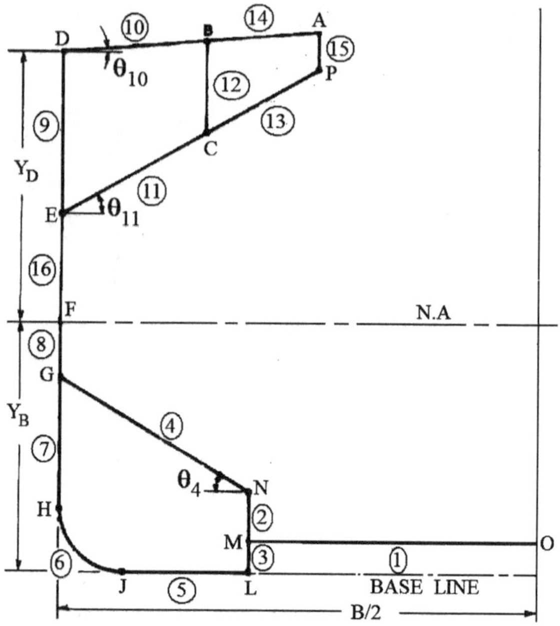

Fig. 11.28 A simplified idealized ship section

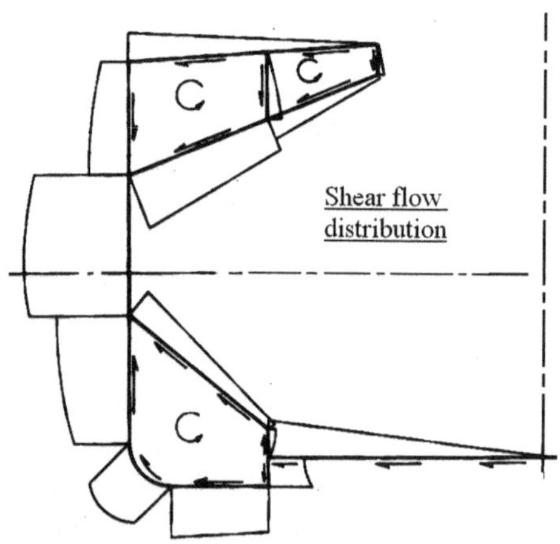

Fig. 11.29 Shear flow distribution over an idealized ship section of a bulk carrier

where t_{je} is the effective plating thickness of' element j; t_j is the original plating thickness of element j; a_{ij} is the area of stiffening member i within element j; l_j is the length of element j.

11.10.3 Shear Flow Distribution

Using a simplified idealization, the shear flow distribution over the ship section is shown in Fig. 11.29.

11.10.4 Shear Stress Distribution

The shear stress at any location on the ship section is calculated as follows

$$\tau = q/t$$

where q is the shear flow; t is the actual plate thickness; τ is the shear stress.

The importance of calculating shear stresses around a ship section stems from the need to determine the maximum allowable shearing force for the ship section.

11.10.5 Shear Flow Distribution Over the Hopper Tank

To simplify the calculations, the complex structure configuration of the double bottom and hopper tank, see Fig. 11.30, is idealized using the concept of effective thickness of plating as given by

$$t_e = t_t + t_B$$

The combined effective thicknesses of the double bottom plating and longitudinals is located at the neutral axis of the double bottom structure, see Fig. 11.31.

Let point O be the centroid of the center line girder, see Fig. 11.31. Due to symmetry of the ship section, the shear flow at point O is *zero*.

Hence

$$q_o = 0.0$$

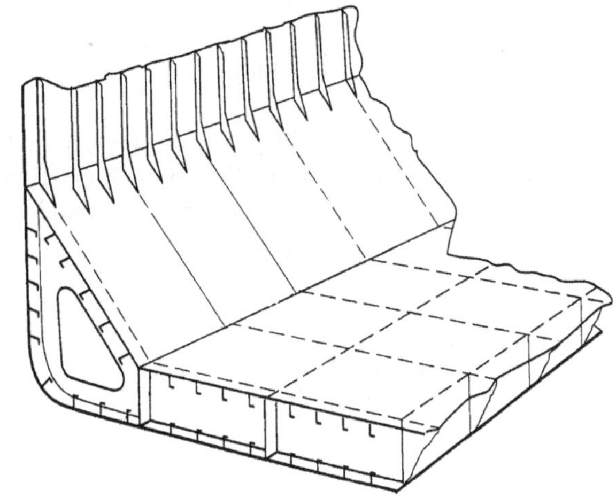

Fig. 11.30 Double bottom and hopper tank structure

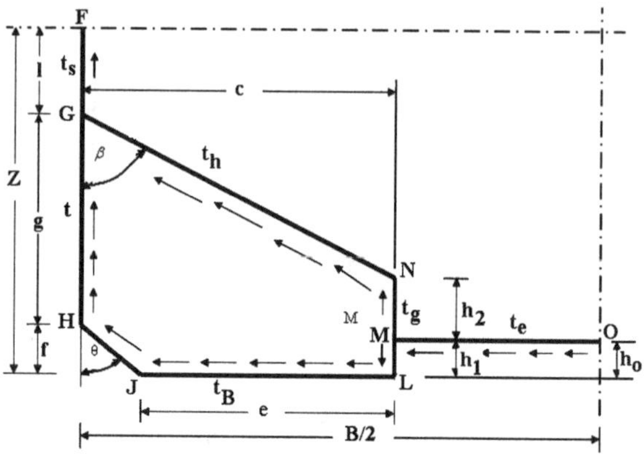

Fig. 11.31 Assumed shear flow over the bottom structure

11.10.5.1 Assumed Shear Flow Distribution Over the Bottom Structure

In order to calculate the shear flow distribution around the hopper tank, the shear flow q_{MN} is assumed to be zero, see Fig. 11.31.

i.e.,

$$q_{MN} = 0.0$$

The shear flow at any point i is given by

$$q_i = F/I \cdot \sum A\bar{y}.$$

Hence

$$\begin{aligned} q_{MO} &= q_0 + A_{OM} \cdot [z - (h_1 + h_O)/2] \\ &= A_{OM} \cdot [z - (h_1 + h_O)/2] \\ q_{NM} &= q_{MO} + q_{MN}. \end{aligned}$$

11.10.5.2 Correcting Shear Flow

The assumed shear flow will cause the hopper tank to distort by an angle θ. A correcting shear flow q_c is applied in the hopper tank to correct this distortion, see Fig. 11.32. This connecting shear flow should produce an angle of twist θ_c equal and opposite to the induced angle of twist θ

$$\theta + \theta_c = 0.0$$

The correcting shear flow is given by

$$q_c = -\oint q \frac{\Delta s}{t} \Big/ \oint \frac{\Delta s}{t}$$

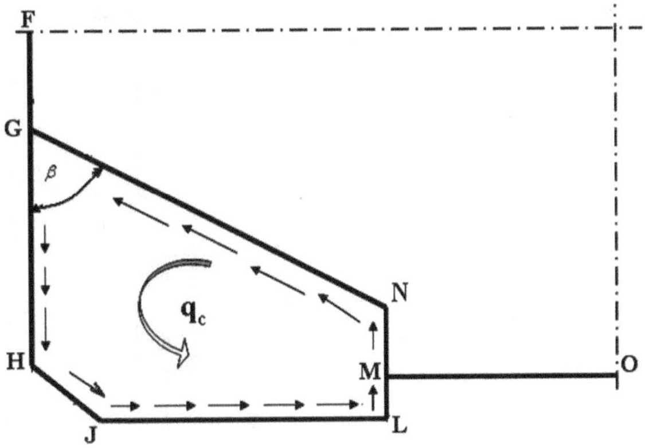

Fig. 11.32 Correcting shear flow in the hopper tank

$$\oint \Delta s/t = h_2/t_g + c/t_h \sin(\beta) + h_1/t_g + e/t_B + f/t_B \cos(\theta) + g/t_l = h/t_g + C/t_h \sin(\beta) + e/t_B + f/t_B \cos(\theta) + g/t_l$$

$$\oint q\Delta s/t = (q_{MO} + 2q_{ML}/3) \cdot h_1/t_g + (q_{LM} + q_{LJ}/2) \cdot e/t_B + (q_{JL} + q_{JH}/2) \cdot f/t_B \cos(\theta) + (q_{HJ} + 2/3 \cdot q_{HG}) \cdot g/t_l - q_{NM} \cdot h_2/3t_g - (q_{NM} + q_{NG}/2) \cdot c/t_h \sin(\beta)$$

The shear flow distribution over the lower part of the ship section of the bulk carrier is therefore given by

$$q_{MO} = A_{OM}[Z - (h_1 + h_O)/2]$$
$$q_{NM} = q_{MO} + q_{MN} + q_C$$
$$q_{GN} = q_{NM} + q_{NG}$$
$$q_{LM} = q_{MO} + q_{ML} - q_C$$
$$q_{JL} = q_{LM} + q_{LJ}$$
$$q_{HJ} = q_{JL} + q_{JH}$$
$$q_{GH} = q_{HJ} + q_{HG}$$
$$q_G = q_{GN} + q_{GH}$$
$$q_{FG} = q_G + q_{GF}$$

11.10.6 Shear Flow Distribution Over the Top Wing Tanks

The calculation of the shear flow over the ship structure above the neutral axis of the ship section could be carried out by assuming a cut at point *B* for cell *1* and another cut at point *E* for cell 2, see Fig. 11.33.

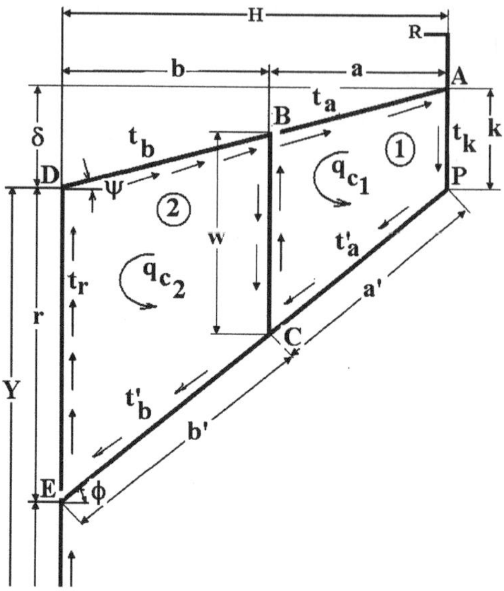

Fig. 11.33 Assumed cuts in cells 1 and 2 and shear flow

11.10.6.1 Assumed Shear Flow

The assumed shear flow distribution over the two cells will induce angles of twist for each cell. As it is assumed that no torsional moments are applied and that the ship section is subjected only to a vertical shear loading, a correcting shear flow is applied to the two cells to nullify the induced angles of twist.

The assumed shear flow is calculated using the following basic equation

$$q_i = F/I \cdot \sum A\bar{y}$$

The induced angle of twist for cell *1* is given by

$$\theta_1 = 1/2A_1G \cdot \oint_1 q_1 ds/t$$

And the induced angle of twist for cell *2* is given by

$$\theta_2 = 1/2A_2G \cdot \oint_2 q_2 ds/t.$$

11.10.6.2 Calculation of the Correcting Shear Flow

Assume the correcting shear flow for cell *1* = q_{c1}, see Fig. 11.34, and the correcting shear flow for cell *2* = q_{c2}.

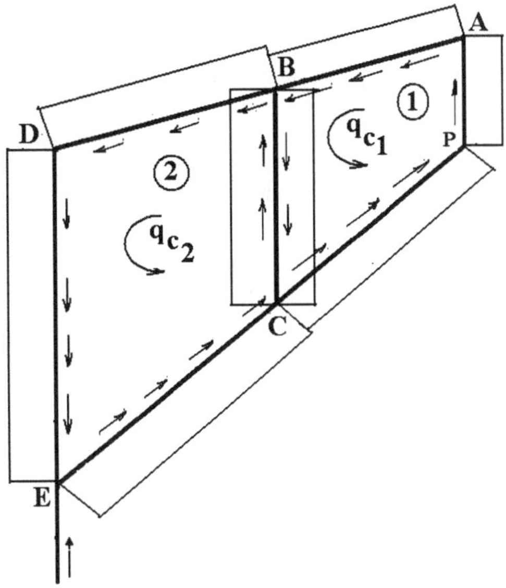

Fig. 11.34 Correcting shear flow

As the magnitude of the angles of twist resulting from the assumed shear flows equals the magnitude of the angles of twist resulting from the correcting shear flows, we have $\theta_1 = \theta_2 = 0.0$ where θ_1 is the angle of twist of cell *1*; θ_2 is the angle of twist of cell *2*.

But

$$\theta_1 = \theta_{q1} - \theta_{c1} \tag{11.1}$$

$$\theta_2 = \theta_{q2} - \theta_{c2} \tag{11.2}$$

where θ_{C1} is the correcting angle of twist for cell *1* and is given by $\theta_{C1} = q_c/2A_1G \cdot \oint_1 \Delta s/t$, θ_{C2} is the correcting angle of twist for cell *2* and is given by $\theta_{C2} = q_c/2A_2G \cdot \oint_2 \Delta s/t$, θ_{q1} is angle of twist resulting from the assumed shear flow in cell *1*, θ_{q2} is the angle of twist resulting from the assumed shear flow in cell 2.

Substituting for the angles of twist in Eqs. (11.1) and (11.2), we get

$$q_{C1} \sum_1 \Delta S/t - q_{C2}(\Delta S/t)_{12} - \oint_1 q_1 \Delta S/t = 0 \tag{11.3}$$

$$q_{C2} \sum_2 \Delta S/t - q_{C1}(\Delta S/t)_{12} - \oint_2 q_2 \Delta S/t = 0. \tag{11.4}$$

Solving Eqs. (11.3) and (11.4), we can determine q_{c1} and q_{c2}.

11.10.6.3 Correct Distribution of the Shear Flow Over the Top Wing Tank

The corrected shear flow at any point in cell *1* is given by

$$(q_i)_1 = q_1 - q_{c1}.$$

The corrected shear flow at any point in cell *2* is given by

$$(q_i)_2 = q_2 - q_{c2}.$$

Part III
Programming Implementation

Chapter 12
Programming Implementation

12.1 Introduction

The aforementioned methodology and its associated numerical examples are modeled as separate subjective modules managed by a main executive program namely *PROP*®. *FORTRAN*® is the programming language in which the modules are written. Attached to the program are *four* standard mathematical subroutines, beyond a few accessories subroutines that function as follows

i. *GAUSELM*: solves a linear system of equations which are required for the calculation of shear center in multi-cell closed section.
ii. *SIMPUN*: calculates the area under a curve using *Simpson's* integration rules for both equal and unequal intervals.
iii. *DPLINT*: determines the polynomial which interpolates a set of discrete data points.
iv. *DPOLVL*: calculates the value of a polynomial together with its first *NDER* derivatives, where the polynomial was produced by a previous call to *DPLINT*.

For any arrangement of rectangular cross sections, e.g., vertical, horizontal, or inclined, *PROP*® calculates their physical properties, e.g., the centre of area, the principal axes together with the corresponding principal moments of inertia and the other sectional area properties, the shear center of closed, opened, and combined cross sections, in addition to the sectorial properties of the thin walled sections. Also, the interested researcher may use the attached standard mathematical subroutines to manipulate the distribution of the sectional and/or sectorial properties together with their first few derivatives (*if necessary*).

The information gathered in this program is expected to be sufficient for the first glance without going into more detailed discussion. However, for further details of the mentioned formulations and their associated computer programs, the interested researcher may consult the appropriate subjective chapters of this book, or he may refer to the listed references at the end of the book. All of the surveyed

M. Shama, *Torsion and Shear Stresses in Ships*,
DOI: 10.1007/978-3-642-14633-6_12, © Springer-Verlag Berlin Heidelberg 2010

formulations and their associated computer programs are computationally fast using the standard *IBM*® compatible computers, without any special requirements of the hardware configuration.

One of the main goals of the *PROP*® program is the easy possibility of addition to and deletion of functional modules as required. The program is not an optimization routine but still considered to belong to the preliminary structural design stage without any economical or optimization consideration.

12.2 Program List

Following is the detailed listing of the *PROP*® program:

```
      PROGRAM PROP
C
C     BEGIN PROLOGUE PROP
C     PURPOSE: DETERMINATION OF THE PHYSICAL PROPERTIES OF A RECTANGULAR
C              SEGMENTAL CLOSED, OPENED, AND COMBINED CROSS SECTION, BY
C              ANALYZING ANY ARRANGEMENT OF RECTANGULAR ELEMENTS FORMING
C              A CROSS-SECTIONAL AREA. THE CALCULATED PHYSICAL PROPERTIES
C              ARE THOSE OF CENTROID, DIRECTION OF THE PRINCIPAL AXES,
C              PRINCIPAL MOMENTS OF INERTIA, LOCATION OF THE SHEAR CENTER
C              WITH RESPECT TO THE ORIGINAL AND PRINCIPAL AXES, AND SECTOR
C              PROPERTIES.
C
C     LIBRARY: STRUCTURAL ANALYSIS
C     TYPE: DOUBLE PRECISION
C     KEYWORDS:
C     AUTHOR: Dr. K.A.Hafez, DEPARTMENT OF NAVAL ARCHITECTURE AND MARINE
C             ENGINEERING, FACULTY OF ENGINEERING, ALEXANDRIA UNIVERSITY
C
C     DESCRIPTION:
C     FORMAL PARAMETERS:
C     DIMENSIONING INFORMATION:
C     REFERENCES:
C     ROUTINES CALLED: PRP, DODO, DADA, DIDI, DRDR, DSTR
C     REVISION HISTORY (YYMMDD): 980101  DATE WRITTEN
C     END PROLOGUE PROP
C
      IMPLICIT REAL*8(A-H,O-Z)
C
C     PARAMETER(NGDOF=154,NNDOF=7,NSTN=21,NSEC=22)
C     PARAMETER(NSEG=,NSTIF=)
C
      DOUBLE PRECISION IDSTIFF,IWSTIFF,MM,KK,KY,KZ,LBP,IYY,IZZ
      COMMON/CONS/PI,N,NL,NJ
      COMMON/GENR/Z1(75),Z2(75),BETA(75),NQ(75),NCB(75)
     &           ,M(75,75),ML(75,75),NC(75,75),ND(75,75),QN(75,75)
      COMMON/ELEM/W(75),D(75),NSTIFF(75),INDELM(75,2),NEL(75,10)
      COMMON/DID/CN(75),CC(75),FA(75),THETR
      COMMON/STIF/ASTIFF(75,10),XSTIFF(75,10),YSTIFF(75,10)
     &           ,IDSTIFF(75,10),IWSTIFF(75,10)
      COMMON/PR/X(75),Y(75),ALP(75),XX(75),YY(75),JN(2,75)
      COMMON/TORWR/CI(21),CJ(21)
      COMMON/SHCOF/KY(21),KZ(21)
      COMMON/SCT/IYY(21),IZZ(21),AST(21),SC(21,2),YA(21),ZA(21),YS(21)
     &           ,ZS(21),ASC(21)
      COMMON/COEF/C(75,75)
      COMMON/DAD/QQ(75,6),QC(2,23),QOS(75)
      COMMON/DOD/ACL(75),YO(21),ZO(21),INDCS(75,4),XXM(75),YYM(75)
     &           ,WM(75),DM(75),RR(75),WW(75,2),AE(5)
      COMMON/DRD/ASTIFFN(75,10),XSTIFFN(75,10),YSTIFFN(75,10)
     &           ,NSTIFFN(75)
C
C    &      ,THETN(75),THETF(75),PHIN(75),PHIF(75)
C
      COMMON/PRDIM/LBP
      COMMON/MK/MM(154,154),KK(154,154),DMK(154,154)
      COMMON /NUMB/NSTN,NNDF,NDOF
      DIMENSION Z(154,154),EGN(154),PRM(7,22,154),XMODE(7,22,154)
     & ,VEC(22,154),XCORD(22),WR(154),WI(154)
C
C    & ,XMD(7,22,154),XM2D(7,22,154)
C    & ,PC(22),YPP(2),YCORD(22)
C     ,IDS,IWS
C
C     OPEN(3,FILE='\DCKST.IP'  ,STATUS='UNKNOWN')
C     OPEN(5,FILE='\SEC.IP'    ,STATUS='UNKNOWN')
C     OPEN(5,FILE='\TRALLL.IP' ,STATUS='UNKNOWN')
C     OPEN(5,FILE='\SEC1.IP'   ,STATUS='UNKNOWN')
C     OPEN(5,FILE='\SEC2.IP'   ,STATUS='UNKNOWN')
C
      OPEN( 10,FILE='\CHK.OP'    ,STATUS='UNKNOWN')
      OPEN( 50,FILE='\PROP.IP'   ,STATUS='UNKNOWN')
      OPEN( 60,FILE='\SEC.OP'    ,STATUS='UNKNOWN')
      OPEN( 90,FILE='\PROP.OP'   ,STATUS='UNKNOWN')
```

```
      OPEN( 96,FILE='\CHICK.OP' ,STATUS='UNKNOWN')
      OPEN(100,FILE='\DEIGEN.IP',STATUS='UNKNOWN')
C
      WRITE(10,*)'PROP I/P'
C
      CALL PRP
      CALL DODO
      CALL DADA
      CALL DIDI
      CALL DRDR
      CALL DSTR
C
      WRITE(10,*)'PROP O/P'
C
      STOP
      END
******
C     BEGIN PROLOGUE PRP
C     PURPOSE:
C
C     LIBRARY: STRUCTURAL ANALYSIS
C     TYPE: DOUBLE PRECISION
C     KEYWORDS:
C     AUTHOR: Dr. K.A.Hafez, DEPARTMENT OF NAVAL ARCHITECTURE AND MARINE
C             ENGINEERING, FACULTY OF ENGINEERING, ALEXANDRIA UNIVERSITY
C
C     FORMAL PARAMETERS:
C     REFERENCES:
C     ROUTINES CALLED: NON
C     REVISION HISTORY (YYMMDD): 980101  DATE WRITTEN
C     END PROLOGUE PRP
C
      SUBROUTINE PRP
      IMPLICIT REAL*8(A-H,O-Z)
      COMMON/CONS/PI,N,NL,NJ
      COMMON/GENR/Z1(75),Z2(75),BETA(75),NQ(75),NCB(75)
     &          ,M(75,75),ML(75,75),NC(75,75),ND(75,75),QN(75,75)
      COMMON/SCT/IYY(21),IZZ(21),AST(21),SC(21,2),YA(21),ZA(21),YS(21)
     &          ,ZS(21),ASC(21)
      COMMON/ELEM/W(75),D(75),NSTIFF(75),INDELM(75,2),NEL(75,10)
      COMMON/DID/CN(75),CC(75),FA(75),THETR
      COMMON/STIF/ASTIFF(75,10),XSTIFF(75,10),YSTIFF(75,10)
     &           ,IDSTIFF(75,10),IWSTIFF(75,10)
      COMMON/PR/X(75),Y(75),ALP(75),XX(75),YY(75),JN(2,75)
      COMMON/COEF/C(75,75)
      COMMON/SHCOF/KY(21),KZ(21)
      COMMON/DAD/QQ(75,6),QC(2,23),QOS(75)
      COMMON/DOD/ACL(75),YO(21),ZO(21),INDCS(75,4),XXM(75),YYM(75)
     &          ,WM(75),DM(75),RR(75),WW(75,2),AE(5)
      COMMON/DRD/ASTIFFN(75,10),XSTIFFN(75,10),YSTIFFN(75,10)
     &          ,NSTIFFN(75)
C    &     ,THETN(75),THETF(75),PHIN(75),PHIF(75)
      DOUBLE PRECISION IDSTIFF,IWSTIFF,KY,KZ,IYY,IZZ
C         ,IDS,IWS
C
      WRITE(10,*)'PRP IP'
C
      PI=22.0D00/7.0D00
C
      READ (50,603)N,NJ,NL
      WRITE (60,603)N,NJ,NL
  603 FORMAT(3I5)
C
      NL1=NL+1
C
      READ  (50,614)(INDELM(I,1),I=1,N)
      WRITE (60,614)(INDELM(I,1),I=1,N)
      READ  (50,614)(INDELM(I,2),I=1,N)
      WRITE (60,614)(INDELM(I,2),I=1,N)
      READ  (50,614)(NCB(L),L=1,N)
      WRITE (60,614)(NCB(L),L=1,N)
  614 FORMAT(15I5)
C
```

```
      DO 31 I=1,N
            NEL(I,1)=0
            DO 31 J=1,N
                  NC(I,J)=0
                  ND(I,J)=0
                  QN(I,J)=0.0
   31 CONTINUE
C
      READ (50,614)(NQ(I),I=1,N)
      WRITE(60,614)(NQ(I),I=1,N)
      READ (50,614)(NSTIFF(I),I=1,N)
      WRITE(60,614)(NSTIFF(I),I=1,N)
C
      DO 222 I=1,N
             IF(NSTIFF(I).NE.0)THEN
                NSTIF=NSTIFF(I)
                READ (50,605) (ASTIFF(I,J),J=1,NSTIF)
                WRITE(60,605)(ASTIFF(I,J),J=1,NSTIF)
             ENDIF
  222 CONTINUE
C
      DO 125 I=1,N
             IF(NSTIFF(I).NE.0)THEN
                NSTIF=NSTIFF(I)
                READ (50,605) (IDSTIFF(I,J),J=1,NSTIF)
                WRITE(60,605)(IDSTIFF(I,J),J=1,NSTIF)
             ENDIF
  125 CONTINUE
C
      DO 126 I=1,N
             IF(NSTIFF(I).NE.0)THEN
                NSTIF=NSTIFF(I)
                READ (50,605) (IWSTIFF(I,J),J=1,NSTIF)
                WRITE(60,605)(IWSTIFF(I,J),J=1,NSTIF)
             ENDIF
  126 CONTINUE
C
      DO 123 I=1,N
             IF(NSTIFF(I).NE.0)THEN
                NSTIF=NSTIFF(I)
                READ (50,605)(XSTIFF(I,J),J=1,NSTIF)
                WRITE(60,605)(XSTIFF(I,J),J=1,NSTIF)
             ENDIF
  123 CONTINUE
C
      DO 124 I=1,N
             IF(NSTIFF(I).NE.0)THEN
                NSTIF=NSTIFF(I)
                READ (50,605)(YSTIFF(I,J),J=1,NSTIF)
                WRITE(60,605)(YSTIFF(I,J),J=1,NSTIF)
             ENDIF
  124 CONTINUE
C
  605 FORMAT(7F11.6)
             READ  (50,609)(II,XX(I),YY(I),I=1,NJ)
             WRITE (60,609)(I,XX(I),YY(I),I=1,NJ)
  609        FORMAT(I5,2F10.3)
C
      K=0
      DO 10 II=1,N
            READ  (50,606)NODE,JN(1,II),JN(2,II),W(II),D(II)
            WRITE (60,606)NODE,JN(1,II),JN(2,II),W(II),D(II)
  606       FORMAT(3I5,2F10.4)
            NQC=NQ(II)
            NCBB=NCB(II)
            READ  (50,607)(NC(II,L),L=1,NCBB)
            WRITE (60,607)(NC(II,L),L=1,NCBB)
            READ  (50,607)(ND(II,L),L=1,NCBB)
            WRITE (60,607)(ND(II,L),L=1,NCBB)
  607       FORMAT(2I5)
            READ  (50,711)(M(II,L),L=1,NQC)
            WRITE (60,711)(M(II,L),L=1,NQC)
  711       FORMAT(15I5)
```

```
            IF(NCBB.NE.0)THEN
               DO 38 L=1,NCBB
                     I=NC(II,L)
                     QN(II,I)=ND(II,L)
                     NEL(I,1)=NEL(I,1)+1
                     KK=1
                     DO 39 LL=1,N
                     DO 39 MM=1,NCBB
                        IF(NC(LL,MM).EQ.I)THEN
                           KK=KK+1
                           NEL(I,KK)=LL
                        ENDIF
   39                CONTINUE
                     J=NEL(I,1)
                     ML(J,I)=II
                     K=K+1
                     CC(K)=ND(II,L)
   38          CONTINUE
            ENDIF
            I=II
            J1=JN(1,I)
            J2=JN(2,I)
            R=YY(J1)-YY(J2)
            Q=(XX(J1)-XX(J2))
            ALP(I)=DATAN2(R,Q)
            ALP(I)=ALP(I)/(PI/180.0D00)
            X(I)=(XX(J1)+XX(J2))/2.0D00
            Y(I)=(YY(J1)+YY(J2))/2.0D00
   10 CONTINUE
C
      DO 50 I=1,N
            NQC=NQ(I)
            IF(NQC.EQ.0)THEN
               NQC=1
               M(I,1)=0
            ENDIF
   50 CONTINUE
C
      DO 40 II=1,N
            NCBB=NCB(II)
            IF(NCBB.EQ.0)THEN
               NCBB=1
               NC(II,1)=0
               ND(II,1)=0
            ENDIF
   40 CONTINUE
C
      AST(11)=0.0D00
      SUMMX  =0.0D00
      SUMMY  =0.0D00
      DO 20 I=1,N
            A=W(I)*D(I)
            AMX=A*Y(I)
            AMY=A*X(I)
            AST(11)=AST(11)+A
            SUMMX   =SUMMX+AMX
            SUMMY   =SUMMY+AMY
            IF(NSTIFF(I).NE.0)THEN
               ASS =0.0D00
               ASSX=0.0D00
               ASSY=0.0D00
               NSTIF=NSTIFF(I)
               DO 77 J=1,NSTIF
                     ASS =ASS +ASTIFF(I,J)
                     ASSX=ASSX+ASTIFF(I,J)*YSTIFF(I,J)
                     ASSY=ASSY+ASTIFF(I,J)*XSTIFF(I,J)
   77          CONTINUE
               AST(11)=AST(11)+ASS
               SUMMX   =SUMMX   +ASSX
               SUMMY   =SUMMY   +ASSY
            ENDIF
   20 CONTINUE
C
```

```
      YA(11)=SUMMY/AST(11)
      ZA(11)=SUMMX/AST(11)
      SUMIX =0.0D00
      SUMIY =0.0D00
      SUMIXY=0.0D00
      DO 30 I=1,N
            AID=D(I)*(W(I)**3)/12.0D00
            AIW=W(I)*(D(I)**3)/12.0D00
            AIX = AIW*(DCOS(ALP(I)*PI/180.0D00))**2
     &           +AID*(DSIN(ALP(I)*PI/180.0D00))**2
     &           +D(I)*W(I)*(Y(I)-ZA(11))**2
            AIY = AID+AIW-AIX
     &           +D(I)*W(I)*(X(I)-YA(11))**2
            AIXY=-0.5D00*(AIW-AID)*DSIN(ALP(I)*PI/90.D00)
     &           +D(I)*W(I)*(Y(I)-ZA(11))*(X(I)-YA(11))
C
C           AIX=AIX+ D(I)*W(I)*(Y(I)-ZA(11))**2
C           AIY=AIY+D(I)*W(I)*(X(I)-YA(11))**2
C           AIXY=AIXY+ D(I)*W(I)*(Y(I)-ZA(11))*(X(I)-YA(11))
C           WRITE(*,*)'AIX,AIY,AIXY'
C           WRITE(*,*) AIX,AIY,AIXY
C            PAUSE
C
            IF(NSTIFF(I).NE.0)THEN
               AIXS =0.0D00
               AIYS =0.0D00
               AIXYS=0.0D00
               NSTIF=NSTIFF(I)
               DO 188 J=1,NSTIF
                     ARMY=YSTIFF(I,J)-ZA(11)
                     ARMX=XSTIFF(I,J)-YA(11)
                     AIXS=AIXS+ASTIFF(I,J)*ARMY**2
                     AIYS=AIYS+ASTIFF(I,J)*ARMX**2
                     AIXYS=AIXYS+ASTIFF(I,J)*ARMX*ARMY
  188          CONTINUE
               AIX=AIX+AIXS
               AIY=AIY+AIYS
               AIXY=AIXY+AIXYS
            ENDIF
            SUMIX =SUMIX+AIX
            SUMIY =SUMIY+AIY
            SUMIXY=SUMIXY+AIXY
   30 CONTINUE
C
      R=2.0D00*SUMIXY
      Q=SUMIY-SUMIX
      THET2=DATAN2(R,Q)
      IYY(11)=((SUMIX+SUMIY)/2.0D00)
     &       +(SUMIX-SUMIY)*DCOS(THET2)/2.0D00-SUMIXY*DSIN(THET2)
      IZZ(11)=SUMIX+SUMIY-IYY(11)
      THETA  =THET2/(PI/90.0D00)
      WRITE(60,  *)'######################'
      WRITE(60,206)AST(11),YA(11),ZA(11),SUMIX,SUMIY,IYY(11),IZZ(11)
     &          ,THETA
  206 FORMAT(F10.5,/,F10.5,2X,F10.5,/,F10.5,2X,F10.5,/,F10.5,2X,F10.5,/
     &      ,F10.5,2X,F10.5,/,F10.5)
      DO 600 I=1,N
            ALPHR=ALP(I)*(PI/180.0D00)
            BETA(I)=ALP(I)-THETA
            THETR=THETA*PI/180.0D00
            YAA=X(I)-YA(11)+W(I)*DCOS(ALPHR)/2.0D00
            ZAA=Y(I)-ZA(11)+W(I)*DSIN(ALPHR)/2.0D00
            Z1(I)= YAA*DCOS(THETR)+ZAA*DSIN(THETR)
            Z2(I)=-YAA*DSIN(THETR)+ZAA*DCOS(THETR)
      WRITE(60,9876)I,BETA(I)
 9876 FORMAT(I5,5X,F12.8)
  600 CONTINUE
C
      WRITE(10,*)'PRP OP'
C
      RETURN
      END
******
```

```
C       BEGIN PROLOGUE DODO
C       PURPOSE:
C
C       LIBRARY: STRUCTURAL ANALYSIS
C       TYPE: DOUBLE PRECISION
C       KEYWORDS:
C       AUTHOR: Dr. K.A.Hafez, DEPARTMENT OF NAVAL ARCHITECTURE AND MARINE
C               ENGINEERING, FACULTY OF ENGINEERING, ALEXANDRIA UNIVERSITY
C
C       FORMAL PARAMETERS:
C       REFERENCES:
C       ROUTINES CALLED: PRMS, GAUSELM
C       REVISION HISTORY (YYMMDD): 980101  DATE WRITTEN
C       END PROLOGUE DODO
C
        SUBROUTINE DODO
        IMPLICIT REAL*8(A-H,O-Z)
        COMMON/CONS/PI,N,NL,NJ
        COMMON/GENR/Z1(75),Z2(75),BETA(75),NQ(75),NCB(75)
     &             ,M(75,75),ML(75,75),NC(75,75),ND(75,75),QN(75,75)
        COMMON/SCT/IYY(21),IZZ(21),AST(21),SC(21,2),YA(21),ZA(21),YS(21)
     &         ,ZS(21),ASC(21)
        COMMON/ELEM/W(75),D(75),NSTIFF(75),INDELM(75,2),NEL(75,10)
        COMMON/DID/CN(75),CC(75),FA(75),THETR
        COMMON/STIF/ASTIFF(75,10),XSTIFF(75,10),YSTIFF(75,10)
     &             ,IDSTIFF(75,10),IWSTIFF(75,10)
        COMMON/PR/X(75),Y(75),ALP(75),XX(75),YY(75),JN(2,75)
        COMMON/COEF/C(75,75)
        COMMON/SHCOF/KY(21),KZ(21)
        COMMON/DAD/QQ(75,6),QC(2,23),QOS(75)
        DOUBLE PRECISION IDSTIFF,IWSTIFF,KY,KZ,IYY,IZZ
C
        WRITE(10,*)'DODO IP'
C
        IF(NL.EQ.0)THEN
           KCL=1
        ELSEIF(NL.GT.0)THEN
           KCL=2
        ENDIF
        NL1=NL+1
        DO 50 L=1,2
              SC(11,L)=0.0D00
              KY(11)=0.0
              KZ(11)=0.0
              DO 25 I2=1,NL1
              DO 25 J2=1,NL1
                    C(I2,J2)=0.0
   25         CONTINUE
              DO 90 JA=1,KCL
                    SM=0.0D00
                    Q =0.0D00
                    DO 10 I=1,N
                          DO 300 KM=1,6
                                 QQ(I,KM)=0.0
  300                     CONTINUE
                          NQC=NQ(I)
                          Q=0.0D00
                          IF(NQC.NE.0.0)THEN
                             DO 20 K=1,NQC
                                   MM=M(I,K)
                                   CALL PRMS(L,MM,SI,YAA,ZAA,BETAR)
                                   Q=Q+ZAA*D(MM)*W(MM)-(W(MM))**2
     &                                  *DSIN(BETAR)*D(MM)/2.0
                                   IF(NSTIFF(K).NE.0)THEN
                                      QS=0.0
                                      NSTIF=NSTIFF(K)
                                      DO 188 J=1,NSTIF
                                             XSS=XSTIFF(K,J)-YA(11)
                                             YSS=YSTIFF(K,J)-ZA(11)
                                      IF(L.EQ.1)THEN
                                      ARM=-XSS*DSIN(THETR)+YSS*DCOS(THETR)
                                      ELSEIF(L.EQ.2)THEN
                                      ARM= XSS*DCOS(THETR)+YSS*DSIN(THETR)
```

```
                                    ELSE
                                       STOP 'WARNING: INCOMP.DIR.'
                                    ENDIF
                                       QS=QS+ASTIFF(K,J)*ARM/SI
  188                               CONTINUE
                                    Q=Q+QS
                                 ENDIF
   20                      CONTINUE
                           QOS(I)=Q/SI
                        ENDIF
                        CALL PRMS(L,I,SI,YAA,ZAA,BETAR)
                        F=(Q*W(I)+ZAA*D(I)*((W(I))**2)/2.-D(I)*(W(I))**3
     &                     *DSIN(BETAR)/6.)/SI
                        IF(NSTIFF(I).NE.0)THEN
                           FS=0.0
                           NSTIF=NSTIFF(I)
                           DO 199 J=1,NSTIF
                                  XSS=XSTIFF(I,J)-YA(11)
                                  YSS=YSTIFF(I,J)-ZA(11)
                                  IF(L.EQ.1)THEN
                                  ARM=-XSS*DSIN(THETR)+YSS*DCOS(THETR)
                                  ELSEIF(L.EQ.2)THEN
                                  ARM= XSS*DCOS(THETR)+YSS*DSIN(THETR)
                                  ENDIF
                                  FS=FS+ASTIFF(I,J)*W(I)*ARM/SI
  199                      CONTINUE
                           F=F+FS
                        ENDIF
C                       QQ(I,1)=Q/SI
                        IF(JA.EQ.2)THEN
                           DO 140 LL=1,NL
                                  F=F+C(LL,NL1)*QN(I,LL)*W(I)
C                                 QQ(I,1)=QQ(I,1)+C(LL,NL1)*QN(I,LL)
  140                      CONTINUE
                        ENDIF
C
C                       DO 777 JJ=2,6
C                              QQ(I,JJ)=QQ(I,JJ-1)+(ZAA*D(I)*
C    &                                  (W(I)/5.0)-(W(I)/5.0)**2
C    &                                  *DSIN(BETAR)*D(I)/2.0)/SI
C 777                   CONTINUE
C                       ASF=0.5*(W(I)/5.0)*((QQ(I,1))**2+2*(QQ(I,2))**2
C    &                    +2*(QQ(I,3))**2+2*(QQ(I,4))**2+2*(QQ(I,5))**2
C    &                    +(QQ(I,6))**2)
C                       IF(L.EQ.1)THEN
C                         KY(11)=KY(11)+D(I)/ASF
C                       ELSEIF(L.EQ.2)THEN
C                         KZ(11)=KZ(11)+D(I)/ASF
C                       ENDIF
C
                        FA(I)=F/D(I)
                        FX=F*DCOS(BETAR)
                        FY=F*DSIN(BETAR)
                        SM=SM+FX*ZAA-FY*YAA
                        SC(11,L)=-SM
C
C                       CALL PRMS(L,I,SI,YAA,ZAA,BETAR)
C                       QQ(I,6)=F/W(I)
C                       DO 777 JJ=5,1,-1
C                              QQ(I,JJ)=QQ(I,JJ-1)-
C                              (ZAA*D(I)*(W(I)/5.0)-(W(I)/5.0)**2
C    &                                  *DSIN(BETAR)*D(I)/2.0)/SI
C 777                   CONTINUE
C                       AII=0.5*(W(I)/5.0)*((QQ(I,1))**2+2*(QQ(I,2))**2
C    &                    +2*(QQ(I,3))**2+2*(QQ(I,4))**2+2*(QQ(I,5))**2
C    &                    +(QQ(I,6))**2)
C                       IF(L.EQ.1)THEN
C                         KY(11)=KY(11)+D(I)/AII
C                       ELSEIF(L.EQ.2)THEN
C                         KZ(11)=KZ(11)+D(I)/AII
C                       ENDIF
C
      WRITE(60,204)I,F,FX,FY,FA(I),W(I)*D(I)
```

```
  204 FORMAT(5X,I2,7(5X,F7.3))
   10               CONTINUE
      IF(NL.NE.0.AND.JA.NE.2)THEN
         K=0
         DO 60 II=1,NL
               KL=NEL(II,1)
               DO 60 JJ=1,KL
                     KC=ML(JJ,II)
                     DO 58 LL=1,NL
                           CN(LL)=0.0D00
   58                CONTINUE
                     CN(II)=1.0D00
                     NCBKC=NCB(KC)
                     IF(NCBKC.GT.1)THEN
                        KT=NC(KC,2)
                        IF(KT.EQ.II)THEN
                           KT=II-1
                        ENDIF
                        CN(KT)=-1.0D00
                     ENDIF
                     K=K+1
                     CC(K)=QN(KC,II)
                     C(II,NL1)=C(II,NL1)-FA(KC)*CC(K)
                     DO 59 LL=1,NL
                           C(II,LL)=C(II,LL)+CN(LL)*W(KC)/D(KC)
   59                CONTINUE
   60    CONTINUE
         CALL GAUSELM
      ENDIF
      IF((NL.NE.0.AND.JA.EQ.2).OR.(NL.EQ.0.AND.JA.EQ.1))THEN
         DO 7788 LL=1,NL
                QC(L,LL)=C(LL,NL1)
 7788    CONTINUE
         WRITE(60,7000)(L,LL,QC(L,LL),LL=1,NL)
 7000    FORMAT(2I5,F10.5)
      ENDIF
   90 CONTINUE
   50 CONTINUE
C
      WRITE (60,999)AST(11)
  999 FORMAT(5X,'A = ',F10.5)
C     SC (,1): VERTICAL POSITION OF SHEAR CENTER DUE TO VERTICAL SHEAR FOR
      YS(11)=YA(11)+SC(11,1)*DCOS(THETR)-SC(11,2)*DSIN(THETR)
C
C     SC (,1): VERTICAL POSITION OF SHEAR CENTER DUE TO HORIZONTAL SHEAR F
      ZS(11)=ZA(11)+SC(11,1)*DSIN(THETR)+SC(11,2)*DCOS(THETR)
C
      WRITE(60,*)'CHECK:  THETR,YA(11),ZA(11)'
      WRITE(60,957)THETR,YA(11),ZA(11)
  957 FORMAT(3F10.4)
      WRITE(60,*)'###############>>>>>>>>>>>>>>>>>>>>>>>>>>>>>>'
      WRITE(60,100)(SC(11,L),L=1,2),YS(11),ZS(11)
  100 FORMAT(F10.5,2X,F10.5,2X,/,F10.5,2X,F10.5,2X)
C
      WRITE(10,*)'DODO OP'
C
      RETURN
      END
******
C     BEGIN PROLOGUE DADA
C     PURPOSE:
C
C     LIBRARY: STRUCTURAL ANALYSIS
C     TYPE: DOUBLE PRECISION
C     KEYWORDS:
C     AUTHOR: Dr. K.A.Hafez, DEPARTMENT OF NAVAL ARCHITECTURE AND MARINE
C             ENGINEERING, FACULTY OF ENGINEERING, ALEXANDRIA UNIVERSITY
C
C     FORMAL PARAMETERS:
C     KY(): SHEAR COEFFICIENT DUE TO FORCE IN THE HORIZONTAL DIRECTION, Y
C     KZ(): SHEAR COEFFICIENT DUE TO FORCE IN THE VERTICAL DIRECTION, Z
C
C     REFERENCES:
```

```
C      ROUTINES CALLED: PRMS
C      REVISION HISTORY (YYMMDD): 980101  DATE WRITTEN
C      END PROLOGUE DADA
C
       SUBROUTINE DADA
       IMPLICIT REAL*8(A-H,O-Z)
       COMMON/CONS/PI,N,NL,NJ
       COMMON/GENR/Z1(75),Z2(75),BETA(75),NQ(75),NCB(75)
     &           ,M(75,75),ML(75,75),NC(75,75),ND(75,75),QN(75,75)
       COMMON/SCT/IYY(21),IZZ(21),AST(21),SC(21,2),YA(21),ZA(21),YS(21)
     &           ,ZS(21),ASC(21)
       COMMON/ELEM/W(75),D(75),NSTIFF(75),INDELM(75,2),NEL(75,10)
       COMMON/DID/CN(75),CC(75),FA(75),THETR
       COMMON/STIF/ASTIFF(75,10),XSTIFF(75,10),YSTIFF(75,10)
     &           ,IDSTIFF(75,10),IWSTIFF(75,10)
       COMMON/PR/X(75),Y(75),ALP(75),XX(75),YY(75),JN(2,75)
       COMMON/COEF/C(75,75)
       COMMON/SHCOF/KY(21),KZ(21)
       COMMON/DAD/QQ(75,6),QC(2,23),QOS(75)
       DOUBLE PRECISION IDSTIFF,IWSTIFF,KY,KZ,IYY,IZZ
C
       WRITE(10,*)'DADA IP'
C
       WRITE(60,*)'DIRECT DET. OF KY, KZ; EXC.THE INF.OF THE ATT.STIFF.'
C
C               ASP=0.0D00
C               DO 387 I=1,N
C                      ASP=ASP+W(I)/D(I)
C  387          CONTINUE
C        DO 4444 L=1,2
C                 H=0.0
C               KY(11)=0.0D00
C                 KZ(11)=0.0D00
C               DO 1111 I=1,N
C                       CALL PRMS(L,I,SI,XA,YA,BETAR)
C                       IF(L.EQ.1)THEN
C                          YA=Y(I)-YA
C                       ELSEIF(L.EQ.2)THEN
C                          YA=X(I)-YA
C                       ENDIF
C                        H=H+(W(I)/D(I)/SI**2)*(D(I)*W(I)*YA)**2
C                       H=H+(W(I)/D(I))*(D(I)*W(I)*YA)**2
C                        H=H+(D(I)*W(I)*YA)**2
C          H=H+1.0D00/ASP/(1.0D00/SI**2)/(W(I)/D(I))*(D(I)*W(I)*YA)**2
C 1111          CONTINUE
C               H=SI**2/H
C                H=1.0D00/ASP/(1.0D00/SI**2)/H
C                H=1.0D00/H
C               IF(L.EQ.1)THEN
C                  WRITE(60,*)'SHEAR COEFF. DUE TO FORCE IN U DIR., KY'
C                  KY(11)=H/AST(11)
C                          WRITE(*,*)KY(11)
C               ELSEIF(L.EQ.2)THEN
C                  WRITE(60,*)'SHEAR COEFF. DUE TO FORCE IN V DIR., KZ'
C                    KZ(11)=H/AST(11)
C                  WRITE(60,*)KZ(11)
C               ENDIF
C 4444 CONTINUE
C
       WRITE(60,*)'################'
C
C        WRITE(60,*)'DETAILED DET. OF KX, KY; EXC.THE INF.OF THE ATT.STIFF.'
C        DO 444 L=1,2
C              H=0.0
C              TERM1=0.0
C              TERM2=0.0
C              DO 116 I=1,N
C                     QE=0.0
C                     TERMS=0.0
C                     JNC=JN(2,I)
C                     CALL PRMS(L,I,SI,XA,YA,BETAR)
C                     IF(L.EQ.1)THEN
C                        ARM=YY(I)-YA
```

```
C                        ELSEIF(L.EQ.2)THEN
C                           ARM=XX(I)-YA
C                        ENDIF
C                        IF(NSTIFF(I).NE.0)THEN
C                           NSTIF=NSTIFF(I)
C                           DO 399 J=1,NSTIF
C                                  XSS=XSTIFF(I,J)-YA(11)
C                                  YSS=YSTIFF(I,J)-ZA(11)
C                                  IF(L.EQ.1)THEN
C                                  ARM= XSS*DCOS(THETR)+YSS*DSIN(THETR)
C                                  ELSEIF(L.EQ.2)THEN
C                                  ARM=-XSS*DSIN(THETR)+YSS*DCOS(THETR)
C                                  ENDIF
C                                  TERMS=TERMS+ASTIFF(I,J)*ARM
C  399                      CONTINUE
C                        ENDIF
C                        TERM1=TERM1+(W(I)/D(I))*(D(I)*W(I)*YA+TERMS)**2
C                        TERM2=TERM2+(W(I)/D(I))*(D(I)*W(I)*YA+TERMS)
C  116            CONTINUE
C                 TERM1=(1.0D00/SI**2)*TERM1
C                 TERM2=(2.0D00/SI)*TERM2
C                 DO 119 I=1,N
C                        NCBB=NCB(I)
C                        DO 111 J=1,NCBB
C                              NCC=NC(I,J)
C                              QE=QE+QC(L,NCC)*(-1)**(J+1)
C  111                   CONTINUE
C                        WRITE(60,701)L,I,QE,QOS(I),QE+QOS(I)
C  701                   FORMAT(2I3,3F20.10)
C                        QE=QE+QOS(I)
C                        TERM1=TERM1+QE**2
C                        TERM2=TERM2*QE
C  119            CONTINUE
C                        H=1.0D00/(TERM1-TERM2)
C                  IF(L.EQ.1)THEN
C                     WRITE(60,*)'SHEAR COEFF. DUE TO FORCE IN U DIR., KY'
C                     KY(11)=H/AST(11)
C                     WRITE(60,*)KY(11)
C                  ELSEIF(L.EQ.2)THEN
C                     WRITE(60,*)'SHEAR COEFF. DUE TO FORCE IN V DIR., KZ'
C                     KZ(11)=H/AST(11)
C                     WRITE(60,*)KZ(11)
C                  ENDIF
C       H=0.0
C       H=H+1.0D00/((W(I)/D(I))*((1.D00/SI**2)*(D(I)*YA*W(I)
C&         +AS*YA)**2+(QE**2)-(2.0D00/SI)*(D(I)*YA*W(I)+AS*YA)*QE))
C  444 CONTINUE
C
      IF(KY(11).LT.0.3D00.OR.KY(11).GT.0.5D00)THEN
      KY(11)=1.0D00/2.40D00
      ENDIF
      IF(KZ(11).LT.0.3D00.OR.KZ(11).GT.0.5D00)THEN
      KZ(11)=1.0D00/3.26D00
      ENDIF
      WRITE(60,*)'KY,KZ'
      WRITE(60,7000)KY,KZ
 7000 FORMAT(2F10.5)
C
      WRITE(10,*)'DADA OP'
C
      RETURN
      END
******
C     BEGIN PROLOGUE PRMS
C     PURPOSE:
C
C     LIBRARY: STRUCTURAL ANALYSIS
C     TYPE: DOUBLE PRECISION
C     KEYWORDS:
C     AUTHOR: Dr. K.A.Hafez, DEPARTMENT OF NAVAL ARCHITECTURE AND MARINE
C             ENGINEERING, FACULTY OF ENGINEERING, ALEXANDRIA UNIVERSITY
C
C     FORMAL PARAMETERS:
```

```
C       REFERENCES:
C       ROUTINES CALLED: NON
C       REVISION HISTORY (YYMMDD): 980101  DATE WRITTEN
C       END PROLOGUE DADA
C
        SUBROUTINE PRMS(L,I,SI,YAA,ZAA,BETAR)
        IMPLICIT REAL*8(A-H,O-Z)
        COMMON/CONS/PI,N,NL,NJ
        COMMON/GENR/Z1(75),Z2(75),BETA(75),NQ(75),NCB(75)
     &             ,M(75,75),ML(75,75),NC(75,75),ND(75,75),QN(75,75)
        COMMON/SCT/IYY(21),IZZ(21),AST(21),SC(21,2),YA(21),ZA(21),YS(21)
     &             ,ZS(21),ASC(21)
        DOUBLE PRECISION IYY,IZZ
C
        WRITE(10,*)'PRMS IP'
C
        IF(L.EQ.1)THEN
           SI=IYY(11)
           YAA=Z1(I)
           ZAA=Z2(I)
           BET=BETA(I)
        ELSE
           SI=IZZ(11)
           YAA=Z2(I)
           ZAA=Z1(I)
           BET=90.0D00-BETA(I)
        ENDIF
        BETAR=BET*PI/180.0D00
C
        WRITE(10,*)'PRMS OP'
C
        RETURN
        END
******
C       BEGIN PROLOGUE GAUSELM
C       PURPOSE: SOLVING A LINEAR SYSTEM OF EQUATIONS BY USING THE GAUSSIAN
C                ELEMINATION METHOD. THE SOLUTION IS REQUIRED FOR THE CALCUL
C                OF SHEAR CENTER IN MULTI-CELL CLOSED SECTION.
C
C       LIBRARY: STRUCTURAL ANALYSIS
C       TYPE: DOUBLE PRECISION
C       KEYWORDS:
C       AUTHOR: Dr. K.A.Hafez, DEPARTMENT OF NAVAL ARCHITECTURE AND MARINE
C               ENGINEERING, FACULTY OF ENGINEERING, ALEXANDRIA UNIVERSITY
C
C       FORMAL PARAMETERS:
C       REFERENCES:
C       ROUTINES CALLED: NON
C       REVISION HISTORY (YYMMDD): 980101  DATE WRITTEN
C       END PROLOGUE GAUSELM
C
        SUBROUTINE GAUSELM
        IMPLICIT REAL*8(A-H,O-Z)
        COMMON/CONS/PI,N,NL,NJ
        COMMON/COEF/C(75,75)
        DATA EPS/1.0E-10/
C
        WRITE(10,*)'GAUSELM I/P'
C
        M=NL+1
        DETER=1.0D00
        DO 9 K=1,NL
              DETER=DETER*C(K,K)
              AB=DABS(C(K,K))
              IF(AB.GT.EPS)THEN
                 KP1=K+1
                 DO 6 J=KP1,M
                       C(K,J)=C(K,J)/C(K,K)
    6            CONTINUE
                 C(K,K)=1.0D00
                 DO 10 I=1,NL
                       IF(I.NE.K)THEN
                          AB=DABS(C(I,K))
```

```
                        IF(AB.NE.0.0D00)THEN
                           DO 8 J=KP1,M
                               C(I,J)=C(I,J)-C(I,K)*C(K,J)
    8                      CONTINUE
                           C(I,K)=0.0D00
                        ENDIF
                     ENDIF
   10          CONTINUE
            ELSE
               WRITE(60,951)
  951          FORMAT('SMALL PIVOT, MATRIX MAY BE SINGULAR')
               GOTO 99
            ENDIF
    9 CONTINUE
C
      WRITE(10,*)'GAUSELM O/P'
C
   99 RETURN
      END
******
C     BEGIN PROLOGUE DIDI
C     PURPOSE:
C
C     LIBRARY: STRUCTURAL ANALYSIS
C     TYPE: DOUBLE PRECISION
C     KEYWORDS:
C     AUTHOR: Dr. K.A.Hafez, DEPARTMENT OF NAVAL ARCHITECTURE AND MARINE
C             ENGINEERING, FACULTY OF ENGINEERING, ALEXANDRIA UNIVERSITY
C
C     FORMAL PARAMETERS:
C     REFERENCES:
C     ROUTINES CALLED: GAUSELM
C     REVISION HISTORY (YYMMDD): 980101  DATE WRITTEN
C     END PROLOGUE DIDI
C
      SUBROUTINE DIDI
      IMPLICIT REAL*8(A-H,O-Z)
      COMMON/CONS/PI,N,NL,NJ
      COMMON/GENR/Z1(75),Z2(75),BETA(75),NQ(75),NCB(75)
     &           ,M(75,75),ML(75,75),NC(75,75),ND(75,75),QN(75,75)
      COMMON/SCT/IYY(21),IZZ(21),AST(21),SC(21,2),YA(21),ZA(21),YS(21)
     &          ,ZS(21),ASC(21)
      COMMON/ELEM/W(75),D(75),NSTIFF(75),INDELM(75,2),NEL(75,10)
      COMMON/DID/CN(75),CC(75),FA(75),THETR
      COMMON/STIF/ASTIFF(75,10),XSTIFF(75,10),YSTIFF(75,10)
     &           ,IDSTIFF(75,10),IWSTIFF(75,10)
      COMMON/PR/X(75),Y(75),ALP(75),XX(75),YY(75),JN(2,75)
      COMMON/TORWR/CI(21),CJ(21)
      COMMON/COEF/C(75,75)
      COMMON/DOD/ACL(75),YO(21),ZO(21),INDCS(75,4),XXM(75),YYM(75)
     &          ,WM(75),DM(75),RR(75),WW(75,2),AE(5)
      COMMON/DRD/ASTIFFN(75,10),XSTIFFN(75,10),YSTIFFN(75,10)
     &          ,NSTIFFN(75)
C    &     ,THETN(75),THETF(75),PHIN(75),PHIF(75)
      COMMON/PRDIM/LBP
      DOUBLE PRECISION IDSTIFF,IWSTIFF,IYY,IZZ
      DOUBLE PRECISION LBP,LO,L,IT,JL,J,LH(10)
      DATA G/82.8E+6/
C     G IS IN KN/M^2
C     ,NU/0.316/,EPS/1E-5/
C
      WRITE(10,*)'DIDI IP'
C
      DO 456 IND=1,5
             AE(IND)=0.0
  456 CONTINUE
      READ(50,117)NODECK
      WRITE(60,117)NODECK
  117 FORMAT(I5)
      IF(NODECK.GT.0)THEN
         DO 511 JI=1,NODECK
               READ(50,117)IND
               WRITE(60,117)IND
```

```
                 READ(50,91)NH,NHL
                 WRITE(60,91)NH,NHL
   91            FORMAT(2I5)
                 READ(50,92)LBP,LO,BO,IT,JL,AH,CH
                 WRITE(60,92)LBP,LO,BO,IT,JL,AH,CH
                 READ(50,92)(LH(II),II=1,NHL)
                 WRITE(60,92)(LH(II),II=1,NHL)
   92            FORMAT(7F10.4)
                 IF(LO.LE.0.0)THEN
                    LO=0.65*LBP
                 ENDIF
C
                 L=0.0
                 DO 19 II=1,NHL
                       L=L+LH(II)
   19            CONTINUE
                 L=L/NH
                 BH=BO/2.0D00-CH
                 MH=AH/BH
                 NH=L/BH
                 K=JL/IT
                 Q=IT*(NH+12*MH*K)/(BH**2*(NH**2*(MH**3-6*MH**2+9*MH-4)-
     &              12*MH*NH*K*(4*MH**3+9*MH**2-12*MH+5)-72*MH**2*K*
     &              (MH**3+6*MH**2-6*MH+2)))
                 P=Q*(NH-6*MH*K)*(1-MH)
                 R=Q/MH*(N*(MH**2-3*MH+2)+6*MH*K*(3*MH**3+11*MH**2-12*MH
     &              +4))
                 TE=31.5D00*(NH-1)*BH*(P+R)/LO/BO*(1-2.63D00*(BO**2/LO))
                 AE(IND)=BO*TE/6.0D00
  511      CONTINUE
      ENDIF
      DO 25 I2=1,NL1
            DO 25 J2=1,NL1
                  C(I2,J2)=0.0
   25 CONTINUE
      NL1=NL+1
      CJ(11)=0.0
      DO 12 I=1,N
            IF(NCB(I).EQ.0)THEN
               CJ(11)=CJ(11)+(W(I)*D(I)**3)
            ENDIF
   12 CONTINUE
      CJ(11)=CJ(11)*G/3.00D00
      IF(NL.NE.0)THEN
         DO 60 II=1,NL
                NN=NEL(II,1)
                MM=NEL(II,2)
                IF(NN.EQ.4.OR.NN.EQ.3)THEN
                   DO 71 I=2,NN-1
                         KK=NEL(II,I+1)
                         IF(INDELM(MM,1).NE.INDELM(KK,1))THEN
C                               ACL(II)=(W(MM)-D(MM)/NCB(MM)*(NCB(MM)-1))
C    &                                 *(W(KK)-D(KK)/NCB(KK)*(NCB(KK)-1))
                            ACL(II)=W(MM)*W(KK)
                         ENDIF
   71              CONTINUE
                ELSEIF(NN.GT.4)THEN
                   STOP 'POLYGON/UNIDENTIFIED CELL'
                ENDIF
                IF(NN.EQ.3)THEN
                   ACL(II)=0.5D00*ACL(II)
                ENDIF
                KL=NEL(II,1)
                DO 61 JJ=1,KL
                      KC=ML(JJ,II)
                      DO 58 LL=1,NL
                            CN(LL)=0.0
   58                 CONTINUE
                      CN(II)=1.0
                      IF(NCB(KC).GT.1)THEN
                         KT=NC(KC,2)
                         IF(KT.EQ.II)THEN
                              KT=II-1
```

```
                        ENDIF
                        CN(KT)=-1.0
                     ENDIF
                     DO 59 LL=1,NL
                           IF(D(KC).LE.0.0)THEN
                             STOP'ERROR IN DATA TRANSFER BETWEEN SUBS'
                           ENDIF
                           C(II,LL)=CN(LL)*(W(KC)/D(KC))+C(II,LL)
   59                CONTINUE
   61          CONTINUE
               C(II,NL1)=2.0*G*ACL(II)
   60    CONTINUE
      CALL GAUSELM
      WRITE(60,*)'Q"=Q/O" @@@@@@@@@@'
      WRITE(60,*)(C(I,NL1),I=1,NL)
      WRITE(60,*)'CELLULAR AREA@@@@@@@@@@@@@@@@@@@@@@@@@@@@@@@@'
      WRITE(60,*)(ACL(I),I=1,NL)
      WRITE(60,*)'CELLULAR ELEMENTS@@@@@@@@@@@@@@@@@@@@@@@@@@@@@@@@'
      WRITE(60,1000)((NEL(I,J),J=1,5),I=1,NL)
 1000 FORMAT(5I5)
      DO 90 I=1,NL
            CJ(11)=CJ(11)+2*ACL(I)*C(I,NL1)
   90 CONTINUE
      ENDIF
      CJ(11)=CJ(11)/G
      WRITE(60,*)'CJ(11)/G,CJ(11)'
      WRITE(60,*)CJ(11),CJ(11)*G
C
      WRITE(10,*)'DIDI OP'
C
      RETURN
      END
******
C     BEGIN PROLOGUE DRDR
C     PURPOSE:
C
C     LIBRARY: STRUCTURAL ANALYSIS
C     TYPE: DOUBLE PRECISION
C     KEYWORDS:
C     AUTHOR: Dr. K.A.Hafez, DEPARTMENT OF NAVAL ARCHITECTURE AND MARINE
C             ENGINEERING, FACULTY OF ENGINEERING, ALEXANDRIA UNIVERSITY
C
C     FORMAL PARAMETERS:
C     REFERENCES:
C     ROUTINES CALLED: NON
C     REVISION HISTORY (YYMMDD): 980101  DATE WRITTEN
C     END PROLOGUE DRDR
C
      SUBROUTINE DRDR
      IMPLICIT REAL*8(A-H,O-Z)
      COMMON/CONS/PI,N,NL,NJ
      COMMON/GENR/Z1(75),Z2(75),BETA(75),NQ(75),NCB(75)
     &          ,M(75,75),ML(75,75),NC(75,75),ND(75,75),QN(75,75)
      COMMON/SCT/IYY(21),IZZ(21),AST(21),SC(21,2),YA(21),ZA(21),YS(21)
     &          ,ZS(21),ASC(21)
      COMMON/ELEM/W(75),D(75),NSTIFF(75),INDELM(75,2),NEL(75,10)
      COMMON/DID/CN(75),CC(75),FA(75),THETR
      COMMON/STIF/ASTIFF(75,10),XSTIFF(75,10),YSTIFF(75,10)
     &          ,IDSTIFF(75,10),IWSTIFF(75,10)
      COMMON/PR/X(75),Y(75),ALP(75),XX(75),YY(75),JN(2,75)
      COMMON/TORWR/CI(21),CJ(21)
      COMMON/COEF/C(75,75)
      COMMON/DOD/ACL(75),YO(21),ZO(21),INDCS(75,4),XXM(75),YYM(75)
     &          ,WM(75),DM(75),RR(75),WW(75,2),AE(5)
      COMMON/DRD/ASTIFFN(75,10),XSTIFFN(75,10),YSTIFFN(75,10)
     &          ,NSTIFFN(75)
C    &     ,THETN(75),THETF(75),PHIN(75),PHIF(75)
      DOUBLE PRECISION IDSTIFF,IWSTIFF,IYY,IZZ
      DOUBLE PRECISION IWW,J
      DATA E/207E+6/,G/82.8E+6/
C
C     DATA E/21.107E+9/,G/8.44E+9/
C     E IS IN KG/M^2 & G IS IN KG/M^2
```

```
C       E IS IN KN/M^2 & G IS IN KN/M^2
C       ,NU/0.316/,EPS/1E-5/
C
        WRITE(10,*)'DRDR IP'
C
        READ  (50,202)INDSC,NN,NNJ,NUT
        WRITE (60,202)INDSC,NN,NNJ,NUT
  202   FORMAT(4I5)
        READ  (50,333)(II,XXM(I),YYM(I),I=1,NNJ)
        WRITE (60,333)(I,XXM(I),YYM(I),I=1,NNJ)
  333   FORMAT(I5,2F12.5)
        READ  (50,333)(II,WM(I),DM(I),I=1,NN)
        WRITE (60,333)(I,WM(I),DM(I),I=1,NN)
        READ  (50,657)(NSTIFFN(I),I=1,NN)
        WRITE (60,657)(NSTIFFN(I),I=1,NN)
  657   FORMAT(15I5)
C
        DO 272 I=1,NN
               IF(NSTIFFN(I).NE.0)THEN
               NSTIF=NSTIFFN(I)
               READ(50,615) (ASTIFFN(I,J),J=1,NSTIF)
               WRITE(60,615)(ASTIFFN(I,J),J=1,NSTIF)
               ENDIF
  272   CONTINUE
  615   FORMAT(7F11.6)
C
        DO 273 I=1,NN
               IF(NSTIFFN(I).NE.0)THEN
               NSTIF=NSTIFFN(I)
               READ(50,615) (XSTIFFN(I,J),J=1,NSTIF)
               WRITE(60,615)(XSTIFFN(I,J),J=1,NSTIF)
               ENDIF
  273   CONTINUE
C
        DO 274 I=1,NN
               IF(NSTIFFN(I).NE.0)THEN
               NSTIF=NSTIFFN(I)
               READ(50,615) (YSTIFFN(I,J),J=1,NSTIF)
               WRITE(60,615)(YSTIFFN(I,J),J=1,NSTIF)
               ENDIF
  274   CONTINUE
C
        READ  (50,111)((INDCS(I,J),J=1,4),I=1,NN)
  111   FORMAT(5X,4I5)
        READ  (50,3)(YO(J),J=1,NUT)
        WRITE (60,3)(YO(J),J=1,NUT)
        READ  (50,3)(ZO(J),J=1,NUT)
        WRITE (60,3)(ZO(J),J=1,NUT)
    3   FORMAT(9F8.3)
        WW(1,1)=0.0
        IWW    =0.0
C
C       IF(INDSC.EQ.1)THEN
           ASC(11)= 2.0D00*DABS(SIMPUN(ZO,YO,NUT))
C       ENDIF
C
        DO 1 I=1,NN
             J1=INDCS(I,2)
             J2=INDCS(I,3)
             IF(INDCS(I,1).EQ.0)THEN
                RR(I)=ZS(11)-YYM(J1)
             ELSEIF(INDCS(I,1).EQ.1)THEN
                RR(I)=YS(11)-XXM(J1)
             ENDIF
C            RR(I)=(YYM(I)-ZS(11)-XXM(I)+YS(11))*COS(APL(I))
             ASS =0.0
C            IF(NSTIFFN(I).NE.0)THEN
C               NSTIF=NSTIFFN(I)
C               DO 77 J=1,NSTIF
C                     ASS =ASS+ASTIFFN(I,J)
C   77          CONTINUE
C            ENDIF
             WW(I,2)=WW(I,1)    +RR(I)*WM(I)+   ASS+AE(I)
```

```
C
           IF(INDSC.EQ.1)THEN
              WW(I,2)=WW(I,2)+(CJ(11)*G/2.0D00/ASC(11))*(W(I)/D(I))
           ENDIF
C
           WW(I+1,1)=WW(I,2)
    1 CONTINUE
C
C       WWO=0.0
C       DO 13 I=1,NN
C             WWO=WWO+WW(I)*D(I)*W(I)
C   13 CONTINUE
C       DO 14 I=1,NN
C             WW(I,1)=WW(I,1)+(1.0/ASC(11))*WWO
C             WW(I,2)=WW(I,2)+(1.0/ASC(11))*WWO
C   14 CONTINUE
C
      DO 122 I=1,NN
           IWW=IWW+DM(I)*WM(I)*(WW(I,1)**2+WW(I,2)**2+WW(I,1)*WW(I,2))
  122 CONTINUE
C
      IWW=(2.0D00/3.0D00)*IWW
      CI(11)=IWW
C
      WRITE (60,*)'@@@@@@@@@@@@@@@@@@@@@'
      WRITE (60,*)'IWW,CI(11)'
      WRITE (60,*)CI(11),CI(11)*E
      WRITE (60,*)'ASC(11)'
      WRITE (60,*)ASC(11)
      WRITE (60,*)'INDSC'
      WRITE (60,*)INDSC
      WRITE (60,*)'CJ(11)'
      WRITE (60,*)CJ(11)
      WRITE (60,*)'AE'
      WRITE (60,459)(AE(I),I=1,NN)
  459 FORMAT(F10.5)
      WRITE (60,*)'@@@@@@@@@@@@@@@@@@@@@'
C
      WRITE (10,*)'DRDR OP'
C
      RETURN
      END
******
C     BEGIN PROLOGUE DSTR
C     PURPOSE:
C
C     LIBRARY: STRUCTURAL ANALYSIS
C     TYPE: DOUBLE PRECISION
C     KEYWORDS:
C     AUTHOR: Dr. K.A.Hafez, DEPARTMENT OF NAVAL ARCHITECTURE AND MARINE
C                ENGINEERING, FACULTY OF ENGINEERING, ALEXANDRIA UNIVERSITY
C
C     FORMAL PARAMETERS:
C     SC(,): CENTER OF AREA TO SHEAR CENTER DISTANCE
C     ZS (): BOTTOM TO SHEAR CENTER DISTANCE
C     ZA (): BOTTOM TO CENTER OF AREA DISTANCE
C     AST(): CROSS-SECTIONAL STEEL AREA
C     ASC(): CROSS-SECTIONAL IMMERSED AREA
C
C     REFERENCES:
C     ROUTINES CALLED: NON
C     REVISION HISTORY (YYMMDD): 980101  DATE WRITTEN
C     END PROLOGUE DSTR
C
      SUBROUTINE DSTR
      IMPLICIT REAL*8(A-H,O-Z)
      COMMON/CONS/PI,N,NL,NJ
      COMMON/GENR/Z1(75),Z2(75),BETA(75),NQ(75),NCB(75)
     &            ,M(75,75),ML(75,75),NC(75,75),ND(75,75),QN(75,75)
      COMMON/ELEM/W(75),D(75),NSTIFF(75),INDELM(75,2),NEL(75,10)
      COMMON/DID/CN(75),CC(75),FA(75),THETR
      COMMON/STIF/ASTIFF(75,10),XSTIFF(75,10),YSTIFF(75,10)
     &            ,IDSTIFF(75,10),IWSTIFF(75,10)
```

```
      COMMON/PR/X(75),Y(75),ALP(75),XX(75),YY(75),JN(2,75)
      COMMON/TORWR/CI(21),CJ(21)
      COMMON/SHCOF/KY(21),KZ(21)
      COMMON/SCT/IYY(21),IZZ(21),AST(21),SC(21,2),YA(21),ZA(21),YS(21)
     &          ,ZS(21),ASC(21)
      COMMON/COEF/C(75,75)
      COMMON/DOD/ACL(75),YO(21),ZO(21),INDCS(75,4),XXM(75),YYM(75)
     &          ,WM(75),DM(75),RR(75),WW(75,2),AE(5)
      COMMON/DIST/XIWW(21),XJW(21),ZSS(21),XA(21),XIXX(21),XIYY(21)
      COMMON/DAD/QQ(75,6),QC(2,23),QOS(75)
      DOUBLE PRECISION IDSTIFF,IWSTIFF,KY,KZ,IYY,IZZ
C
      WRITE(10,*)'DSTR IP'
C
      WRITE(60,*)'CHECKING THE PROPERTY DISTRIBUTION PARAMETERS'
C
C     DO 444 I=1,21
C           READ  (50,389)DB,DD,DH,DW
C           WRITE (60,389)DB,DD,DH,DW
C 389       FORMAT(4F10.4)
C           XIWW(I)=DD**3*DB**2*DW*(2*DB*DH+3.0D00*DD*DW)/12.0D00/
C    &              (DB*DH+6.0D00*DD*DW)
C           XJW(I)=1.0D00/3.0D00*(DB*DH**3+2*DD*DW**3)
C           ZSS(I)=DD**2*DW/(2*DD*DW+DB*DH)+3*DD**2*DW/(DB*DH+6.0D00*DD*D
C           XA(I)=2*DD*DW+DB*DH
C           XIXX(I)=1.0D00/6.0D00*DD**3*DW+DD**3*DW*DB*DH/(4*DD*DW+2*DB*D
C           XIYY(I)=DB**3*DH/12.0D00+1.0D00/2.0D00*DB**2*DD*DW
C 444 CONTINUE
C
C     DO 465 I=1,21
C           CJ(I)=CJ(11)*XJW(I)/XJW(11)
C           CI(I)=CI(11)*XIWW(I)/XIWW(11)
C           SC(I,1)=0.0
C           SC(I,2)=SC(11,2)*ZS(I)/ZS(11)
C           ZS(I)=ZS(11)*ZSS(I)/ZSS(11)
C           AST(I)=AST(11)*XA(I)/XA(11)
C           KY(I)=KY(11)*XA(I)/XA(11)
C           KZ(I)=KZ(11)*XA(I)/XA(11)
C           IYY(I)=IYY(11)*XIXX(I)/XIXX(11)
C           IZZ(I)=IZZ(11)*XIYY(I)/XIYY(11)
C 465 CONTINUE
C
      DO 396 I=1,21
C           IF(I.NE.11)THEN
            CJ(I)  =CJ(11)
            CI(I)  =CI(11)
            SC(I,1)=0.0
            SC(I,2)=SC(11,2)
            ZS(I)  =ZS(11)
            ZA(I)  =ZA(11)
            AST(I) =AST(11)
            ASC(I) =ASC(11)
            KY(I)  =KY(11)
            KZ(I)  =KZ(11)
            IYY(I) =IYY(11)
            IZZ(I) =IZZ(11)
C           ENDIF
  396 CONTINUE
C
      WRITE (60,*)'CALC.OF THE PROP.DIST.PARA.FOR THE INDIV.SHIP SECS.'
      WRITE (60,675)
  675 FORMAT(13X,'CJ',13X,'CI',13X,'SC(2)',10X,'KY',13X,'KZ'
     &      ,13X,'AST',13X,'IV',13X,'IW',11X,'ZS(I)',11X,'ZDK')
      WRITE (60,897)(I,CJ(I),CI(I),SC(I,2),KY(I),KZ(I)
     &              ,AST(I),IYY(I),IZZ(I),ZS(I)
     &              ,(IYY(I)/(21.5D00-ZA(I))),I=1,21)
  897 FORMAT(I4,10E15.5)
C
      WRITE(10,*)'DSTR OP'
C
      RETURN
      END
******
```

```
C      BEGIN PROLOGUE SIMPUN
C      PURPOSE: INTEGRATION OF AN AREA OVER AN EQUAL AND/OR UNEQUAL INTERVA
C               BY USING SIMPSON'S RULE
C
C      LIBRARY: MATHEMATICS
C      TYPE: DOUBLE PRECISION
C      KEYWORDS:
C      AUTHOR: Dr. K.A.Hafez, DEPARTMENT OF NAVAL ARCHITECTURE AND MARINE
C              ENGINEERING, FACULTY OF ENGINEERING, ALEXANDRIA UNIVERSITY
C
C      FORMAL PARAMETERS:
C      REFERENCES:
C      ROUTINES CALLED: NON
C      REVISION HISTORY (YYMMDD): 980101  DATE WRITTEN
C      END PROLOGUE SIMPUN
C
       FUNCTION SIMPUN(X,Y,N)
       IMPLICIT REAL*8(A-H,O-Z)
       DIMENSION X(50),Y(50)
C
       WRITE(10,*)'SIMPUN IN'
C
         WRITE(60,*)N
         WRITE(60,39)(X(J),J=1,N)
         WRITE(60,39)(Y(J),J=1,N)
       IF(N-2)7,5,4
    7     S=0.0D00
          GOTO 6
    5     S=(Y(1)+Y(2))*(X(2)-X(1))/2.0D00
          GOTO 6
    4     M=N-1
       IF(((X(1)-X(3)).EQ.0.0).OR.((X(2)-X(3)).EQ.0.0))THEN
       PAUSE 'WARNING<SIMPUN>: INV. DEN., CHECK X(1)-X(3) AND X(2)-X(3)'
       ENDIF
       S=(X(2)-X(1))/6.D00*(Y(1)*((X(2)-X(3))/(X(1)-X(3))+2.D00)+
     &    Y(2)*((X(1)-X(3))/(X(2)-X(3))+2.D00)-
     &    Y(3)*(X(2)-X(1))**2/((X(1)-X(3))*(X(2)-X(3))))
       LB=2
       IF(N-3)8,8,9
    9  CONTINUE
       IF(((X(2)-X(4)).EQ.0.0).OR.((X(3)-X(4)).EQ.0.0))THEN
       PAUSE 'WARNING<SIMPUN>: INV. DEN., CHECK X(2)-X(4) AND X(3)-X(4)'
       ENDIF
       S=S+(X(3)-X(2))/6.0D00*(Y(2)*((X(3)-X(4))/(X(2)-X(4))+2.D00)+
     &       Y(3)*((X(2)-X(4))/(X(3)-X(4))+2.D00)-Y(4)*(X(3)-
     &       X(2))**2/((X(2)-X(4))*(X(3)-X(4))))
       LB=3
    8  DO 1 K=LB,M
            WRITE(60,*)'K,X(K)'
            WRITE(60,*)K,X(K)
            XDIFF=DABS(X(K+1)-X(1))
            XDIFF1=DABS(X(K)-X(1))
            IF(XDIFF-XDIFF1)3,11,11
    3          WRITE(60,2)K,X(K)
    2          FORMAT('WARNING<SIMPUN>: NON MONOTONE X',I4,D12.4)
               GOTO 7
   11  CONTINUE
       IF(((X(K)-X(K-1)).EQ.0.0).OR.((X(K+1)-X(K-1)).EQ.0.0))THEN
       PAUSE 'WARNING<SIMPUN>: INV. DEN., CHECK X(2)-X(4) AND X(3)-X(4)'
       ENDIF
       S=S+(X(K+1)-X(K))/6.0D00*(Y(K)*((X(K+1)-X(K-1))/(X(K)-X(K-1))+
     &      2.D00)+Y(K+1)*((X(K)-X(K-1))/(X(K+1)-X(K-1))+2.D00)-
     &      Y(K-1)*(X(K+1)-X(K))**2/((X(K)-X(K-1))*(X(K+1)-X(K-1))))
    1  CONTINUE
    6  SIMPUN=S
       WRITE(60,39)(X(J),J=1,N)
       WRITE(60,39)(Y(J),J=1,N)
   39  FORMAT(9F8.3)
       WRITE(60,*)S
C
       WRITE(10,*)'SIMPUN OUT'
C
       RETURN
```

```
      END
******
******
******
      SUBROUTINE DPLINT (N, X, Y, C)
      IMPLICIT REAL*8(A-H,O-Z)
C
C     BEGIN PROLOGUE DPLINT
C     PURPOSE PRODUCE THE POLYNOMIAL WHICH INTERPOLATES A SET OF DISCRETE
C             DATA POINTS.
C     LIBRARY: SLATEC
C     CATEGORY: E1B
C     TYPE: DOUBLE PRECISION (POLINT-S, DPLINT-D)
C     KEYWORDS: POLYNOMIAL INTERPOLATION
C     AUTHOR: HUDDLESTON, R. E., (SNLL)
C     DESCRIPTION:
C
C     ABSTRACT:
C               SUBROUTINE DPLINT IS DESIGNED TO PRODUCE THE POLYNOMIAL WH
C               INTERPOLATES THE DATA  (X(I),Y(I)), I=1,...,N.  DPLINT SET
C               INFORMATION IN THE ARRAY C WHICH CAN BE USED BY SUBROUTINE
C               TO EVALUATE THE POLYNOMIAL AND ITS DERIVATIVES AND BY SUBR
C               DPOLCF TO PRODUCE THE COEFFICIENTS.
C
C     FORMAL PARAMETERS:
C     ALL TYPE REAL VARIABLES ARE DOUBLE PRECISION
C               N: THE NUMBER OF DATA POINTS  (N .GE. 1)
C               X: THE ARRAY OF ABSCISSAS (ALL OF WHICH MUST BE DISTINCT)
C               Y: THE ARRAY OF ORDINATES
C               C: AN ARRAY OF INFORMATION USED BY SUBROUTINES
C
C     DIMENSIONING INFORMATION:
C               ARRAYS X,Y, AND C MUST BE DIMENSIONED AT LEAST N IN THE CA
C               PROGRAM.
C
C     REFERENCES:
C               L. F. SHAMPINE, S. M. DAVENPORT AND R. E. HUDDLESTON,
C               CURVE FITTING BY POLYNOMIALS IN ONE VARIABLE, REPORT
C               SLA-74-0270, SANDIA LABORATORIES, JUNE 1974.
C
C     ROUTINES CALLED: XERMSG
C     REVISION HISTORY (YYMMDD):
C               740601  DATE WRITTEN
C               891006  COSMETIC CHANGES TO PROLOGUE.  (WRB)
C               891006  REVISION DATE FROM VERSION 3.2
C               891214  PROLOGUE CONVERTED TO VERSION 4.0 FORMAT.  (BAB)
C               900315  CALLS TO XERROR CHANGED TO CALLS TO XERMSG.  (THJ)
C               920501  REFORMATTED THE REFERENCES SECTION.  (WRB)
C     END PROLOGUE  DPLINT
C
      INTEGER I,K,KM1,N
      DIMENSION C(*),X(*),Y(*)
C
      WRITE(10,*)'DPLINT I/P'
C
C     FIRST EXECUTABLE STATEMENT  DPLINT
      IF (N .LE. 0) GO TO 91
      C(1)=Y(1)
      IF(N .EQ. 1) RETURN
      DO 10010 K=2,N
      C(K)=Y(K)
      KM1=K-1
      DO 10010 I=1,KM1
C
C     CHECK FOR DISTINCT X VALUES
      DIF = X(I)-X(K)
      IF (DIF .EQ. 0.0) GO TO 92
      C(K) = (C(I)-C(K))/DIF
10010 CONTINUE
      RETURN
   91 CALL XERMSG ('SLATEC', 'DPLINT', 'N IS ZERO OR NEGATIVE.', 2, 1)
      RETURN
   92 CALL XERMSG ('SLATEC', 'DPLINT',
```

```
     +    'THE ABSCISSAS ARE NOT DISTINCT.', 2, 1)
C
      WRITE(10,*)'DPLINT O/P'
C
      RETURN
      END
******
      SUBROUTINE XERMSG (LIBRAR, SUBROU, MESSG, NERR, LEVEL)
      IMPLICIT REAL*8(A-H,O-Z)
C
C     BEGIN PROLOGUE OF XERMSG
C     PURPOSE: PROCESS ERROR MESSAGES FOR SLATEC AND OTHER LIBRARIES.
C     LIBRARY: SLATEC (XERROR)
C     CATEGORY: R3C
C     TYPE: ALL (XERMSG-A)
C     KEYWORDS: ERROR MESSAGE, XERROR
C
C     AUTHOR: FONG, KIRBY, (NMFECC AT LLNL)
C
C     DESCRIPTION:
C     XERMSG PROCESSES A DIAGNOSTIC MESSAGE IN A MANNER DETERMINED BY THE
C     VALUE OF LEVEL AND THE CURRENT VALUE OF THE LIBRARY ERROR CONTROL
C     FLAG, KONTRL.  SEE SUBROUTINE XSETF FOR DETAILS.
C
C     LIBRAR   A CHARACTER CONSTANT (OR CHARACTER VARIABLE) WITH THE NAME
C              OF THE LIBRARY.  THIS WILL BE 'SLATEC' FOR THE SLATEC
C              COMMON MATH LIBRARY.  THE ERROR HANDLING PACKAGE IS
C              GENERAL ENOUGH TO BE USED BY MANY LIBRARIES
C              SIMULTANEOUSLY, SO IT IS DESIRABLE FOR THE ROUTINE THAT
C              DETECTS AND REPORTS AN ERROR TO IDENTIFY THE LIBRARY NAME
C              AS WELL AS THE ROUTINE NAME.
C
C     SUBROU   A CHARACTER CONSTANT (OR CHARACTER VARIABLE) WITH THE NAME
C              OF THE ROUTINE THAT DETECTED THE ERROR.  USUALLY IT IS THE
C              NAME OF THE ROUTINE THAT IS CALLING XERMSG.  THERE ARE
C              SOME INSTANCES WHERE A USER CALLABLE LIBRARY ROUTINE CALLS
C              LOWER LEVEL SUBSIDIARY ROUTINES WHERE THE ERROR IS
C              DETECTED.  IN SUCH CASES IT MAY BE MORE INFORMATIVE TO
C              SUPPLY THE NAME OF THE ROUTINE THE USER CALLED RATHER THAN
C              THE NAME OF THE SUBSIDIARY ROUTINE THAT DETECTED THE
C              ERROR.
C
C    MESSG     A CHARACTER CONSTANT (OR CHARACTER VARIABLE) WITH THE TEXT
C              OF THE ERROR OR WARNING MESSAGE.  IN THE EXAMPLE BELOW,
C              THE MESSAGE IS A CHARACTER CONSTANT THAT CONTAINS A
C              GENERIC MESSAGE.
C
C                    CALL XERMSG ('SLATEC', 'MMPY',
C                   *'THE ORDER OF THE MATRIX EXCEEDS THE ROW DIMENSION',
C                   *3, 1)
C
C              IT IS POSSIBLE (AND IS SOMETIMES DESIRABLE) TO GENERATE A
C              SPECIFIC MESSAGE--E.G., ONE THAT CONTAINS ACTUAL NUMERIC
C              VALUES.  SPECIFIC NUMERIC VALUES CAN BE CONVERTED INTO
C              CHARACTER STRINGS USING FORMATTED WRITE STATEMENTS INTO
C              CHARACTER VARIABLES.  THIS IS CALLED STANDARD FORTRAN
C              INTERNAL FILE I/O AND IS EXEMPLIFIED IN THE FIRST THREE
C              LINES OF THE FOLLOWING EXAMPLE.  YOU CAN ALSO CATENATE
C              SUBSTRINGS OF CHARACTERS TO CONSTRUCT THE ERROR MESSAGE.
C              HERE IS AN EXAMPLE SHOWING THE USE OF BOTH WRITING TO
C              AN INTERNAL FILE AND CATENATING CHARACTER STRINGS.
C
C                    CHARACTER*5 CHARN, CHARL
C                    WRITE (CHARN,10) N
C                    WRITE (CHARL,10) LDA
C                 10 FORMAT(I5)
C                    CALL XERMSG ('SLATEC', 'MMPY', 'THE ORDER'//CHARN//
C                   *   ' OF THE MATRIX EXCEEDS ITS ROW DIMENSION OF'//
C                   *   CHARL, 3, 1)
C
C              THERE ARE TWO SUBTLETIES WORTH MENTIONING.  ONE IS THAT
C              THE // FOR CHARACTER CATENATION IS USED TO CONSTRUCT THE
C              ERROR MESSAGE SO THAT NO SINGLE CHARACTER CONSTANT IS
```

```
C           CONTINUED TO THE NEXT LINE.  THIS AVOIDS CONFUSION AS TO
C           WHETHER THERE ARE TRAILING BLANKS AT THE END OF THE LINE.
C           THE SECOND IS THAT BY CATENATING THE PARTS OF THE MESSAGE
C           AS AN ACTUAL ARGUMENT RATHER THAN ENCODING THE ENTIRE
C           MESSAGE INTO ONE LARGE CHARACTER VARIABLE, WE AVOID
C           HAVING TO KNOW HOW LONG THE MESSAGE WILL BE IN ORDER TO
C           DECLARE AN ADEQUATE LENGTH FOR THAT LARGE CHARACTER
C           VARIABLE.  XERMSG CALLS XERPRN TO PRINT THE MESSAGE USING
C           MULTIPLE LINES IF NECESSARY.  IF THE MESSAGE IS VERY LONG,
C           XERPRN WILL BREAK IT INTO PIECES OF 72 CHARACTERS (AS
C           REQUESTED BY XERMSG) FOR PRINTING ON MULTIPLE LINES.
C           ALSO, XERMSG ASKS XERPRN TO PREFIX EACH LINE WITH ' *  '
C           SO THAT THE TOTAL LINE LENGTH COULD BE 76 CHARACTERS.
C           NOTE ALSO THAT XERPRN SCANS THE ERROR MESSAGE BACKWARDS
C           TO IGNORE TRAILING BLANKS.  ANOTHER FEATURE IS THAT
C           THE SUBSTRING '$$' IS TREATED AS A NEW LINE SENTINEL
C           BY XERPRN.  IF YOU WANT TO CONSTRUCT A MULTILINE
C           MESSAGE WITHOUT HAVING TO COUNT OUT MULTIPLES OF 72
C           CHARACTERS, JUST USE '$$' AS A SEPARATOR.  '$$'
C           OBVIOUSLY MUST OCCUR WITHIN 72 CHARACTERS OF THE
C           START OF EACH LINE TO HAVE ITS INTENDED EFFECT SINCE
C           XERPRN IS ASKED TO WRAP AROUND AT 72 CHARACTERS IN
C           ADDITION TO LOOKING FOR '$$'.
C
C     NERR  AN INTEGER VALUE THAT IS CHOSEN BY THE LIBRARY ROUTINE'S
C           AUTHOR.  IT MUST BE IN THE RANGE -99 TO 999 (THREE
C           PRINTABLE DIGITS).  EACH DISTINCT ERROR SHOULD HAVE ITS
C           OWN ERROR NUMBER.  THESE ERROR NUMBERS SHOULD BE DESCRIBED
C           IN THE MACHINE READABLE DOCUMENTATION FOR THE ROUTINE.
C           THE ERROR NUMBERS NEED BE UNIQUE ONLY WITHIN EACH ROUTINE,
C           SO IT IS REASONABLE FOR EACH ROUTINE TO START ENUMERATING
C           ERRORS FROM 1 AND PROCEEDING TO THE NEXT INTEGER.
C
C     LEVEL AN INTEGER VALUE IN THE RANGE 0 TO 2 THAT INDICATES THE
C           LEVEL (SEVERITY) OF THE ERROR.  THEIR MEANINGS ARE
C
C          -1  A WARNING MESSAGE.  THIS IS USED IF IT IS NOT CLEAR
C              THAT THERE REALLY IS AN ERROR, BUT THE USER'S ATTENTION
C              MAY BE NEEDED.  AN ATTEMPT IS MADE TO ONLY PRINT THIS
C              MESSAGE ONCE.
C
C           0  A WARNING MESSAGE.  THIS IS USED IF IT IS NOT CLEAR
C              THAT THERE REALLY IS AN ERROR, BUT THE USER'S ATTENTION
C              MAY BE NEEDED.
C
C           1  A RECOVERABLE ERROR.  THIS IS USED EVEN IF THE ERROR IS
C              SO SERIOUS THAT THE ROUTINE CANNOT RETURN ANY USEFUL
C              ANSWER.  IF THE USER HAS TOLD THE ERROR PACKAGE TO
C              RETURN AFTER RECOVERABLE ERRORS, THEN XERMSG WILL
C              RETURN TO THE LIBRARY ROUTINE WHICH CAN THEN RETURN TO
C              THE USER'S ROUTINE.  THE USER MAY ALSO PERMIT THE ERROR
C              PACKAGE TO TERMINATE THE PROGRAM UPON ENCOUNTERING A
C              RECOVERABLE ERROR.
C
C           2  A FATAL ERROR.  XERMSG WILL NOT RETURN TO ITS CALLER
C              AFTER IT RECEIVES A FATAL ERROR.  THIS LEVEL SHOULD
C              HARDLY EVER BE USED; IT IS MUCH BETTER TO ALLOW THE
C              USER A CHANCE TO RECOVER.  AN EXAMPLE OF ONE OF THE FEW
C              CASES IN WHICH IT IS PERMISSIBLE TO DECLARE A LEVEL 2
C              ERROR IS A REVERSE COMMUNICATION LIBRARY ROUTINE THAT
C              IS LIKELY TO BE CALLED REPEATEDLY UNTIL IT INTEGRATES
C              ACROSS SOME INTERVAL.  IF THERE IS A SERIOUS ERROR IN
C              THE INPUT SUCH THAT ANOTHER STEP CANNOT BE TAKEN AND
C              THE LIBRARY ROUTINE IS CALLED AGAIN WITHOUT THE INPUT
C              ERROR HAVING BEEN CORRECTED BY THE CALLER, THE LIBRARY
C              ROUTINE WILL PROBABLY BE CALLED FOREVER WITH IMPROPER
C              INPUT.  IN THIS CASE, IT IS REASONABLE TO DECLARE THE
C              ERROR TO BE FATAL.
C
C     EACH OF THE ARGUMENTS TO XERMSG IS INPUT; NONE WILL BE MODIFIED BY
C     XERMSG.  A ROUTINE MAY MAKE MULTIPLE CALLS TO XERMSG WITH WARNING
C     LEVEL MESSAGES; HOWEVER, AFTER A CALL TO XERMSG WITH A RECOVERABLE
C     ERROR, THE ROUTINE SHOULD RETURN TO THE USER.  DO NOT TRY TO CALL
```

```
C     XERMSG WITH A SECOND RECOVERABLE ERROR AFTER THE FIRST RECOVERABLE
C     ERROR BECAUSE THE ERROR PACKAGE SAVES THE ERROR NUMBER.  THE USER
C     CAN RETRIEVE THIS ERROR NUMBER BY CALLING ANOTHER ENTRY POINT IN
C     THE ERROR HANDLING PACKAGE AND THEN CLEAR THE ERROR NUMBER WHEN
C     RECOVERING FROM THE ERROR.  CALLING XERMSG IN SUCCESSION CAUSES THE
C     OLD ERROR NUMBER TO BE OVERWRITTEN BY THE LATEST ERROR NUMBER.
C     THIS IS CONSIDERED HARMLESS FOR ERROR NUMBERS ASSOCIATED WITH
C     WARNING MESSAGES BUT MUST NOT BE DONE FOR ERROR NUMBERS OF SERIOUS
C     ERRORS.  AFTER A CALL TO XERMSG WITH A RECOVERABLE ERROR, THE USER
C     MUST BE GIVEN A CHANCE TO CALL NUMXER OR XERCLR TO RETRIEVE OR
C     CLEAR THE ERROR NUMBER.
C     REFERENCES  R. E. JONES AND D. K. KAHANER, XERROR, THE SLATEC
C                   ERROR-HANDLING PACKAGE, SAND82-0800, SANDIA
C                   LABORATORIES, 1982.
C     ROUTINES CALLED  FDUMP, J4SAVE, XERCNT, XERHLT, XERPRN, XERSVE
C     REVISION HISTORY  (YYMMDD)
C     880101  DATE WRITTEN
C     880621  REVISED AS DIRECTED AT SLATEC CML MEETING OF FEBRUARY 1988.
C             THERE ARE TWO BASIC CHANGES.
C             1.  A NEW ROUTINE, XERPRN, IS USED INSTEAD OF XERPRT TO
C                 PRINT MESSAGES.  THIS ROUTINE WILL BREAK LONG MESSAGES
C                 INTO PIECES FOR PRINTING ON MULTIPLE LINES.  '$$' IS
C                 ACCEPTED AS A NEW LINE SENTINEL.  A PREFIX CAN BE
C                 ADDED TO EACH LINE TO BE PRINTED.  XERMSG USES EITHER
C                 ' ***' OR ' *  ' AND LONG MESSAGES ARE BROKEN EVERY
C                 72 CHARACTERS (AT MOST) SO THAT THE MAXIMUM LINE
C                 LENGTH OUTPUT CAN NOW BE AS GREAT AS 76.
C             2.  THE TEXT OF ALL MESSAGES IS NOW IN UPPER CASE SINCE THE
C                 FORTRAN STANDARD DOCUMENT DOES NOT ADMIT THE EXISTENCE
C                 OF LOWER CASE.
C     880708  REVISED AFTER THE SLATEC CML MEETING OF JUNE 29 AND 30.
C             THE PRINCIPAL CHANGES ARE
C             1.  CLARIFY COMMENTS IN THE PROLOGUES
C             2.  RENAME XRPRNT TO XERPRN
C             3.  REWORK HANDLING OF '$$' IN XERPRN TO HANDLE BLANK LINES
C                 SIMILAR TO THE WAY FORMAT STATEMENTS HANDLE THE /
C                 CHARACTER FOR NEW RECORDS.
C     890706  REVISED WITH THE HELP OF FRED FRITSCH AND REG CLEMENS TO
C             CLEAN UP THE CODING.
C     890721  REVISED TO USE NEW FEATURE IN XERPRN TO COUNT CHARACTERS IN
C             PREFIX.
C     891013  REVISED TO CORRECT COMMENTS.
C     891214  PROLOGUE CONVERTED TO VERSION 4.0 FORMAT.  (WRB)
C     900510  CHANGED TEST ON NERR TO BE -9999999 < NERR < 99999999, BUT
C             NERR .NE. 0, AND ON LEVEL TO BE -2 < LEVEL < 3.  ADDED
C             LEVEL=-1 LOGIC, CHANGED CALLS TO XERSAV TO XERSVE, AND
C             XERCTL TO XERCNT.  (RWC)
C     920501  REFORMATTED THE REFERENCES SECTION.  (WRB)
C      END PROLOGUE  XERMSG
C
       CHARACTER*(*) LIBRAR, SUBROU, MESSG
       CHARACTER*8 XLIBR, XSUBR
       CHARACTER*72  TEMP
       CHARACTER*20  LFIRST
C
       WRITE(10,*)'XERMSG I/P'
C
C      FIRST EXECUTABLE STATEMENT  XERMSG
       LKNTRL = J4SAVE (2, 0, .FALSE.)
       MAXMES = J4SAVE (4, 0, .FALSE.)
C
C      LKNTRL IS A LOCAL COPY OF THE CONTROL FLAG KONTRL.
C      MAXMES IS THE MAXIMUM NUMBER OF TIMES ANY PARTICULAR MESSAGE
C      SHOULD BE PRINTED.
C
C      WE PRINT A FATAL ERROR MESSAGE AND TERMINATE FOR AN ERROR IN
C      CALLING XERMSG.  THE ERROR NUMBER SHOULD BE POSITIVE,
C      AND THE LEVEL SHOULD BE BETWEEN 0 AND 2.
       IF (NERR.LT.-9999999 .OR. NERR.GT.99999999 .OR. NERR.EQ.0 .OR.
     *    LEVEL.LT.-1 .OR. LEVEL.GT.2) THEN
          CALL XERPRN (' ***', -1, 'FATAL ERROR IN...$$ ' //
     *       'XERMSG -- INVALID ERROR NUMBER OR LEVEL$$ '//
     *       'JOB ABORT DUE TO FATAL ERROR.', 72)
```

```
            CALL XERSVE (' ', ' ', ' ', 0, 0, 0, KDUMMY)
            CALL XERHLT (' ***XERMSG -- INVALID INPUT')
            RETURN
         ENDIF
C
C        RECORD THE MESSAGE.
         I = J4SAVE (1, NERR, .TRUE.)
         CALL XERSVE (LIBRAR, SUBROU, MESSG, 1, NERR, LEVEL, KOUNT)
C
C        HANDLE PRINT-ONCE WARNING MESSAGES.
         IF (LEVEL.EQ.-1 .AND. KOUNT.GT.1) RETURN
C
C        ALLOW TEMPORARY USER OVERRIDE OF THE CONTROL FLAG.
         XLIBR  = LIBRAR
         XSUBR  = SUBROU
         LFIRST = MESSG
         LERR   = NERR
         LLEVEL = LEVEL

         CALL XERCNT (XLIBR, XSUBR, LFIRST, LERR, LLEVEL, LKNTRL)
C
         LKNTRL = MAX(-2, MIN(2,LKNTRL))
         MKNTRL = ABS(LKNTRL)
C
C        SKIP PRINTING IF THE CONTROL FLAG VALUE AS RESET IN XERCNT IS
C        ZERO AND THE ERROR IS NOT FATAL.
         IF (LEVEL.LT.2 .AND. LKNTRL.EQ.0) GO TO 30
         IF (LEVEL.EQ.0 .AND. KOUNT.GT.MAXMES) GO TO 30
         IF (LEVEL.EQ.1 .AND. KOUNT.GT.MAXMES .AND. MKNTRL.EQ.1) GO TO 30
         IF (LEVEL.EQ.2 .AND. KOUNT.GT.MAX(1,MAXMES)) GO TO 30
C
C        ANNOUNCE THE NAMES OF THE LIBRARY AND SUBROUTINE BY BUILDING A
C        MESSAGE IN CHARACTER VARIABLE TEMP (NOT EXCEEDING 66 CHARACTERS)
C        AND SENDING IT OUT VIA XERPRN.  PRINT ONLY IF CONTROL FLAG
C        IS NOT ZERO.
         IF (LKNTRL .NE. 0) THEN
            TEMP(1:21) = 'MESSAGE FROM ROUTINE '
            I = MIN(LEN(SUBROU), 16)
            TEMP(22:21+I) = SUBROU(1:I)
            TEMP(22+I:33+I) = ' IN LIBRARY '
            LTEMP = 33 + I
            I = MIN(LEN(LIBRAR), 16)
            TEMP(LTEMP+1:LTEMP+I) = LIBRAR (1:I)
            TEMP(LTEMP+I+1:LTEMP+I+1) = '.'
            LTEMP = LTEMP + I + 1
            CALL XERPRN (' ***', -1, TEMP(1:LTEMP), 72)
         ENDIF
C
C        IF LKNTRL IS POSITIVE, PRINT AN INTRODUCTORY LINE BEFORE
C        PRINTING THE MESSAGE.  THE INTRODUCTORY LINE TELLS THE CHOICE
C        FROM EACH OF THE FOLLOWING THREE OPTIONS.
C        1.  LEVEL OF THE MESSAGE
C                'INFORMATIVE MESSAGE'
C                'POTENTIALLY RECOVERABLE ERROR'
C                'FATAL ERROR'
C        2.  WHETHER CONTROL FLAG WILL ALLOW PROGRAM TO CONTINUE
C                'PROG CONTINUES'
C                'PROG ABORTED'
C        3.  WHETHER OR NOT A TRACEBACK WAS REQUESTED.  (THE TRACEBACK
C            MAY NOT BE IMPLEMENTED AT SOME SITES, SO THIS ONLY TELLS
C            WHAT WAS REQUESTED, NOT WHAT WAS DELIVERED.)
C                'TRACEBACK REQUESTED'
C                'TRACEBACK NOT REQUESTED'
C        NOTICE THAT THE LINE INCLUDING FOUR PREFIX CHARACTERS WILL NOT
C        EXCEED 74 CHARACTERS.
C        WE SKIP THE NEXT BLOCK IF THE INTRODUCTORY LINE IS NOT NEEDED.
         IF (LKNTRL .GT. 0) THEN
C
C        THE FIRST PART OF THE MESSAGE TELLS ABOUT THE LEVEL.
            IF (LEVEL .LE. 0) THEN
               TEMP(1:20) = 'INFORMATIVE MESSAGE,'
               LTEMP = 20
            ELSEIF (LEVEL .EQ. 1) THEN
```

```
            TEMP(1:30) = 'POTENTIALLY RECOVERABLE ERROR,'
            LTEMP = 30
         ELSE
            TEMP(1:12) = 'FATAL ERROR,'
            LTEMP = 12
         ENDIF
C
C     THEN WHETHER THE PROGRAM WILL CONTINUE.
         IF ((MKNTRL.EQ.2 .AND. LEVEL.GE.1) .OR.
     *        (MKNTRL.EQ.1 .AND. LEVEL.EQ.2)) THEN
            TEMP(LTEMP+1:LTEMP+14) = ' PROG ABORTED,'
            LTEMP = LTEMP + 14
         ELSE
            TEMP(LTEMP+1:LTEMP+16) = ' PROG CONTINUES,'
            LTEMP = LTEMP + 16
         ENDIF
C
C     FINALLY TELL WHETHER THERE SHOULD BE A TRACEBACK.
         IF (LKNTRL .GT. 0) THEN
            TEMP(LTEMP+1:LTEMP+20) = ' TRACEBACK REQUESTED'
            LTEMP = LTEMP + 20
         ELSE
            TEMP(LTEMP+1:LTEMP+24) = ' TRACEBACK NOT REQUESTED'
            LTEMP = LTEMP + 24
         ENDIF
         CALL XERPRN (' ***', -1, TEMP(1:LTEMP), 72)
      ENDIF
C
C     NOW SEND OUT THE MESSAGE.
      CALL XERPRN (' *  ', -1, MESSG, 72)
C
C     IF LKNTRL IS POSITIVE, WRITE THE ERROR NUMBER AND REQUEST A
C        TRACEBACK.
      IF (LKNTRL .GT. 0) THEN
         WRITE (TEMP, '(''ERROR NUMBER = '', I8)') NERR
         DO 10 I=16,22
            IF (TEMP(I:I) .NE. ' ') GO TO 20
   10    CONTINUE
C
   20    CALL XERPRN (' *  ', -1, TEMP(1:15) // TEMP(I:23), 72)
         CALL FDUMP
      ENDIF
C
C     IF LKNTRL IS NOT ZERO, PRINT A BLANK LINE AND AN END OF MESSAGE.
      IF (LKNTRL .NE. 0) THEN
         CALL XERPRN (' *  ', -1, ' ', 72)
         CALL XERPRN (' ***', -1, 'END OF MESSAGE', 72)
         CALL XERPRN ('    ',  0, ' ', 72)
      ENDIF
C
C     IF THE ERROR IS NOT FATAL OR THE ERROR IS RECOVERABLE AND THE
C     CONTROL FLAG IS SET FOR RECOVERY, THEN RETURN.
   30 IF (LEVEL.LE.0 .OR. (LEVEL.EQ.1 .AND. MKNTRL.LE.1)) RETURN
C
C     THE PROGRAM WILL BE STOPPED DUE TO AN UNRECOVERED ERROR OR A
C     FATAL ERROR.  PRINT THE REASON FOR THE ABORT AND THE ERROR
C     SUMMARY IF THE CONTROL FLAG AND THE MAXIMUM ERROR COUNT PERMIT.
      IF (LKNTRL.GT.0 .AND. KOUNT.LT.MAX(1,MAXMES)) THEN
         IF (LEVEL .EQ. 1) THEN
            CALL XERPRN
     *         (' ***', -1, 'JOB ABORT DUE TO UNRECOVERED ERROR.', 72)
         ELSE
            CALL XERPRN(' ***', -1, 'JOB ABORT DUE TO FATAL ERROR.', 72)
         ENDIF
         CALL XERSVE (' ', ' ', ' ', -1, 0, 0, KDUMMY)
         CALL XERHLT (' ')
      ELSE
         CALL XERHLT (MESSG)
      ENDIF
C
      WRITE(10,*)'XERMSG O/P'
C
      RETURN
```

```
      END
******
      SUBROUTINE FDUMP
      IMPLICIT REAL*8(A-H,O-Z)
C
C     BEGIN PROLOGUE FDUMP
C     DATE WRITTEN: 790801 (YYMMDD)
C     REVISION DATE: 820801 (YYMMDD)
C     CATEGORY NO.: Z
C     KEYWORDS: ERROR, XERROR PACKAGE
C     AUTHOR: JONES, R. E., (SNLA)
C     PURPOSE: SYMBOLIC DUMP (SHOULD BE LOCALLY WRITTEN)
C
C     DESCRIPTION:
C     NOTE: MACHINE DEPENDENT ROUTINE
C     FDUMP IS INTENDED TO BE REPLACED BY A LOCALLY WRITTEN
C     VERSION WHICH PRODUCES A SYMBOLIC DUMP. FAILING THIS,
C     IT SHOULD BE REPLACED BY A VERSION WHICH PRINTS THE
C     SUBPROGRAM NESTING LIST. NOTE THAT THIS DUMP MUST BE
C     PRINTED ON EACH OF UP TO FIVE FILES, AS INDICATED BY THE
C     XGETUA ROUTINE.
C
C     WRITTEN BY RON JONES, WITH SLATEC COMMON MATH LIBRARY SUBCOMMITTEE
C     LATEST REVISION ---  23 MAY 1979
C
C     ROUTINES CALLED  (NONE)
C     END PROLOGUE  FDUMP
C
      WRITE(10,*)'FDUMP I/P'
C
C     ***FIRST EXECUTABLE STATEMENT  FDUMP
C
      WRITE(10,*)'FDUMP O/P'
C
      RETURN
      END
******
      FUNCTION J4SAVE(IWHICH,IVALUE,ISET)
C
C     ABSTRACT:
C     J4SAVE SAVES AND RECALLS SEVERAL GLOBAL VARIABLES NEEDED
C     BY THE LIBRARY ERROR HANDLING ROUTINES.
C
C     DESCRIPTION OF PARAMETERS:
C     INPUT:
C        IWHICH - INDEX OF ITEM DESIRED.
C                 = 1 REFERS TO CURRENT ERROR NUMBER.
C                 = 2 REFERS TO CURRENT ERROR CONTROL FLAG.
C                 = 3 REFERS TO CURRENT UNIT NUMBER TO WHICH ERROR
C                     MESSAGES ARE TO BE SENT.  (0 MEANS USE STANDARD.)
C                 = 4 REFERS TO THE MAXIMUM NUMBER OF TIMES ANY
C                     MESSAGE IS TO BE PRINTED (AS SET BY XERMAX).
C                 = 5 REFERS TO THE TOTAL NUMBER OF UNITS TO WHICH
C                     EACH ERROR MESSAGE IS TO BE WRITTEN.
C                 = 6 REFERS TO THE 2ND UNIT FOR ERROR MESSAGES
C                 = 7 REFERS TO THE 3RD UNIT FOR ERROR MESSAGES
C                 = 8 REFERS TO THE 4TH UNIT FOR ERROR MESSAGES
C                 = 9 REFERS TO THE 5TH UNIT FOR ERROR MESSAGES
C        IVALUE - THE VALUE TO BE SET FOR THE IWHICH-TH PARAMETER,
C                 IF ISET IS .TRUE. .
C        ISET   - IF ISET=.TRUE., THE IWHICH-TH PARAMETER WILL BE
C                 GIVEN THE VALUE, IVALUE.  IF ISET=.FALSE., THE
C                 IWHICH-TH PARAMETER WILL BE UNCHANGED, AND IVALUE
C                 IS A DUMMY PARAMETER.
C     OUTPUT:
C        THE (OLD) VALUE OF THE IWHICH-TH PARAMETER WILL BE RETURNED
C        IN THE FUNCTION VALUE, J4SAVE.
C
C     WRITTEN BY RON JONES, WITH SLATEC COMMON MATH LIBRARY SUBCOMMITTEE
C     ADAPTED FROM BELL LABORATORIES PORT LIBRARY ERROR HANDLER
C     LATEST REVISION ---  23 MAY 1979
C
      LOGICAL ISET
```

```
      INTEGER IPARAM(9)
      DATA IPARAM(1),IPARAM(2),IPARAM(3),IPARAM(4)/0,1,0,10/
      DATA IPARAM(5)/1/
      DATA IPARAM(6),IPARAM(7),IPARAM(8),IPARAM(9)/0,0,0,0/
C
      WRITE(10,*)'J4SAVE I/P'
C
      J4SAVE = IPARAM(IWHICH)
      IF (ISET) IPARAM(IWHICH) = IVALUE
C
      WRITE(10,*)'J4SAVE O/P'
C
      RETURN
      END
******
      SUBROUTINE XERCNT (LIBRAR, SUBROU, MESSG, NERR, LEVEL, KONTRL)
      IMPLICIT REAL*8(A-H,O-Z)
C
C     BEGIN PROLOGUE  XERCNT
C     PURPOSE: ALLOW USER CONTROL OVER HANDLING OF ERRORS.
C     LIBRARY: SLATEC (XERROR)
C     CATEGORY  R3C
C     TYPE ALL (XERCNT-A)
C     KEYWORDS: ERROR, XERROR
C     AUTHOR: JONES, R. E., (SNLA)
C
C     DESCRIPTION:
C     ABSTRACT:
C        ALLOWS USER CONTROL OVER HANDLING OF INDIVIDUAL ERRORS.
C        JUST AFTER EACH MESSAGE IS RECORDED, BUT BEFORE IT IS
C        PROCESSED ANY FURTHER (I.E., BEFORE IT IS PRINTED OR
C        A DECISION TO ABORT IS MADE), A CALL IS MADE TO XERCNT.
C        IF THE USER HAS PROVIDED HIS OWN VERSION OF XERCNT, HE
C        CAN THEN OVERRIDE THE VALUE OF KONTROL USED IN PROCESSING
C        THIS MESSAGE BY REDEFINING ITS VALUE.
C        KONTRL MAY BE SET TO ANY VALUE FROM -2 TO 2.
C        THE MEANINGS FOR KONTRL ARE THE SAME AS IN XSETF, EXCEPT
C        THAT THE VALUE OF KONTRL CHANGES ONLY FOR THIS MESSAGE.
C        IF KONTRL IS SET TO A VALUE OUTSIDE THE RANGE FROM -2 TO 2,
C        IT WILL BE MOVED BACK INTO THAT RANGE.
C
C     DESCRIPTION OF PARAMETERS:
C     INPUT:
C        LIBRAR - THE LIBRARY THAT THE ROUTINE IS IN.
C        SUBROU - THE SUBROUTINE THAT XERMSG IS BEING CALLED FROM
C        MESSG  - THE FIRST 20 CHARACTERS OF THE ERROR MESSAGE.
C        NERR   - SAME AS IN THE CALL TO XERMSG.
C        LEVEL  - SAME AS IN THE CALL TO XERMSG.
C        KONTRL - THE CURRENT VALUE OF THE CONTROL FLAG AS SET
C                 BY A CALL TO XSETF.
C
C     OUTPUT:
C        KONTRL - THE NEW VALUE OF KONTRL.  IF KONTRL IS NOT
C                 DEFINED, IT WILL REMAIN AT ITS ORIGINAL VALUE.
C                 THIS CHANGED VALUE OF CONTROL AFFECTS ONLY
C                 THE CURRENT OCCURRENCE OF THE CURRENT MESSAGE.
C
C     REFERENCES: R. E. JONES AND D. K. KAHANER, XERROR, THE SLATEC
C                 ERROR-HANDLING PACKAGE, SAND82-0800, SANDIA
C                 LABORATORIES, 1982.
C     ROUTINES CALLED: (NONE)
C     REVISION HISTORY:  (YYMMDD)
C     790801  DATE WRITTEN
C     861211  REVISION DATE FROM VERSION 3.2
C     891214  PROLOGUE CONVERTED TO VERSION 4.0 FORMAT.  (BAB)
C     900206  ROUTINE CHANGED FROM USER-CALLABLE TO SUBSIDIARY.  (WRB)
C     900510  CHANGED CALLING SEQUENCE TO INCLUDE LIBRARY AND SUBROUTINE
C             NAMES, CHANGED ROUTINE NAME FROM XERCTL TO XERCNT.  (RWC)
C     920501  REFORMATTED THE REFERENCES SECTION.  (WRB)
C     END PROLOGUE  XERCNT
C
      CHARACTER*(*)LIBRAR,SUBROU,MESSG
C
```

```
      WRITE(10,*)'XERCNT I/P'
C
C***FIRST EXECUTABLE STATEMENT XERCNT
C
      WRITE(10,*)'XERCNT O/P'
C
      RETURN
      END
******
*      SUBROUTINE XERHLT (MESSG)
*      IMPLICIT REAL*8(A-H,O-Z)
C***BEGIN PROLOGUE  XERHLT
C***SUBSIDIARY
C***PURPOSE  ABORT PROGRAM EXECUTION AND PRINT ERROR MESSAGE.
C***LIBRARY   SLATEC (XERROR)
C***CATEGORY  R3C
C***TYPE      ALL (XERHLT-A)
C***KEYWORDS  ABORT PROGRAM EXECUTION, ERROR, XERROR
C***AUTHOR  JONES, R. E., (SNLA)
C***DESCRIPTION
C
C     ABSTRACT
C        ***NOTE*** MACHINE DEPENDENT ROUTINE
C        XERHLT ABORTS THE EXECUTION OF THE PROGRAM.
C        THE ERROR MESSAGE CAUSING THE ABORT IS GIVEN IN THE CALLING
C        SEQUENCE, IN CASE ONE NEEDS IT FOR PRINTING ON A DAYFILE,
C        FOR EXAMPLE.
C
C     DESCRIPTION OF PARAMETERS
C        MESSG IS AS IN XERMSG.
C
C***REFERENCES  R. E. JONES AND D. K. KAHANER, XERROR, THE SLATEC
C                 ERROR-HANDLING PACKAGE, SAND82-0800, SANDIA
C                 LABORATORIES, 1982.
C***ROUTINES CALLED  (NONE)
C***REVISION HISTORY  (YYMMDD)
C   790801  DATE WRITTEN
C   861211  REVISION DATE FROM VERSION 3.2
C   891214  PROLOGUE CONVERTED TO VERSION 4.0 FORMAT.  (BAB)
C   900206  ROUTINE CHANGED FROM USER-CALLABLE TO SUBSIDIARY.  (WRB)
C   900510  CHANGED CALLING SEQUENCE TO DELETE LENGTH OF CHARACTER
C           AND CHANGED ROUTINE NAME FROM XERABT TO XERHLT.  (RWC)
C   920501  REFORMATTED THE REFERENCES SECTION.  (WRB)
C***END PROLOGUE  XERHLT
*      CHARACTER*(*) MESSG
C
C     FIRST EXECUTABLE STATEMENT  XERHLT
      WRITE(10,*)'XERHLT I/P'
C
      WRITE(10,*)'XERHLT O/P'
C
      STOP
      END
******
      SUBROUTINE XERPRN (PREFIX, NPREF, MESSG, NWRAP)
      IMPLICIT REAL*8(A-H,O-Z)
C
C     BEGIN PROLOGUE OF XERPRN
C     PURPOSE: PRINT ERROR MESSAGES PROCESSED BY XERMSG.
C     LIBRARY: SLATEC (XERROR)
C     CATEGORY: R3C
C     TYPE: ALL (XERPRN-A)
C     KEYWORDS:  ERROR MESSAGES, PRINTING, XERROR
C     AUTHOR:  FONG, KIRBY, (NMFECC AT LLNL)
C
C     DESCRIPTION:
C     THIS ROUTINE SENDS ONE OR MORE LINES TO EACH OF THE (UP TO FIVE)
C     LOGICAL UNITS TO WHICH ERROR MESSAGES ARE TO BE SENT.  THIS ROUTINE
C     IS CALLED SEVERAL TIMES BY XERMSG, SOMETIMES WITH A SINGLE LINE TO
C     PRINT AND SOMETIMES WITH A (POTENTIALLY VERY LONG) MESSAGE THAT MAY
C     WRAP AROUND INTO MULTIPLE LINES.
C
C     PREFIX Input argument of type CHARACTER.  This argument contains
```

```
C             characters to be put at the beginning of each line before
C             the body of the message.  No more than 16 characters of
C             PREFIX will be used.
C
C      NPREF  Input argument of type INTEGER.  This argument is the number
C             of characters to use from PREFIX.  If it is negative, the
C             intrinsic function LEN is used to determine its length.  If
C             it is zero, PREFIX is not used.  If it exceeds 16 or if
C             LEN(PREFIX) exceeds 16, only the first 16 characters will be
C             used.  If NPREF is positive and the length of PREFIX is less
C             than NPREF, a copy of PREFIX extended with blanks to length
C             NPREF will be used.
C
C      MESSG  Input argument of type CHARACTER.  This is the text of a
C             message to be printed.  If it is a long message, it will be
C             broken into pieces for printing on multiple lines.  Each line
C             will start with the appropriate prefix and be followed by a
C             piece of the message.  NWRAP is the number of characters per
C             piece; that is, after each NWRAP characters, we break and
C             start a new line.  In addition the characters '$$' embedded
C             in MESSG are a sentinel for a new line.  The counting of
C             characters up to NWRAP starts over for each new line.  The
C             value of NWRAP typically used by XERMSG is 72 since many
C             older error messages in the SLATEC Library are laid out to
C             rely on wrap-around every 72 characters.
C
C      NWRAP  Input argument of type INTEGER.  This gives the maximum size
C             piece into which to break MESSG for printing on multiple
C             lines.  An embedded '$$' ends a line, and the count restarts
C             at the following character.  If a line break does not occur
C             on a blank (it would split a word) that word is moved to the
C             next line.  Values of NWRAP less than 16 will be treated as
C             16.  Values of NWRAP greater than 132 will be treated as 132.
C             The actual line length will be NPREF + NWRAP after NPREF has
C             been adjusted to fall between 0 and 16 and NWRAP has been
C             adjusted to fall between 16 and 132.
C
C      REFERENCES  R. E. JONES AND D. K. KAHANER, XERROR, THE SLATEC
C                  ERROR-HANDLING PACKAGE, SAND82-0800, SANDIA
C                  LABORATORIES, 1982.
C      ROUTINES CALLED: I1MACH, XGETUA
C      REVISION HISTORY  (YYMMDD)
C      880621 DATE WRITTEN
C      880708 REVISED AFTER THE SLATEC CML SUBCOMMITTEE MEETING OF
C             JUNE 29 AND 30 TO CHANGE THE NAME TO XERPRN AND TO REWORK
C             THE HANDLING OF THE NEW LINE SENTINEL TO BEHAVE LIKE THE
C             SLASH CHARACTER IN FORMAT STATEMENTS.
C      890706 REVISED WITH THE HELP OF FRED FRITSCH AND REG CLEMENS TO
C             STREAMLINE THE CODING AND FIX A BUG THAT CAUSED EXTRA BLANK
C             LINES TO BE PRINTED.
C      890721 REVISED TO ADD A NEW FEATURE.  A NEGATIVE VALUE OF NPREF
C             CAUSES LEN(PREFIX) TO BE USED AS THE LENGTH.
C      891013 REVISED TO CORRECT ERROR IN CALCULATING PREFIX LENGTH.
C      891214 PROLOGUE CONVERTED TO VERSION 4.0 FORMAT.  (WRB)
C      900510 ADDED CODE TO BREAK MESSAGES BETWEEN WORDS.  (RWC)
C      920501 REFORMATTED THE REFERENCES SECTION.  (WRB)
C      END PROLOGUE  XERPRN
C
       CHARACTER*(*) PREFIX, MESSG
       INTEGER NPREF, NWRAP
       CHARACTER*148 CBUFF
       INTEGER IU(5), NUNIT
       CHARACTER*2 NEWLIN
       PARAMETER (NEWLIN = '$$')
C
       WRITE(10,*)'XERPRN I/P'
C
C      FIRST EXECUTABLE STATEMENT  XERPRN
       CALL XGETUA(IU,NUNIT)
C
C      A ZERO VALUE FOR A LOGICAL UNIT NUMBER MEANS TO USE THE STANDARD
C      ERROR MESSAGE UNIT INSTEAD.  I1MACH(4) RETRIEVES THE STANDARD
C      ERROR MESSAGE UNIT.
```

```
      N=I1MACH(4)
      DO 10 I=1,NUNIT
         IF(IU(I).EQ.0)IU(I)=N
   10 CONTINUE
C
C     LPREF IS THE LENGTH OF THE PREFIX.  THE PREFIX IS PLACED AT THE
C     BEGINNING OF CBUFF, THE CHARACTER BUFFER, AND KEPT THERE DURING
C     THE REST OF THIS ROUTINE.
      IF(NPREF.LT.0)THEN
         LPREF=LEN(PREFIX)
      ELSE
         LPREF=NPREF
      ENDIF
      LPREF=MIN(16, LPREF)
      IF(LPREF.NE.0)CBUFF(1:LPREF)=PREFIX
C
C     LWRAP IS THE MAXIMUM NUMBER OF CHARACTERS WE WANT TO TAKE AT ONE
C     TIME FROM MESSG TO PRINT ON ONE LINE.
      LWRAP=MAX(16,MIN(132,NWRAP))
C
C     SET LENMSG TO THE LENGTH OF MESSG, IGNORE ANY TRAILING BLANKS.
      LENMSG = LEN(MESSG)
      N = LENMSG
      DO 20 I=1,N
         IF(MESSG(LENMSG:LENMSG).NE.' ')GOTO 30
         LENMSG=LENMSG-1
   20 CONTINUE
   30 CONTINUE
C
C     IF THE MESSAGE IS ALL BLANKS, THEN PRINT ONE BLANK LINE.
      IF(LENMSG.EQ.0)THEN
         CBUFF(LPREF+1:LPREF+1)=' '
         DO 40 I=1,NUNIT
            WRITE(IU(I),'(A)')CBUFF(1:LPREF+1)
   40    CONTINUE
         RETURN
      ENDIF
C
C     SET NEXTC TO THE POSITION IN MESSG WHERE THE NEXT SUBSTRING
C     STARTS.  FROM THIS POSITION WE SCAN FOR THE NEW LINE SENTINEL.
C     WHEN NEXTC EXCEEDS LENMSG, THERE IS NO MORE TO PRINT.
C     WE LOOP BACK TO LABEL 50 UNTIL ALL PIECES HAVE BEEN PRINTED.
C
C     WE LOOK FOR THE NEXT OCCURRENCE OF THE NEW LINE SENTINEL.  THE
C     INDEX INTRINSIC FUNCTION RETURNS ZERO IF THERE IS NO OCCURRENCE
C     OR IF THE LENGTH OF THE FIRST ARGUMENT IS LESS THAN THE LENGTH
C     OF THE SECOND ARGUMENT.
C
C     THERE ARE SEVERAL CASES WHICH SHOULD BE CHECKED FOR IN THE
C     FOLLOWING ORDER.  WE ARE ATTEMPTING TO SET LPIECE TO THE NUMBER
C     OF CHARACTERS THAT SHOULD BE TAKEN FROM MESSG STARTING AT
C     POSITION NEXTC.
C
C     LPIECE.EQ.0 THE NEW LINE SENTINEL DOES NOT OCCUR IN THE
C                 REMAINDER OF THE CHARACTER STRING.  LPIECE
C                 SHOULD BE SET TO LWRAP OR LENMSG+1-NEXTC,
C                 WHICHEVER IS LESS.
C
C     LPIECE.EQ.1 THE NEW LINE SENTINEL STARTS AT MESSG(NEXTC:
C                 NEXTC).  LPIECE IS EFFECTIVELY ZERO, AND WE
C                 PRINT NOTHING TO AVOID PRODUCING UNNECESSARY
C                 BLANK LINES.  THIS TAKES CARE OF THE SITUATION
C                 WHERE THE LIBRARY ROUTINE HAS A MESSAGE OF
C                 EXACTLY 72 CHARACTERS FOLLOWED BY A NEW LINE
C                 SENTINEL FOLLOWED BY MORE CHARACTERS.  NEXTC
C                 SHOULD BE INCREMENTED BY 2.
C
C     LPIECE .GT. LWRAP+1  REDUCE LPIECE TO LWRAP.
C
C     ELSE        THIS LAST CASE MEANS 2 .LE. LPIECE .LE. LWRAP+1
C                 RESET LPIECE = LPIECE-1.  NOTE THAT THIS
C                 PROPERLY HANDLES THE END CASE WHERE LPIECE .EQ.
C                 LWRAP+1.  THAT IS, THE SENTINEL FALLS EXACTLY
```

```
C                 AT THE END OF A LINE.
C
      NEXTC = 1
   50 LPIECE = INDEX(MESSG(NEXTC:LENMSG), NEWLIN)
      IF(LPIECE.EQ.0)THEN
C
C     THERE WAS NO NEW LINE SENTINEL FOUND.
      IDELTA=0
      LPIECE=MIN(LWRAP, LENMSG+1-NEXTC)
      IF(LPIECE.LT.LENMSG+1-NEXTC)THEN
         DO 52 I=LPIECE+1,2,-1
            IF(MESSG(NEXTC+I-1:NEXTC+I-1) .EQ. ' ')THEN
              LPIECE = I-1
              IDELTA = 1
              GOTO 54
            ENDIF
   52    CONTINUE
      ENDIF
   54 CBUFF(LPREF+1:LPREF+LPIECE)=MESSG(NEXTC:NEXTC+LPIECE-1)
      NEXTC=NEXTC+LPIECE+IDELTA
      ELSEIF(LPIECE.EQ.1)THEN
C
C     WE HAVE A NEW LINE SENTINEL AT MESSG(NEXTC:NEXTC+1).
C     DON'T PRINT A BLANK LINE.
      NEXTC=NEXTC+2
      GOTO 50
      ELSEIF(LPIECE.GT.LWRAP+1)THEN
C
C     LPIECE SHOULD BE SET DOWN TO LWRAP.
      IDELTA=0
      LPIECE=LWRAP
      DO 56 I=LPIECE+1,2,-1
         IF(MESSG(NEXTC+I-1:NEXTC+I-1) .EQ. ' ')THEN
           LPIECE = I-1
           IDELTA = 1
           GOTO 58
         ENDIF
   56 CONTINUE
   58 CBUFF(LPREF+1:LPREF+LPIECE) = MESSG(NEXTC:NEXTC+LPIECE-1)
      NEXTC = NEXTC + LPIECE + IDELTA
      ELSE
C
C     IF WE ARRIVE HERE, IT MEANS 2 .LE. LPIECE .LE. LWRAP+1.
C     WE SHOULD DECREMENT LPIECE BY ONE.
      LPIECE = LPIECE - 1
      CBUFF(LPREF+1:LPREF+LPIECE) = MESSG(NEXTC:NEXTC+LPIECE-1)
      NEXTC  = NEXTC + LPIECE + 2
      ENDIF
C
C     PRINT
      DO 60 I=1,NUNIT
      WRITE(IU(I),'(A)')CBUFF(1:LPREF+LPIECE)
   60 CONTINUE
C
      IF(NEXTC.LE.LENMSG)GOTO 50
C
      WRITE(10,*)'XERPRN O/P'
C
      RETURN
      END
******
      SUBROUTINE XERSVE(LIBRAR,SUBROU,MESSG,KFLAG,NERR,LEVEL,ICOUNT)
      IMPLICIT REAL*8(A-H,O-Z)
C
C     BEGIN PROLOGUE OF XERSVE
C     PURPOSE: RECORD THAT AN ERROR HAS OCCURRED.
C     LIBRARY: SLATEC (XERROR)
C     CATEGORY: R3
C     TYPE: ALL (XERSVE-A)
C     KEYWORDS: ERROR, XERROR
C     AUTHOR: JONES, R. E., (SNLA)
C
C     USAGE:
```

```
C          INTEGER KFLAG, NERR, LEVEL, ICOUNT
C          CHARACTER * (LEN) LIBRAR, SUBROU, MESSG
C
C          CALL XERSVE (LIBRAR, SUBROU, MESSG, KFLAG, NERR, LEVEL, ICOUNT)
C
C       ARGUMENTS:
C          LIBRAR :IN    IS THE LIBRARY THAT THE MESSAGE IS FROM.
C          SUBROU :IN    IS THE SUBROUTINE THAT THE MESSAGE IS FROM.
C          MESSG  :IN    IS THE MESSAGE TO BE SAVED.
C          KFLAG  :IN    INDICATES THE ACTION TO BE PERFORMED.
C                        WHEN KFLAG > 0, THE MESSAGE IN MESSG IS SAVED.
C                        WHEN KFLAG=0 THE TABLES WILL BE DUMPED AND
C                        CLEARED.
C                        WHEN KFLAG < 0, THE TABLES WILL BE DUMPED AND
C                        NOT CLEARED.
C          NERR   :IN    IS THE ERROR NUMBER.
C          LEVEL  :IN    IS THE ERROR SEVERITY.
C          ICOUNT :OUT   THE NUMBER OF TIMES THIS MESSAGE HAS BEEN SEEN,
C                        OR ZERO IF THE TABLE HAS OVERFLOWED AND DOES NOT
C                        CONTAIN THIS MESSAGE SPECIFICALLY.  WHEN KFLAG=0,
C                        ICOUNT WILL NOT BE ALTERED.
C
C       DESCRIPTION:
C          RECORD THAT THIS ERROR OCCURRED AND POSSIBLY DUMP AND CLEAR THE
C          TABLES.
C
C       REFERENCES  R. E. JONES AND D. K. KAHANER, XERROR, THE SLATEC
C                   ERROR-HANDLING PACKAGE, SAND82-0800, SANDIA
C                   LABORATORIES, 1982.
C       ROUTINES CALLED  I1MACH, XGETUA
C       REVISION HISTORY  (YYMMDD)
C       800319  DATE WRITTEN
C       861211  REVISION DATE FROM VERSION 3.2
C       891214  PROLOGUE CONVERTED TO VERSION 4.0 FORMAT.  (BAB)
C       900413  ROUTINE MODIFIED TO REMOVE REFERENCE TO KFLAG.  (WRB)
C       900510  CHANGED TO ADD LIBRARY NAME AND SUBROUTINE TO CALLING
C               SEQUENCE, USE IF-THEN-ELSE, MAKE NUMBER OF SAVED ENTRIES
C               EASILY CHANGEABLE, CHANGED ROUTINE NAME FROM XERSAV TO
C               XERSVE.  (RWC)
C       910626  ADDED LIBTAB AND SUBTAB TO SAVE STATEMENT.  (BKS)
C       920501  REFORMATTED THE REFERENCES SECTION.  (WRB)
C       END PROLOGUE OF XERSVE
C
        PARAMETER (LENTAB=10)
        INTEGER LUN(5)
        CHARACTER*(*) LIBRAR, SUBROU, MESSG
        CHARACTER*8  LIBTAB(LENTAB), SUBTAB(LENTAB), LIB, SUB
        CHARACTER*20 MESTAB(LENTAB), MES
        DIMENSION NERTAB(LENTAB), LEVTAB(LENTAB), KOUNT(LENTAB)
        SAVE LIBTAB, SUBTAB, MESTAB, NERTAB, LEVTAB, KOUNT, KOUNTX, NMSG
        DATA KOUNTX/0/, NMSG/0/
C
        WRITE(10,*)'XERSVE I/P'
C
C       FIRST EXECUTABLE STATEMENT  XERSVE
        IF(KFLAG.LE.0)THEN
C
C       DUMP THE TABLE.
C
        IF(NMSG.EQ.0)RETURN
C
C       PRINT TO EACH UNIT.
        CALL XGETUA (LUN, NUNIT)
        DO 20 KUNIT = 1,NUNIT
        IUNIT = LUN(KUNIT)
        IF (IUNIT.EQ.0) IUNIT = I1MACH(4)
C
C       PRINT THE TABLE HEADER.
        WRITE (IUNIT,9000)
C
C       PRINT BODY OF TABLE.
        DO 10 I = 1,NMSG
        WRITE(IUNIT,9010)LIBTAB(I),SUBTAB(I),MESTAB(I),
```

```
     *                     NERTAB(I),LEVTAB(I),KOUNT (I)
   10 CONTINUE
C
C     PRINT NUMBER OF OTHER ERRORS.
      IF(KOUNTX.NE.0) WRITE(IUNIT,9020)KOUNTX
      WRITE (IUNIT,9030)
   20 CONTINUE
C
C     CLEAR THE ERROR TABLES.
      IF(KFLAG.EQ.0)THEN
      NMSG = 0
      KOUNTX = 0
      ENDIF
      ELSE
C
C     PROCESS A MESSAGE...
C     SEARCH FOR THIS MESSG, OR ELSE AN EMPTY SLOT FOR THIS MESSG,
C     OR ELSE DETERMINE THAT THE ERROR TABLE IS FULL.
      LIB = LIBRAR
      SUB = SUBROU
      MES = MESSG
      DO 30 I = 1,NMSG
      IF(LIB.EQ.LIBTAB(I).AND.SUB.EQ.SUBTAB(I).AND.
     *   MES.EQ.MESTAB(I).AND.NERR.EQ.NERTAB(I).AND.
     * LEVEL.EQ.LEVTAB(I))THEN
      KOUNT(I) = KOUNT(I) + 1
      ICOUNT = KOUNT(I)
      RETURN
      ENDIF
   30 CONTINUE
C
      IF (NMSG.LT.LENTAB) THEN
C
C     EMPTY SLOT FOUND FOR NEW MESSAGE.
      NMSG = NMSG + 1
      LIBTAB(I) = LIB
      SUBTAB(I) = SUB
      MESTAB(I) = MES
      NERTAB(I) = NERR
      LEVTAB(I) = LEVEL
      KOUNT (I) = 1
      ICOUNT    = 1
      ELSE
C
C     TABLE IS FULL.
      KOUNTX = KOUNTX+1
      ICOUNT = 0
      ENDIF
      ENDIF
C
C     FORMATS
 9000 FORMAT ('0          ERROR MESSAGE SUMMARY' /
     +   ' LIBRARY    SUBROUTINE MESSAGE START              NERR',
     +   '       LEVEL      COUNT')
 9010 FORMAT (1X,A,3X,A,3X,A,3I10)
 9020 FORMAT ('0OTHER ERRORS NOT INDIVIDUALLY TABULATED = ', I10)
 9030 FORMAT (1X)
C
      WRITE(10,*)'XERSVE O/P'
C
      RETURN
      END
******
      INTEGER FUNCTION I1MACH(I)
      IMPLICIT REAL*8(A-H,O-Z)
C
C     I/O UNIT NUMBERS.
C     I1MACH( 1) = THE STANDARD INPUT UNIT.
C     I1MACH( 2) = THE STANDARD OUTPUT UNIT.
C     I1MACH( 3) = THE STANDARD PUNCH UNIT.
C     I1MACH( 4) = THE STANDARD ERROR MESSAGE UNIT.
C
C     WORDS.
```

```
C       I1MACH( 5) = THE NUMBER OF BITS PER INTEGER STORAGE UNIT.
C       I1MACH( 6) = THE NUMBER OF CHARACTERS PER CHARACTER STORAGE UNIT.
C                    FOR FORTRAN 77, THIS IS ALWAYS 1.  FOR FORTRAN 66,
C                    CHARACTER STORAGE UNIT = INTEGER STORAGE UNIT.
C       INTEGERS:
C       ASSUME INTEGERS ARE REPRESENTED IN THE S-DIGIT, BASE-A FORM
C       SIGN ( X(S-1)*A**(S-1) + ... + X(1)*A + X(0) )
C       WHERE 0 .LE. X(I) .LT. A FOR I=0,...,S-1.
C       I1MACH( 7) = A, THE BASE.
C       I1MACH( 8) = S, THE NUMBER OF BASE-A DIGITS.
C       I1MACH( 9) = A**S - 1, THE LARGEST MAGNITUDE.
C
C       FLOATING-POINT NUMBERS:
C       ASSUME FLOATING-POINT NUMBERS ARE REPRESENTED IN THE T-DIGIT,
C       BASE-B FORM:
C       SIGN (B**E)*( (X(1)/B) + ... + (X(T)/B**T) )
C       WHERE 0 .LE. X(I) .LT. B FOR I=1,...,T,
C       0 .LT. X(1), AND EMIN .LE. E .LE. EMAX.
C       I1MACH(10) = B, THE BASE.
C
C       SINGLE-PRECISION:
C       I1MACH(11) = T, THE NUMBER OF BASE-B DIGITS.
C       I1MACH(12) = EMIN, THE SMALLEST EXPONENT E.
C       I1MACH(13) = EMAX, THE LARGEST EXPONENT E.
C
C       DOUBLE-PRECISION:
C       I1MACH(14) = T, THE NUMBER OF BASE-B DIGITS.
C       I1MACH(15) = EMIN, THE SMALLEST EXPONENT E.
C       I1MACH(16) = EMAX, THE LARGEST EXPONENT E.
C
C       TO ALTER THIS FUNCTION FOR A PARTICULAR ENVIRONMENT,
C       THE DESIRED SET OF DATA STATEMENTS SHOULD BE ACTIVATED BY
C       REMOVING THE C FROM COLUMN 1.  ALSO, THE VALUES OF
C       I1MACH(1) - I1MACH(4) SHOULD BE CHECKED FOR CONSISTENCY
C       WITH THE LOCAL OPERATING SYSTEM.  FOR FORTRAN 77, YOU MAY WISH
C       TO ADJUST THE DATA STATEMENT SO IMACH(6) IS SET TO 1, AND
C       THEN TO COMMENT OUT THE EXECUTABLE TEST ON I .EQ. 6 BELOW.
C
C       FOR IEEE-ARITHMETIC MACHINES (BINARY STANDARD), THE FIRST
C       SET OF CONSTANTS BELOW SHOULD BE APPROPRIATE, EXCEPT PERHAPS
C       FOR IMACH(1) - IMACH(4).
C
C       COMMENTS JUST BEFORE THE END STATEMENT (LINES STARTING WITH *)
C       GIVE C SOURCE FOR I1MACH.
C
        INTEGER IMACH(16), OUTPUT, SANITY, SMALL(2), I
C/6S
C/7S
        SAVE IMACH, SANITY
C/
C        REAL RMACH
C
        EQUIVALENCE (IMACH(4),OUTPUT), (RMACH,SMALL(1))
C
C       MACHINE CONSTANTS FOR IEEE ARITHMETIC MACHINES, SUCH AS THE AT&T
C       3B SERIES, MOTOROLA 68000 BASED MACHINES (E.G. SUN 3 AND AT&T
C       PC 7300), AND 8087 BASED MICROS (E.G. IBM PC AND AT&T 6300).
C        DATA IMACH( 1) /    5 /
C        DATA IMACH( 2) /    6 /
C        DATA IMACH( 3) /    7 /
C        DATA IMACH( 4) /    6 /
C        DATA IMACH( 5) /   32 /
C        DATA IMACH( 6) /    4 /
C        DATA IMACH( 7) /    2 /
C        DATA IMACH( 8) /   31 /
C        DATA IMACH( 9) / 2147483647 /
C        DATA IMACH(10) /    2 /
C        DATA IMACH(11) /   24 /
C        DATA IMACH(12) / -125 /
C        DATA IMACH(13) /  128 /
C        DATA IMACH(14) /   53 /
C        DATA IMACH(15) / -1021 /
C        DATA IMACH(16) /  1024 /, SANITY/987/
```

```
C
C     MACHINE CONSTANTS FOR AMDAHL MACHINES.
C
C      DATA IMACH( 1) /    5 /
C      DATA IMACH( 2) /    6 /
C      DATA IMACH( 3) /    7 /
C      DATA IMACH( 4) /    6 /
C      DATA IMACH( 5) /   32 /
C      DATA IMACH( 6) /    4 /
C      DATA IMACH( 7) /    2 /
C      DATA IMACH( 8) /   31 /
C      DATA IMACH( 9) / 2147483647 /
C      DATA IMACH(10) /   16 /
C      DATA IMACH(11) /    6 /
C      DATA IMACH(12) /  -64 /
C      DATA IMACH(13) /   63 /
C      DATA IMACH(14) /   14 /
C      DATA IMACH(15) /  -64 /
C      DATA IMACH(16) /   63 /, SANITY/987/
C
C     MACHINE CONSTANTS FOR THE BURROUGHS 1700 SYSTEM.
C
C      DATA IMACH( 1) /    7 /
C      DATA IMACH( 2) /    2 /
C      DATA IMACH( 3) /    2 /
C      DATA IMACH( 4) /    2 /
C      DATA IMACH( 5) /   36 /
C      DATA IMACH( 6) /    4 /
C      DATA IMACH( 7) /    2 /
C      DATA IMACH( 8) /   33 /
C      DATA IMACH( 9) / Z1FFFFFFFF /
C      DATA IMACH(10) /    2 /
C      DATA IMACH(11) /   24 /
C      DATA IMACH(12) / -256 /
C      DATA IMACH(13) /  255 /
C      DATA IMACH(14) /   60 /
C      DATA IMACH(15) / -256 /
C      DATA IMACH(16) /  255 /, SANITY/987/
C
C     MACHINE CONSTANTS FOR THE BURROUGHS 5700 SYSTEM.
C
C      DATA IMACH( 1) /    5 /
C      DATA IMACH( 2) /    6 /
C      DATA IMACH( 3) /    7 /
C      DATA IMACH( 4) /    6 /
C      DATA IMACH( 5) /   48 /
C      DATA IMACH( 6) /    6 /
C      DATA IMACH( 7) /    2 /
C      DATA IMACH( 8) /   39 /
C      DATA IMACH( 9) / O0007777777777777 /
C      DATA IMACH(10) /    8 /
C      DATA IMACH(11) /   13 /
C      DATA IMACH(12) /  -50 /
C      DATA IMACH(13) /   76 /
C      DATA IMACH(14) /   26 /
C      DATA IMACH(15) /  -50 /
C      DATA IMACH(16) /   76 /, SANITY/987/
C
C     MACHINE CONSTANTS FOR THE BURROUGHS 6700/7700 SYSTEMS.
C
C      DATA IMACH( 1) /    5 /
C      DATA IMACH( 2) /    6 /
C      DATA IMACH( 3) /    7 /
C      DATA IMACH( 4) /    6 /
C      DATA IMACH( 5) /   48 /
C      DATA IMACH( 6) /    6 /
C      DATA IMACH( 7) /    2 /
C      DATA IMACH( 8) /   39 /
C      DATA IMACH( 9) / O0007777777777777 /
C      DATA IMACH(10) /    8 /
C      DATA IMACH(11) /   13 /
C      DATA IMACH(12) /  -50 /
C      DATA IMACH(13) /   76 /
```

```
C       DATA IMACH(14) /  26 /
C       DATA IMACH(15) / -32754 /
C       DATA IMACH(16) /  32780 /, SANITY/987/
C
C     MACHINE CONSTANTS FOR FTN4 ON THE CDC 6000/7000 SERIES.
C
C       DATA IMACH( 1) /    5 /
C       DATA IMACH( 2) /    6 /
C       DATA IMACH( 3) /    7 /
C       DATA IMACH( 4) /    6 /
C       DATA IMACH( 5) /   60 /
C       DATA IMACH( 6) /   10 /
C       DATA IMACH( 7) /    2 /
C       DATA IMACH( 8) /   48 /
C       DATA IMACH( 9) / 00007777777777777777B /
C       DATA IMACH(10) /    2 /
C       DATA IMACH(11) /   47 /
C       DATA IMACH(12) / -929 /
C       DATA IMACH(13) / 1070 /
C       DATA IMACH(14) /   94 /
C       DATA IMACH(15) / -929 /
C       DATA IMACH(16) / 1069 /, SANITY/987/
C
C     MACHINE CONSTANTS FOR FTN5 ON THE CDC 6000/7000 SERIES.
C
C       DATA IMACH( 1) /    5 /
C       DATA IMACH( 2) /    6 /
C       DATA IMACH( 3) /    7 /
C       DATA IMACH( 4) /    6 /
C       DATA IMACH( 5) /   60 /
C       DATA IMACH( 6) /   10 /
C       DATA IMACH( 7) /    2 /
C       DATA IMACH( 8) /   48 /
C       DATA IMACH( 9) / O"00007777777777777777" /
C       DATA IMACH(10) /    2 /
C       DATA IMACH(11) /   47 /
C       DATA IMACH(12) / -929 /
C       DATA IMACH(13) / 1070 /
C       DATA IMACH(14) /   94 /
C       DATA IMACH(15) / -929 /
C       DATA IMACH(16) / 1069 /, SANITY/987/
C
C     MACHINE CONSTANTS FOR CONVEX C-1.
C
C       DATA IMACH( 1) /    5 /
C       DATA IMACH( 2) /    6 /
C       DATA IMACH( 3) /    7 /
C       DATA IMACH( 4) /    6 /
C       DATA IMACH( 5) /   32 /
C       DATA IMACH( 6) /    4 /
C       DATA IMACH( 7) /    2 /
C       DATA IMACH( 8) /   31 /
C       DATA IMACH( 9) / 2147483647 /
C       DATA IMACH(10) /    2 /
C       DATA IMACH(11) /   24 /
C       DATA IMACH(12) / -128 /
C       DATA IMACH(13) /  127 /
C       DATA IMACH(14) /   53 /
C       DATA IMACH(15) /-1024 /
C       DATA IMACH(16) / 1023 /, SANITY/987/
C
C     MACHINE CONSTANTS FOR THE CRAY 1, XMP, 2, AND 3.
C
C       DATA IMACH( 1) /     5 /
C       DATA IMACH( 2) /     6 /
C       DATA IMACH( 3) /   102 /
C       DATA IMACH( 4) /     6 /
C       DATA IMACH( 5) /    64 /
C       DATA IMACH( 6) /     8 /
C       DATA IMACH( 7) /     2 /
C       DATA IMACH( 8) /    63 /
C       DATA IMACH( 9) /  777777777777777777777B /
C       DATA IMACH(10) /     2 /
```

```
C       DATA IMACH(11) /    47 /
C       DATA IMACH(12) / -8189 /
C       DATA IMACH(13) /  8190 /
C       DATA IMACH(14) /    94 /
C       DATA IMACH(15) / -8099 /
C       DATA IMACH(16) /  8190 /, SANITY/987/
C
C     MACHINE CONSTANTS FOR THE DATA GENERAL ECLIPSE S/200.
C
C       DATA IMACH( 1) /    11 /
C       DATA IMACH( 2) /    12 /
C       DATA IMACH( 3) /     8 /
C       DATA IMACH( 4) /    10 /
C       DATA IMACH( 5) /    16 /
C       DATA IMACH( 6) /     2 /
C       DATA IMACH( 7) /     2 /
C       DATA IMACH( 8) /    15 /
C       DATA IMACH( 9) /32767 /
C       DATA IMACH(10) /    16 /
C       DATA IMACH(11) /     6 /
C       DATA IMACH(12) /   -64 /
C       DATA IMACH(13) /    63 /
C       DATA IMACH(14) /    14 /
C       DATA IMACH(15) /   -64 /
C       DATA IMACH(16) /    63 /, SANITY/987/
C
C     MACHINE CONSTANTS FOR THE HARRIS SLASH 6 AND SLASH 7.
C
C       DATA IMACH( 1) /        5 /
C       DATA IMACH( 2) /        6 /
C       DATA IMACH( 3) /        0 /
C       DATA IMACH( 4) /        6 /
C       DATA IMACH( 5) /       24 /
C       DATA IMACH( 6) /        3 /
C       DATA IMACH( 7) /        2 /
C       DATA IMACH( 8) /       23 /
C       DATA IMACH( 9) /  8388607 /
C       DATA IMACH(10) /        2 /
C       DATA IMACH(11) /       23 /
C       DATA IMACH(12) /     -127 /
C       DATA IMACH(13) /      127 /
C       DATA IMACH(14) /       38 /
C       DATA IMACH(15) /     -127 /
C       DATA IMACH(16) /      127 /, SANITY/987/
C
C     MACHINE CONSTANTS FOR THE HONEYWELL DPS 8/70 SERIES.
C
C       DATA IMACH( 1) /    5 /
C       DATA IMACH( 2) /    6 /
C       DATA IMACH( 3) /   43 /
C       DATA IMACH( 4) /    6 /
C       DATA IMACH( 5) /   36 /
C       DATA IMACH( 6) /    4 /
C       DATA IMACH( 7) /    2 /
C       DATA IMACH( 8) /   35 /
C       DATA IMACH( 9) / O377777777777 /
C       DATA IMACH(10) /    2 /
C       DATA IMACH(11) /   27 /
C       DATA IMACH(12) / -127 /
C       DATA IMACH(13) /  127 /
C       DATA IMACH(14) /   63 /
C       DATA IMACH(15) / -127 /
C       DATA IMACH(16) /  127 /, SANITY/987/
C
C     MACHINE CONSTANTS FOR THE IBM 360/370 SERIES,
C     THE XEROX SIGMA 5/7/9 AND THE SEL SYSTEMS 85/86.
C
C       DATA IMACH( 1) /   5 /
C       DATA IMACH( 2) /   6 /
C       DATA IMACH( 3) /   7 /
C       DATA IMACH( 4) /   6 /
C       DATA IMACH( 5) /  32 /
C       DATA IMACH( 6) /   4 /
```

```
C       DATA IMACH( 7) /   2 /
C       DATA IMACH( 8) /  31 /
C       DATA IMACH( 9) / Z7FFFFFFF /
C       DATA IMACH(10) /  16 /
C       DATA IMACH(11) /   6 /
C       DATA IMACH(12) / -64 /
C       DATA IMACH(13) /  63 /
C       DATA IMACH(14) /  14 /
C       DATA IMACH(15) / -64 /
C       DATA IMACH(16) /  63 /, SANITY/987/
C
C     MACHINE CONSTANTS FOR THE INTERDATA 8/32
C     WITH THE UNIX SYSTEM FORTRAN 77 COMPILER.
C
C     FOR THE INTERDATA FORTRAN VII COMPILER REPLACE
C     THE Z'S SPECIFYING HEX CONSTANTS WITH Y'S.
C
C       DATA IMACH( 1) /   5 /
C       DATA IMACH( 2) /   6 /
C       DATA IMACH( 3) /   6 /
C       DATA IMACH( 4) /   6 /
C       DATA IMACH( 5) /  32 /
C       DATA IMACH( 6) /   4 /
C       DATA IMACH( 7) /   2 /
C       DATA IMACH( 8) /  31 /
C       DATA IMACH( 9) / Z'7FFFFFFF' /
C       DATA IMACH(10) /  16 /
C       DATA IMACH(11) /   6 /
C       DATA IMACH(12) / -64 /
C       DATA IMACH(13) /  62 /
C       DATA IMACH(14) /  14 /
C       DATA IMACH(15) / -64 /
C       DATA IMACH(16) /  62 /, SANITY/987/
C
C     MACHINE CONSTANTS FOR THE PDP-10 (KA PROCESSOR).
C
C       DATA IMACH( 1) /    5 /
C       DATA IMACH( 2) /    6 /
C       DATA IMACH( 3) /    7 /
C       DATA IMACH( 4) /    6 /
C       DATA IMACH( 5) /   36 /
C       DATA IMACH( 6) /    5 /
C       DATA IMACH( 7) /    2 /
C       DATA IMACH( 8) /   35 /
C       DATA IMACH( 9) / "377777777777 /
C       DATA IMACH(10) /    2 /
C       DATA IMACH(11) /   27 /
C       DATA IMACH(12) / -128 /
C       DATA IMACH(13) /  127 /
C       DATA IMACH(14) /   54 /
C       DATA IMACH(15) / -101 /
C       DATA IMACH(16) /  127 /, SANITY/987/
C
C     MACHINE CONSTANTS FOR THE PDP-10 (KI PROCESSOR).
C
C       DATA IMACH( 1) /    5 /
C       DATA IMACH( 2) /    6 /
C       DATA IMACH( 3) /    7 /
C       DATA IMACH( 4) /    6 /
C       DATA IMACH( 5) /   36 /
C       DATA IMACH( 6) /    5 /
C       DATA IMACH( 7) /    2 /
C       DATA IMACH( 8) /   35 /
C       DATA IMACH( 9) / "377777777777 /
C       DATA IMACH(10) /    2 /
C       DATA IMACH(11) /   27 /
C       DATA IMACH(12) / -128 /
C       DATA IMACH(13) /  127 /
C       DATA IMACH(14) /   62 /
C       DATA IMACH(15) / -128 /
C       DATA IMACH(16) /  127 /, SANITY/987/
C
C     MACHINE CONSTANTS FOR PDP-11 FORTRANS SUPPORTING
```

```
C     32-BIT INTEGER ARITHMETIC.
C
C      DATA IMACH( 1) /    5 /
C      DATA IMACH( 2) /    6 /
C      DATA IMACH( 3) /    7 /
C      DATA IMACH( 4) /    6 /
C      DATA IMACH( 5) /   32 /
C      DATA IMACH( 6) /    4 /
C      DATA IMACH( 7) /    2 /
C      DATA IMACH( 8) /   31 /
C      DATA IMACH( 9) / 2147483647 /
C      DATA IMACH(10) /    2 /
C      DATA IMACH(11) /   24 /
C      DATA IMACH(12) / -127 /
C      DATA IMACH(13) /  127 /
C      DATA IMACH(14) /   56 /
C      DATA IMACH(15) / -127 /
C      DATA IMACH(16) /  127 /, SANITY/987/
C
C     MACHINE CONSTANTS FOR PDP-11 FORTRANS SUPPORTING
C     16-BIT INTEGER ARITHMETIC.
C
C      DATA IMACH( 1) /    5 /
C      DATA IMACH( 2) /    6 /
C      DATA IMACH( 3) /    7 /
C      DATA IMACH( 4) /    6 /
C      DATA IMACH( 5) /   16 /
C      DATA IMACH( 6) /    2 /
C      DATA IMACH( 7) /    2 /
C      DATA IMACH( 8) /   15 /
C      DATA IMACH( 9) / 32767 /
C      DATA IMACH(10) /    2 /
C      DATA IMACH(11) /   24 /
C      DATA IMACH(12) / -127 /
C      DATA IMACH(13) /  127 /
C      DATA IMACH(14) /   56 /
C      DATA IMACH(15) / -127 /
C      DATA IMACH(16) /  127 /, SANITY/987/
C
C     MACHINE CONSTANTS FOR THE PRIME 50 SERIES SYSTEMS
C     WTIH 32-BIT INTEGERS AND 64V MODE INSTRUCTIONS,
C     SUPPLIED BY IGOR BRAY.
C
C      DATA IMACH( 1) /            1 /
C      DATA IMACH( 2) /            1 /
C      DATA IMACH( 3) /            2 /
C      DATA IMACH( 4) /            1 /
C      DATA IMACH( 5) /           32 /
C      DATA IMACH( 6) /            4 /
C      DATA IMACH( 7) /            2 /
C      DATA IMACH( 8) /           31 /
C      DATA IMACH( 9) / :17777777777 /
C      DATA IMACH(10) /            2 /
C      DATA IMACH(11) /           23 /
C      DATA IMACH(12) /         -127 /
C      DATA IMACH(13) /         +127 /
C      DATA IMACH(14) /           47 /
C      DATA IMACH(15) /       -32895 /
C      DATA IMACH(16) /       +32637 /, SANITY/987/
C
C     MACHINE CONSTANTS FOR THE SEQUENT BALANCE 8000.
C
C      DATA IMACH( 1) /     0 /
C      DATA IMACH( 2) /     0 /
C      DATA IMACH( 3) /     7 /
C      DATA IMACH( 4) /     0 /
C      DATA IMACH( 5) /    32 /
C      DATA IMACH( 6) /     1 /
C      DATA IMACH( 7) /     2 /
C      DATA IMACH( 8) /    31 /
C      DATA IMACH( 9) /  2147483647 /
C      DATA IMACH(10) /     2 /
C      DATA IMACH(11) /    24 /
```

```
C      DATA IMACH(12) /  -125 /
C      DATA IMACH(13) /   128 /
C      DATA IMACH(14) /    53 /
C      DATA IMACH(15) / -1021 /
C      DATA IMACH(16) /  1024 /, SANITY/987/
C
C     MACHINE CONSTANTS FOR THE UNIVAC 1100 SERIES.
C
C     NOTE THAT THE PUNCH UNIT, I1MACH(3), HAS BEEN SET TO 7
C     WHICH IS APPROPRIATE FOR THE UNIVAC-FOR SYSTEM.
C     IF YOU HAVE THE UNIVAC-FTN SYSTEM, SET IT TO 1.
C
C      DATA IMACH( 1) /    5 /
C      DATA IMACH( 2) /    6 /
C      DATA IMACH( 3) /    7 /
C      DATA IMACH( 4) /    6 /
C      DATA IMACH( 5) /   36 /
C      DATA IMACH( 6) /    6 /
C      DATA IMACH( 7) /    2 /
C      DATA IMACH( 8) /   35 /
C      DATA IMACH( 9) / O377777777777 /
C      DATA IMACH(10) /    2 /
C      DATA IMACH(11) /   27 /
C      DATA IMACH(12) / -128 /
C      DATA IMACH(13) /  127 /
C      DATA IMACH(14) /   60 /
C      DATA IMACH(15) /-1024 /
C      DATA IMACH(16) / 1023 /, SANITY/987/
C
C     MACHINE CONSTANTS FOR VAX.
C
C      DATA IMACH( 1) /    5 /
C      DATA IMACH( 2) /    6 /
C      DATA IMACH( 3) /    7 /
C      DATA IMACH( 4) /    6 /
C      DATA IMACH( 5) /   32 /
C      DATA IMACH( 6) /    4 /
C      DATA IMACH( 7) /    2 /
C      DATA IMACH( 8) /   31 /
C      DATA IMACH( 9) / 2147483647 /
C      DATA IMACH(10) /    2 /
C      DATA IMACH(11) /   24 /
C      DATA IMACH(12) / -127 /
C      DATA IMACH(13) /  127 /
C      DATA IMACH(14) /   56 /
C      DATA IMACH(15) / -127 /
C      DATA IMACH(16) /  127 /, SANITY/987/
C
C  ***  ISSUE STOP 775 IF ALL DATA STATEMENTS ARE COMMENTED...
C
C     FORMATS.
C
      WRITE(10,*)'X1MACH I/P'
C
      IF (SANITY .NE. 987) THEN
*        *** CHECK FOR AUTODOUBLE ***
         SMALL(2) = 0
         RMACH = 1E13
         IF (SMALL(2) .NE. 0) THEN
*           *** AUTODOUBLED ***
            IF (      (SMALL(1) .EQ. 1117925532
     *           .AND. SMALL(2) .EQ. -448790528)
     *       .OR.     (SMALL(2) .EQ. 1117925532
     *           .AND. SMALL(1) .EQ. -448790528)) THEN
*               *** IEEE ***
               IMACH(10) = 2
               IMACH(14) = 53
               IMACH(15) = -1021
               IMACH(16) = 1024
            ELSE IF ( SMALL(1) .EQ. -2065213935
     *          .AND. SMALL(2) .EQ. 10752) THEN
*               *** VAX WITH D_FLOATING ***
               IMACH(10) = 2
```

```
                 IMACH(14) = 56
                 IMACH(15) = -127
                 IMACH(16) = 127
              ELSE IF ( SMALL(1) .EQ. 1267827943
     *            .AND. SMALL(2) .EQ. 704643072) THEN
*                 *** IBM MAINFRAME ***
                 IMACH(10) = 16
                 IMACH(14) = 14
                 IMACH(15) =-64
                 IMACH(16) = 63
              ELSE
                 WRITE(*,*)'ADJUST AUTODOUBLED I1MACH BY UNCOMMENTING'
                 WRITE(*,*)'DATA STATEMENTS APPROPRIATE FOR YOUR MACHINE'
                 WRITE(*,*)'AND SETTING IMACH(I) = IMACH(I+3) FOR'
                 WRITE(*,*)'I = 11, 12, AND 13.'
                 STOP 777
                 END IF
              IMACH(11) = IMACH(14)
              IMACH(12) = IMACH(15)
              IMACH(13) = IMACH(16)
           ELSE
              RMACH = 1234567.
              IF (SMALL(1) .EQ. 1234613304) THEN
*                 *** IEEE ***
                 IMACH(10) = 2
                 IMACH(11) = 24
                 IMACH(12) = -125
                 IMACH(13) = 128
                 IMACH(14) = 53
                 IMACH(15) = -1021
                 IMACH(16) = 1024
                 SANITY = 987
              ELSE IF (SMALL(1) .EQ. -1271379306) THEN
*                 *** VAX ***
                 IMACH(10) = 2
                 IMACH(11) = 24
                 IMACH(12) = -127
                 IMACH(13) = 127
                 IMACH(14) = 56
                 IMACH(15) = -127
                 IMACH(16) = 127
                 SANITY = 987
              ELSE IF (SMALL(1) .EQ. 1175639687) THEN
*                 *** IBM ***
                 IMACH(10) = 16
                 IMACH(11) = 6
                 IMACH(12) = -64
                 IMACH(13) = 63
                 IMACH(14) = 14
                 IMACH(15) = -64
                 IMACH(16) = 63
                 SANITY = 987
              ELSE
                 WRITE(*,*)'ADJUST I1MACH BY UNCOMMENTING'
                 WRITE(*,*)'DATA STATEMENTS APPROPRIATE FOR YOUR MACHINE.'
                 STOP 777
                 END IF
              END IF
           IMACH( 1) = 5
           IMACH( 2) = 6
           IMACH( 3) = 7
           IMACH( 4) = 6
           IMACH( 5) = 32
           IMACH( 6) = 4
           IMACH( 7) = 2
           IMACH( 8) = 31
           IMACH( 9) = 2147483647
           SANITY = 987
           END IF
        IF (I .LT. 1  .OR.  I .GT. 16) GO TO 30
C
        I1MACH = IMACH(I)
C/6S
```

```
C/7S
      IF (I .EQ. 6) I1MACH = 1
C/
      RETURN
C
 30   WRITE(*,*) 'I1MACH(I): I =',I,' IS OUT OF BOUNDS.'
C
*     CALL FDUMP
C
      WRITE(10,*)'X1MACH O/P'
C
      STOP
      END
******
      SUBROUTINE XGETUA(IUNIT,N)
      IMPLICIT REAL*8(A-H,O-Z)
C
C     ABSTRACT
C        XGETUA MAY BE CALLED TO DETERMINE THE UNIT NUMBER OR NUMBERS
C        TO WHICH ERROR MESSAGES ARE BEING SENT.
C        THESE UNIT NUMBERS MAY HAVE BEEN SET BY A CALL TO XSETUN,
C        OR A CALL TO XSETUA, OR MAY BE A DEFAULT VALUE.
C
C     DESCRIPTION OF PARAMETERS
C      --OUTPUT--
C        IUNIT - AN ARRAY OF ONE TO FIVE UNIT NUMBERS, DEPENDING
C                ON THE VALUE OF N.  A VALUE OF ZERO REFERS TO THE
C                DEFAULT UNIT, AS DEFINED BY THE I1MACH MACHINE
C                CONSTANT ROUTINE.  ONLY IUNIT(1),...,IUNIT(N) ARE
C                DEFINED BY XGETUA.  THE VALUES OF IUNIT(N+1),...,
C                IUNIT(5) ARE NOT DEFINED (FOR N.LT.5) OR ALTERED
C                IN ANY WAY BY XGETUA.
C        N     - THE NUMBER OF UNITS TO WHICH COPIES OF THE
C                ERROR MESSAGES ARE BEING SENT.  N WILL BE IN THE
C                RANGE FROM 1 TO 5.
C
C     WRITTEN BY RON JONES, WITH SLATEC COMMON MATH LIBRARY SUBCOMMITTEE
C
      DIMENSION IUNIT(5)
C
      WRITE(10,*)'XGETUA I/P'
C
      N = J4SAVE(5,0,.FALSE.)
      DO 30 I=1,N
         INDEX = I+4
         IF (I.EQ.1) INDEX = 3
         IUNIT(I) = J4SAVE(INDEX,0,.FALSE.)
   30 CONTINUE
C
      WRITE(10,*)'XGETUA O/P'
C
      RETURN
      END
******
      SUBROUTINE DPOLVL (NDER, XX, YFIT, YP, N, X, C)
      IMPLICIT REAL*8(A-H,O-Z)
C
C     BEGIN PROLOGUE DPOLVL
C     PURPOSE: CALCULATE THE VALUE OF A POLYNOMIAL AND ITS FIRST NDER
C              DERIVATIVES WHERE THE POLYNOMIAL WAS PRODUCED BY A PREVIOUS
C              CALL TO DPLINT.
C     LIBRARY: SLATEC
C     CATEGORY: E3
C     TYPE: DOUBLE PRECISION (POLYVL-S, DPOLVL-D)
C     KEYWORDS: POLYNOMIAL EVALUATION
C     AUTHOR: HUDDLESTON, R. E., (SNLL)
C
C     DESCRIPTION:
C     ABSTRACT:
C                SUBROUTINE DPOLVL CALCULATES THE VALUE OF THE POLYNOMIAL A
C                ITS FIRST NDER DERIVATIVES WHERE THE POLYNOMIAL WAS PRODUC
C                A PREVIOUS CALL TO DPLINT.
C                THE VARIABLE N AND THE ARRAYS X AND C MUST NOT BE ALTERED
```

```
C              BETWEEN THE CALL TO DPLINT AND THE CALL TO DPOLVL.
C
C      DIMENSIONING INFORMATION:
C      YP   MUST BE DIMENSIONED BY AT LEAST NDER
C      X    MUST BE DIMENSIONED BY AT LEAST N (SEE THE ABSTRACT )
C      C    MUST BE DIMENSIONED BY AT LEAST N (SEE THE ABSTRACT )
C      WORK MUST BE DIMENSIONED BY AT LEAST 2*N IF NDER IS .GT. 0.
C
C      NOTE:
C      IF NDER=0, NEITHER YP NOR WORK NEED TO BE DIMENSIONED VARIABLES.
C      IF NDER=1, YP DOES NOT NEED TO BE A DIMENSIONED VARIABLE.
C
C      INPUT PARAMETERS:
C      ALL TYPE REAL VARIABLES ARE DOUBLE PRECISION
C      NDER: THE NUMBER OF DERIVATIVES TO BE EVALUATED
C      XX: THE ARGUMENT AT WHICH THE POLYNOMIAL AND ITS DERIVATIVES
C          ARE TO BE EVALUATED.
C      N, X, AND C MUST NOT BE ALTERED BETWEEN THE CALL TO DPLINT AND
C      THE CALL TO DPOLVL.
C
C      OUTPUT PARAMETERS:
C      ALL TYPE REAL VARIABLES ARE DOUBLE PRECISION
C      YFIT: THE VALUE OF THE POLYNOMIAL AT XX
C      YP: THE DERIVATIVES OF THE POLYNOMIAL AT XX.  THE DERIVATIVE OF
C          ORDER J AT XX IS STORED IN  YP(J) , J = 1,...,NDER.
C
C      IERR: OUTPUT ERROR FLAG WITH THE FOLLOWING POSSIBLE VALUES.
C          = 1  INDICATES NORMAL EXECUTION
C
C      STORAGE PARAMETERS:
C      WORK: THIS IS AN ARRAY TO PROVIDE INTERNAL WORKING STORAGE FOR
C            DPOLVL.  IT MUST BE DIMENSIONED BY AT LEAST 2*N IF NDER IS
C            .GT. 0.  IF NDER=0, WORK DOES NOT NEED TO BE A DIMENSIONED
C            VARIABLE.
C
C      REFERENCES: L. F. SHAMPINE, S. M. DAVENPORT AND R. E. HUDDLESTON,
C                  CURVE FITTING BY POLYNOMIALS IN ONE VARIABLE, REPORT
C                  SLA-74-0270, SANDIA LABORATORIES, JUNE 1974.
C      ROUTINES CALLED:  (NONE)
C      REVISION HISTORY: (YYMMDD)
C      740601  DATE WRITTEN
C      890531  CHANGED ALL SPECIFIC INTRINSICS TO GENERIC.  (WRB)
C      891006  COSMETIC CHANGES TO PROLOGUE.  (WRB)
C      891006  REVISION DATE FROM VERSION 3.2
C      891214  PROLOGUE CONVERTED TO VERSION 4.0 FORMAT.  (BAB)
C      920501  REFORMATTED THE REFERENCES SECTION.  (WRB)
C      END PROLOGUE  DPOLVL
C
       INTEGER I,IERR,IM1,IZERO,K,KM1,KM1PI,KM2PN,KM2PNI,M,MM,N,NDR,NDER,
     * NMKP1,NPKM1
       DIMENSION  C(*),X(*),YP(*),WORK(44)
C      ,XK,XX,YFIT,FAC,PIONE,PITWO,PONE,PTWO
C
       WRITE(10,*)'DPOLVL I/P'
C
C      FIRST EXECUTABLE STATEMENT  DPOLVL
       IERR=1
       IF(NDER.GT.0)GOTO 10020
C
C      CODING FOR THE CASE NDER = 0
       PIONE=1.0D0
       PONE=C(1)
       YFIT=PONE
       IF(N.EQ.1)RETURN
       DO 10010 K=2,N
       PITWO=(XX-X(K-1))*PIONE
       PIONE=PITWO
       PTWO=PONE+PITWO*C(K)
       PONE=PTWO
10010  CONTINUE
       YFIT=PTWO
       RETURN
C      END OF NDER = 0 CASE
```

```
C
10020 CONTINUE
         IF (N.GT.1) GO TO 10040
      YFIT=C(1)
C
C     CODING FOR THE CASE  N=1 AND NDER .GT. 0
      DO 10030 K=1,NDER
      YP(K)=0.0D0
10030 CONTINUE
      RETURN
C     END OF THE CASE  N = 1 AND  NDER .GT. 0
C
10040 CONTINUE
         IF (NDER.LT.N) GO TO 10050
C
C     SET FLAGS FOR NUMBER OF DERIVATIVES AND FOR DERIVATIVES
C     IN EXCESS OF THE DEGREE (N-1) OF THE POLYNOMIAL.
      IZERO=1
      NDR=N-1
      GOTO 10060
10050 CONTINUE
      IZERO=0
      NDR=NDER
10060 CONTINUE
      M=NDR+1
      MM=M
C
C     START OF THE CASE NDER .GT. 0  AND N .GT. 1
C     THE POLYNOMIAL AND ITS DERIVATIVES WILL BE EVALUATED AT XX
      DO 10070 K=1,NDR
      YP(K)=C(K+1)
10070 CONTINUE
C
C     THE FOLLOWING SECTION OF CODE IS EASIER TO READ IF ONE
C     BREAKS WORK INTO TWO ARRAYS W AND V. THE CODE WOULD THEN
C     READ
C     W(1) = 1.
C     PONE = C(1)
C     DO   K = 2,N
C     V(K-1) =  XX - X(K-1)
C     W(K)   =  V(K-1)*W(K-1)
C     PTWO   =  PONE + W(K)*C(K)
C     PONE   =  PWO
C
C               YFIT = PTWO
C
      WORK(1)=1.0D0
      PONE=C(1)
      DO 10080 K=2,N
      KM1=K-1
      NPKM1=N+K-1
      WORK(NPKM1)=XX-X(KM1)
      WORK(K)=WORK(NPKM1)*WORK(KM1)
      PTWO=PONE+WORK(K)*C(K)
      PONE=PTWO
10080 CONTINUE
      YFIT=PTWO
C
C     AT THIS POINT THE POLYNOMIAL HAS BEEN EVALUATED AND INFORMATION
C     FOR THE DERIVATIVE EVALUATIONS HAVE BEEN STORED IN THE ARRAY
C     WORK
      IF(N.EQ.2)GOTO 10110
      IF(M.EQ.N)MM=NDR
C
C     EVALUATE THE DERIVATIVES AT XX
C                  ******  DO K=2,MM    (FOR MOST CASES, MM = NDER + 1)
C                  *  ******  DO I=2,N-K+1
C                  *  *         W(I) = V(K-2+I)*W(I-1) + W(I)
C                  *  *         YP(K-1) = YP(K-1) + W(I)*C(K-1+I)
C                  ******  CONTINUE
C
      DO 10090 K=2,MM
      NMKP1=N-K+1
```

```
      KM1=K-1
      KM2PN=K-2+N
      DO 10090 I=2,NMKP1
      KM2PNI=KM2PN+I
      IM1=I-1
      KM1PI=KM1+I
      WORK(I)=WORK(KM2PNI)*WORK(IM1)+WORK(I)
      YP(KM1)=YP(KM1)+WORK(I)*C(KM1PI)
10090 CONTINUE
         IF (NDR.EQ.1) GO TO 10110
      FAC=1.0D0
      DO 10100 K=2,NDR
      XK=K
      FAC=XK*FAC
      YP(K)=FAC*YP(K)
10100 CONTINUE
C     END OF DERIVATIVE EVALUATIONS
C
10110 CONTINUE
      IF (IZERO.EQ.0) RETURN
C
C     SET EXCESS DERIVATIVES TO ZERO.
      DO 10120 K=N,NDER
      YP(K)=0.0D0
10120 CONTINUE
C
      WRITE(10,*)'DPOLVL O/P'
C
      RETURN
      END
******
```

12.3 Solved Problems

(1) Figure 12.1 shows a cross section having 3 closed cells.

The centroid and the principal axes of the whole cross section are calculated using the program *PROP*®.and tabulated in Table 12.1.

The unknown redundant shear flows Q_1, Q_2, and Q_3 at the 3 artificial cuts of the 3 closed cells are obtained simultaneously and having the values: $Q_1 = -0.0025$, $Q_2 = -0.056$, $Q_3 = -0.130$. Figure 12.2 shows the shear forces in the open cells.

The shear flow due to cells openings are shown in Fig. 12.3, whereas the corresponding shear forces are shown in Fig. 12.4.

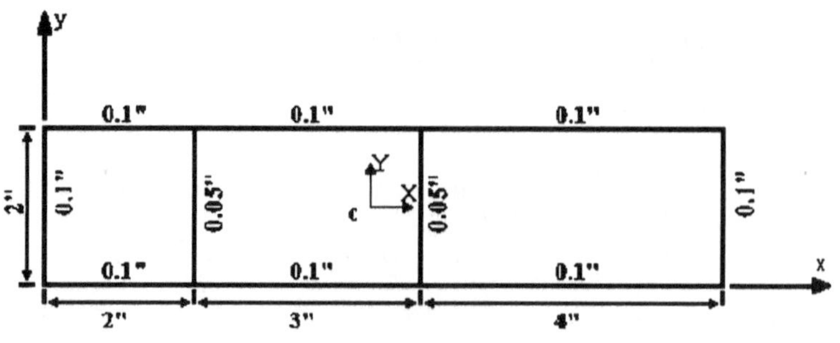

Fig. 12.1 Cross section of a 3 closed cells

Table 12.1 Geometrical properties of the 3 closed cells section

Geometrical constants	
Area	$A = 2.400\ \text{in}^2$
Centroid	$x_c = 4.417$ in, $y_c = 1.000$ in
Principal moments of inertia	$I_X = 2.002\ \text{in}^4$, $I_Y = 20.8884\ \text{in}^4$
Direction of principal axes	$\Theta = 0.000°$

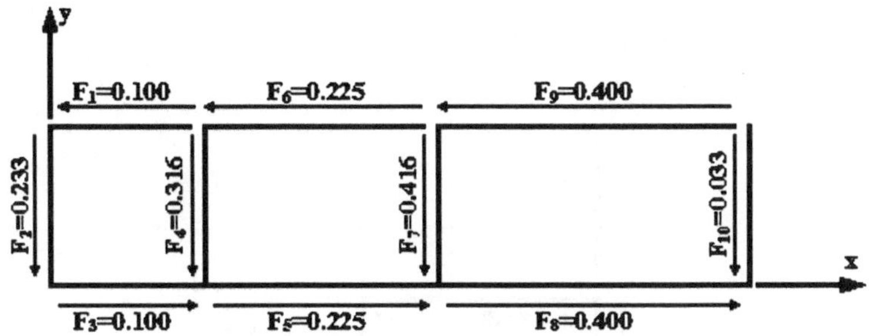

Fig. 12.2 Shear forces in open cells

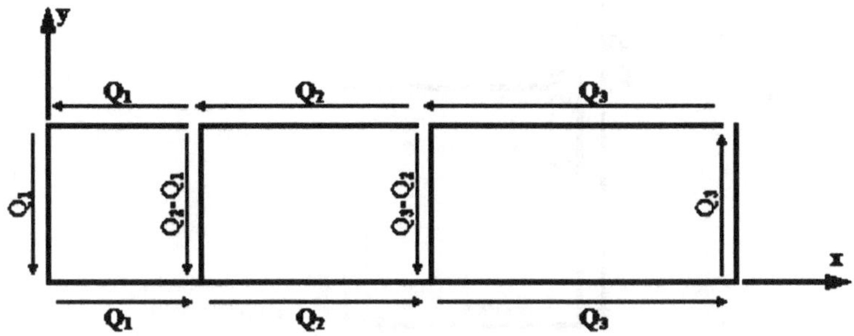

Fig. 12.3 Unknown shear flow due to cells openings

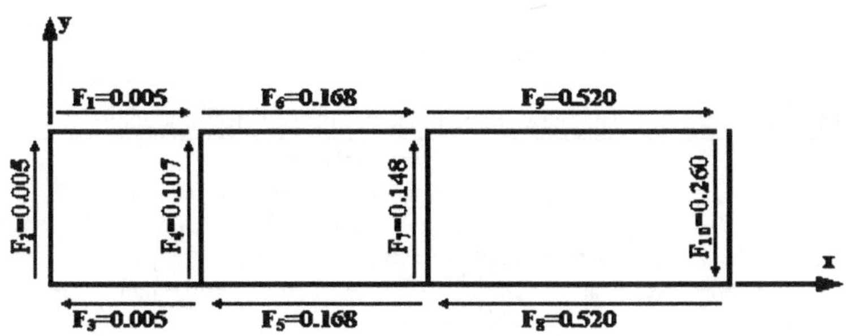

Fig. 12.4 Shear forces due to the unknown shear flows

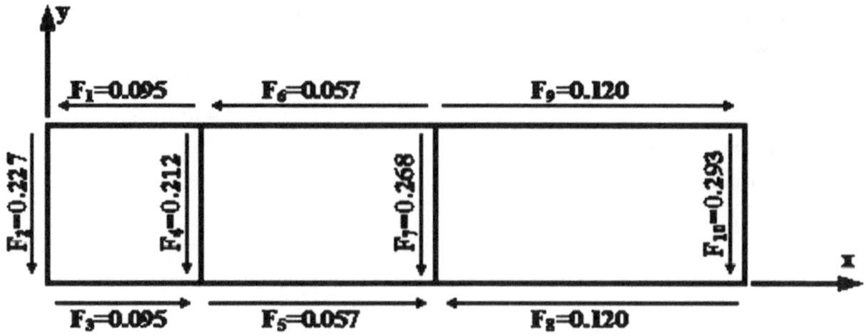

Fig. 12.5 Final shear forces in the closed cells

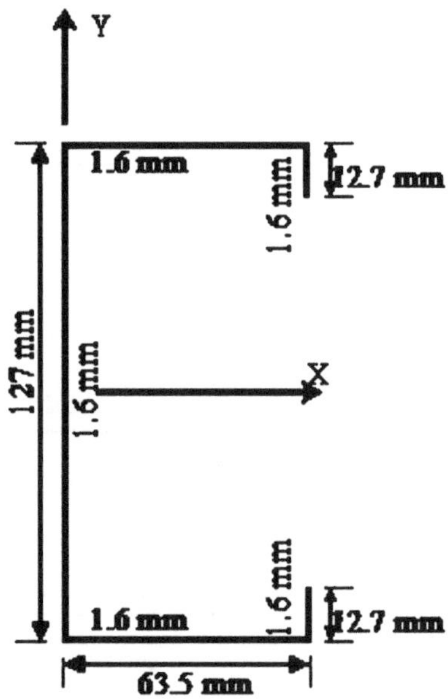

Fig. 12.6 C-channel section

Table 12.2 Geometrical, shear, and warping properties of the C-rolled section

Area	$A = 447.04\ \text{mm}^2$
Centroid	$x_c = 20.205$ mm, $y_c = 0.0$ mm
Principal moments of inertia	$I_{xx} = 1.226 \times 10^6\ \text{mm}^4$, $I_{yy} = 0.255 \times 10^6\ \text{mm}^4$
Polar moment of inertia (Torsion constant)	$J = 0.381 \times 10^3\ \text{mm}^4$
Shear center position (wrt the arb. selected coord. sys.)	$x_s = -29.60$ mm, $y_s = 0.0$ mm
Warping constant	$I_\omega = 829 \times 10^6\ \text{mm}^6$

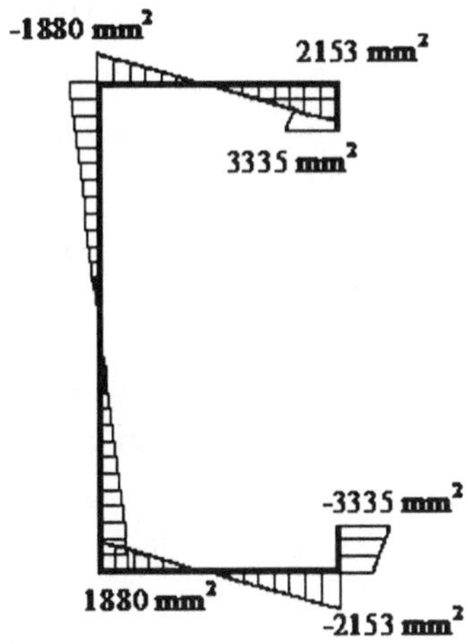

Fig. 12.7 Distribution of the principal sectorial coordinate, $\omega(s)$

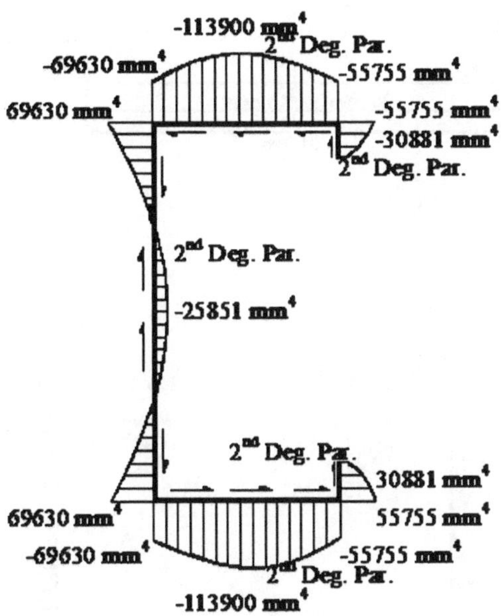

Fig. 12.8 Distribution and direction of the sectorial shear functions $S_\omega(s)$

Table 12.3 Geometrical, shear, and warping properties of a container ship

Area	$A = 2.21\ \text{mm}^2$
Centroid	$x_c = 0.0$ m, $y_c = 6.24$ m
Principal moments of inertia	$I_{xx} = 100.46\ \text{m}^4$, $I_{yy} = 331.83\ \text{m}^4$
Polar moment of inertia (Torsion constant)	$J = 5.6685\ \text{m}^4$
Shear center position (wrt the arb. selected coord. sys. x, y)	$x_s = 0.0$ m, $y_s = 8.30$ m
Warping constant	$I_\omega = 21851\ \text{mm}^6$

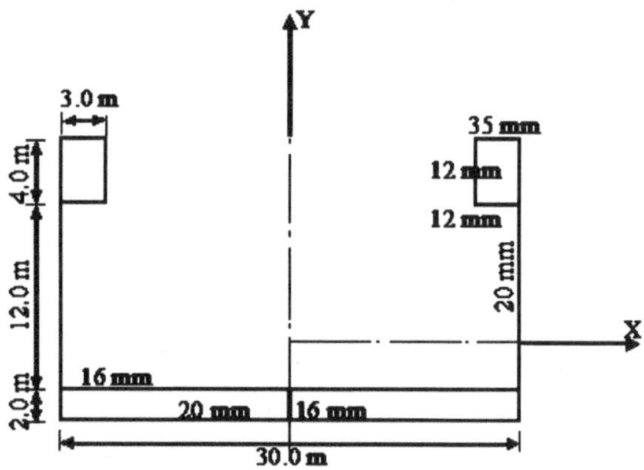

Fig. 12.9 Midship section of a container ship

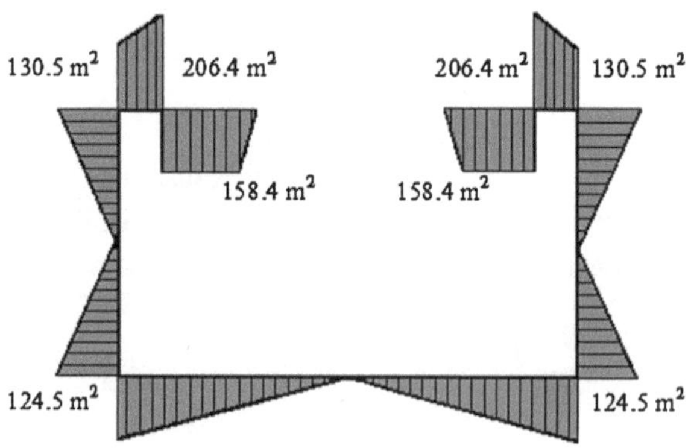

Fig. 12.10 Sectorial area of a container ship

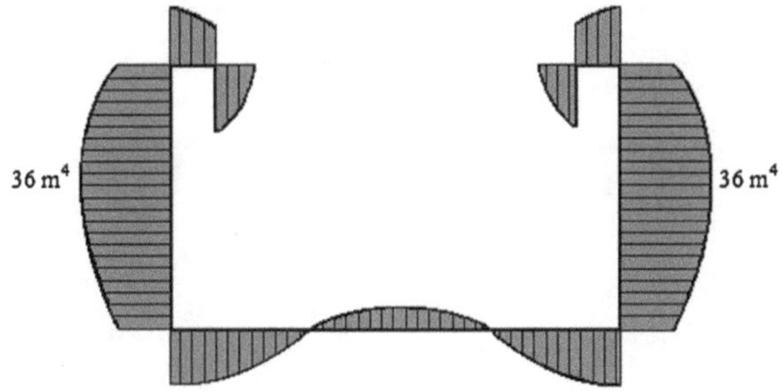

Fig. 12.11 Sectorial moment of a container ship

Finally, the final shear forces in the given 3 closed cells section are shown in Fig. 12.5.

The location of the shear center with respect to the principal axes is (−0.080″, 0.000).

(2) Figure 12.6 shows an open cross section "C rolled section".

The geometrical, shear, and warping properties of the C-cross section are calculated using the program *PROP*® and tabulated in Table 12.2.

Figures 12.7 and 12.8 shows the distribution and direction of the principal sectorial coordinate $\omega(s)$, and the sectorial shear function $S_\omega(s)$ respectively.

(3) Figure 12.9 shows a typical open midship section of a container ship.

The geometrical, shear, and warping properties of the cross section are calculated using the program *PROP*® and tabulated in Table 12.3.

Figures 12.10 and 12.11 shows the distribution and direction of the sectorial area, and the sectorial moment of area of the given container ship respectively.

Chapter 13
Problems

1. The catamaran vessel shown in Fig. 13.1 has a length L = 40 m. The vessel section is subjected to a clockwise twisting moment T = 900 tonnes.m and $G = 8.08 \times 10^6$ tonnes/m^2. Calculate:

   i. The torsion shear stresses at points A and B.
   ii. The angle of twist φ.

2. The catamaran vessel shown in Fig. 13.2 is subjected to a 15 tonnes vertical shear force. Calculate the shear stresses at point A.
3. The catamaran ship section shown in Fig. 13.3 is subjected to a vertical shear force F = 10 tonnes and a twisting moment T = 50 tonnes.m, calculate the shear stresses at points A and P and also the maximum shear stress due to F and T. The particulars of the ship section are: L = 40 m, B = 10 m, D = 7 m, T = 2.5 m, H = 3 m, d = 4 m, w = 5 m, $t_D = t_B = t_S = 7$ mm, $t_L = t_K = 10$ mm.
4. Explain how you determine the torsion shear stresses in the two-cell section shown in Fig. 13.4. The section is subjected to a torsion moment T.
5. The ship section shown in Fig. 13.5 is subjected to a clockwise torque of 50 t.m and a vertical shear force of 400 tonnes. Calculate:

   i. The maximum shear stress in the side shell plating due to the vertical shear force.
   ii. The torsion constant of the ship section.
   iii. The angle of twist θ of the ship section.
   iv. The torsion stress in: *Side shell plating*, *Deck plating*, *Tank top plating*.
   v. The total shear stress in the side shell plating, illustrate your answer with a sketch showing the stress distribution over the plate thickness.

6. The catamaran vessel shown in Fig. 13.6 has a length L = 40 m. the vessel section is subjected to a vertical shear force F = 20 tonnes and a twisting moment T = 90 tonnes.m. Calculate the shear stresses at points A and B.

M. Shama, *Torsion and Shear Stresses in Ships*,
DOI: 10.1007/978-3-642-14633-6_13, © Springer-Verlag Berlin Heidelberg 2010

Fig. 13.1

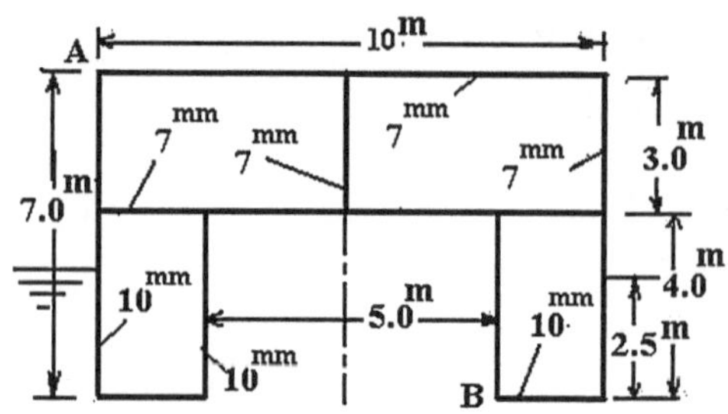

Fig. 13.2

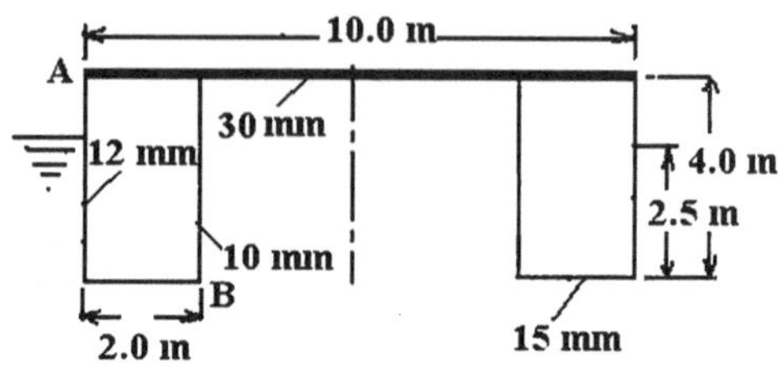

Fig. 13.3

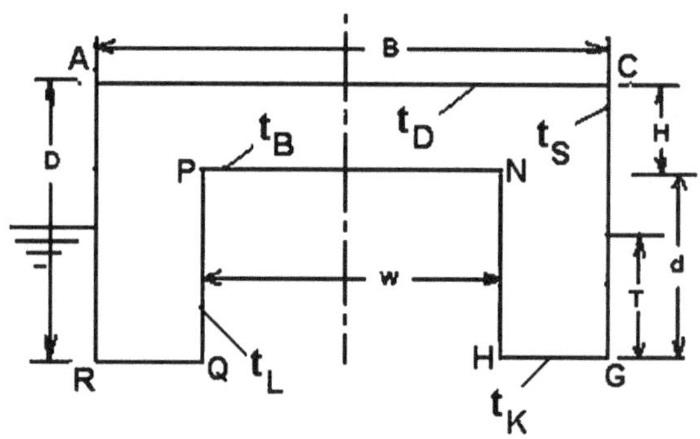

Fig. 13.4

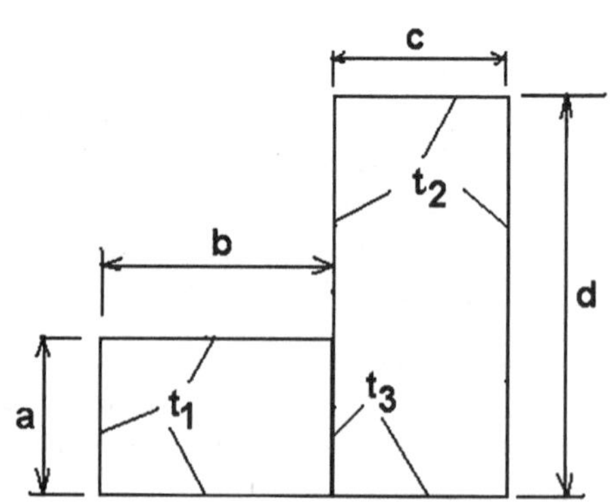

Fig. 13.5

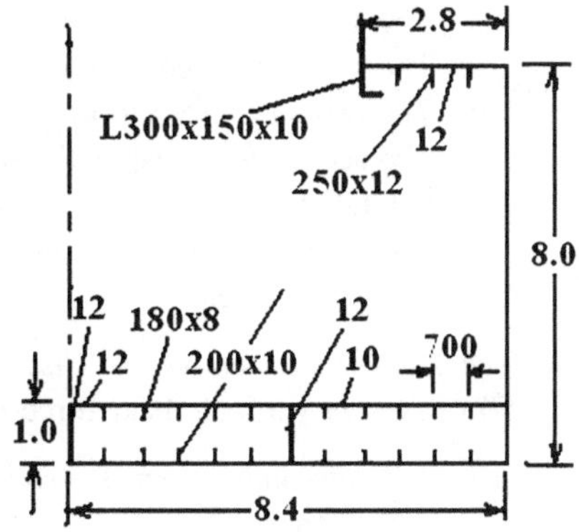

Fig. 13.6

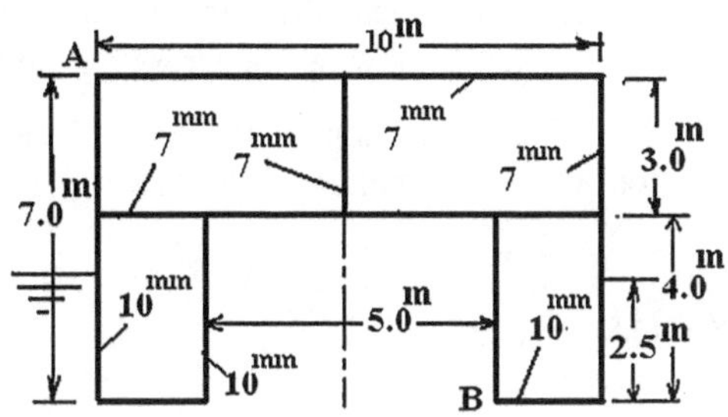

Fig. 13.7

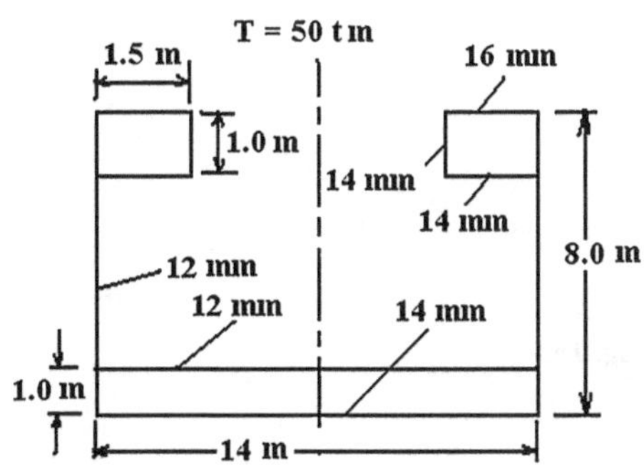

7. The midship section of the general cargo ship shown in Fig. 13.7 has the following particulars: L = 118 m, B = 16.8 m, D = 8.0 m, T = 5.2 m, $C_b = 0.7$, V = 15 knots, hold length = 18.2 m, side frame spacing = 650 mm, hatch opening = 13 × 11.2 m, the deck loading p = 2.5 tonnes/m^2. Make all necessary assumptions to calculate:

i. The still-water bending moment Ms.
ii. The wave induced bending moment Mw.
iii. The primary stress in the tanktop plating of the bottom side girder.
iv. The secondary stress in the bottom side girder at its end at the transverse bulkhead.
v. The shear flow distribution over the ship section, the vertical shear force F = 120 tonnes.

8. The ship section shown in Fig. 13.8 is subjected to a clockwise torque of 50 t.m and a vertical shear force of 400 tonnes. Calculate:

    i. The maximum shear stress in the side shell plating due to the vertical shear force
    ii. The torsion constant of the ship section
    iii. The angle of twist θ of the ship section
    iv. The torsion stress in: *Side shell, Deck,* and *Tank top platings.*
    v. The total shear stress in the side shell plating, illustrate your answer with a sketch showing the stress distribution over the plate thickness

9. The midship section of the general cargo ship shown in Fig. 13.9 has the following particulars: L = 118 m, B = 16.8 m, D = 8.0 m, T = 5.2 m,

Fig. 13.8

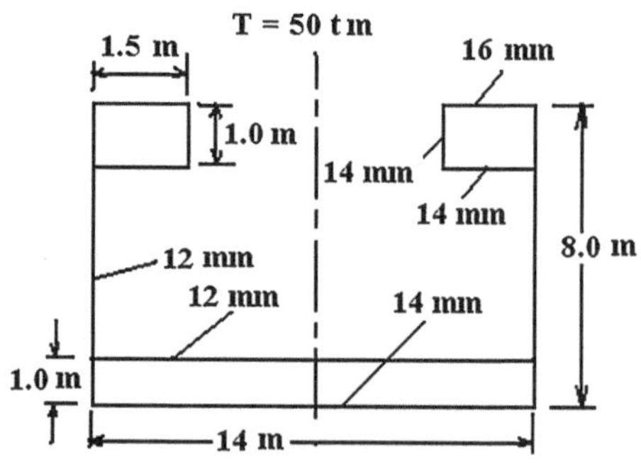

Fig. 13.9

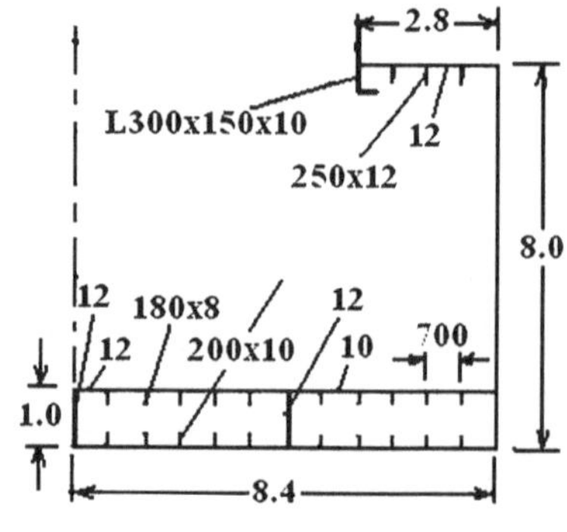

$C_b = 0.7$, $V = 15$ knots, hold length = 18.2 m, Side frame spacing = 650 mm, Hatch opening = 13 × 11.2 m. The deck loading $p = 2.5$ tonnes/m^2. The vertical shear force $F = 120$ tonnes. Make all necessary assumptions to calculate the shear flow distribution over the ship section.

10. The catamaran ship section shown in Fig. 13.10 is subjected to a vertical shear force $F = 10$ tonnes and a twisting moment $T = 50$ tonnes.m, calculate the shear stresses at points A and P and also the maximum shear stress due to F and T. The particulars of the ship section are: $L = 40$ m, $B = 10$ m, $D = 7$ m, $T = 2.5$ m, $H = 3$ m, $d = 4$ m, $w = 5$ m, $t_D = t_B = t_S = 7$ mm, $t_L = t_K = 10$ mm.
11. The idealized barge section shown in Fig. 13.11 has the following particulars: $L = 60$ m, $B = 12$ m, $D = 6$ m, $a = 2$ m, $t_B = 10$ mm, $t_D = 12$ mm, $t_S = 10$ mm. The ship section is subjected to a hull girder bending stress at the

Fig. 13.10

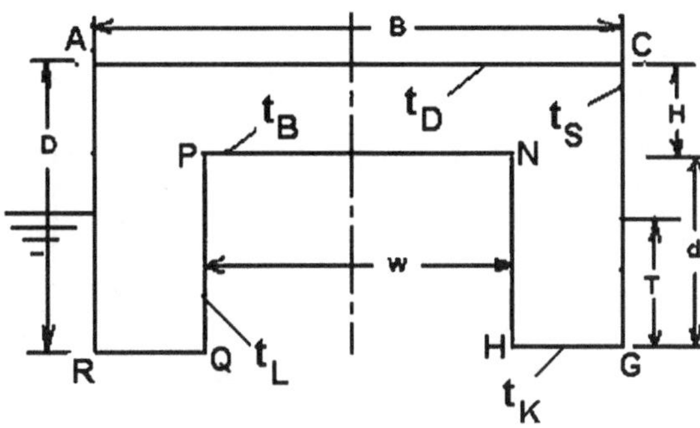

Fig. 13.11

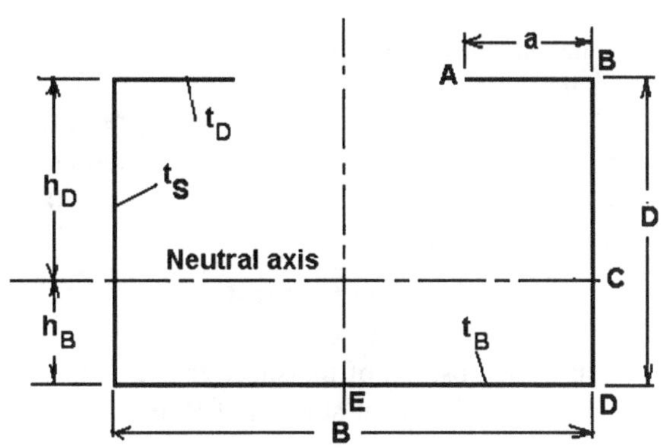

Fig. 13.12

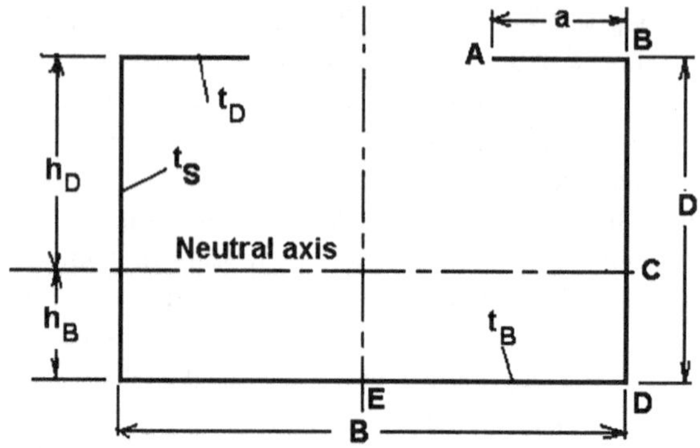

Fig. 13.13

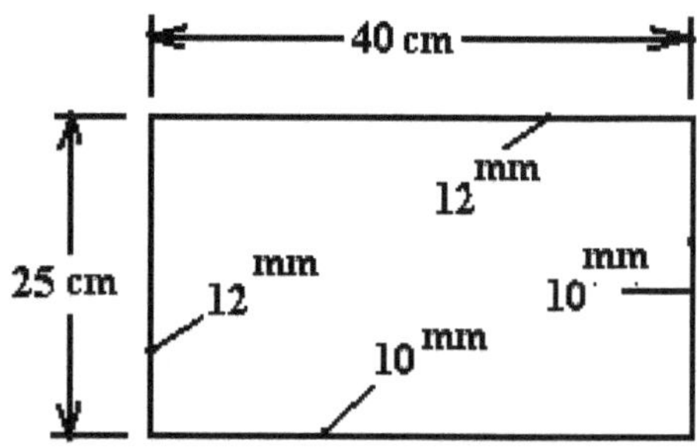

deck = 0.6 tonnes/cm^2 and a vertical shear force F = 20 tonnes and a torsion moment 30 tm. Make all necessary assumptions to calculate:

i. The shear stresses at B and C.
ii. The equivalent stress at point B.
iii. The torsion stress at point C.

12. The idealized barge section shown in Fig. 13.12 has the following particulars: L = 60 m, B = 12 m, D = 6 m, a = 2 m, t_B = 10 mm, t_D = 12 mm, t_S = 10 mm. The ship section is subjected to a hull girder bending stress at the deck = 0.8 tonnes/cm^2 and a horizontal shear force F = 20 tonnes and a torsion moment 30 tonnes.m. Make all necessary assumptions to calculate:

i. The torsion stress at B, C and D.
ii. The shear stresses at B, C, D and E.
iii. The position of the shear centre of the section.
iv. The equivalent stresses at B and D.

13. Calculate the shear flow distribution over the closed box girder shown in Fig. 13.13. The box girder is subjected to a vertical shear force F = 5.0 tonnes acting at the shear centre of the section.

References

de Wilde (1967) Structural problems in ships with large hatch openings. Int Shipbuild Prog 14(149):7

Evans JH (1983) Ship structural design concepts. Cornell Maritime Press, Cambridge

Frick W (1995) Structural design of container and multi-purpose vessels. Germanischer Lloyds of Shipping. 1st Pan American Symposium on Marine structures

Timoshenko S (1960) Strength of materials, part II, D. van Nostrand, Prentice Hall

Pilky WD, Pilky OH (1974) Mechanics of solids. Quantum Publishers

Shama MA (2006) Buckling of ship structure. Faculty of Engineering, Alexandria University

Shama MA (1976) Stress analysis and design of fabricated asymmetrical sections. Schiffstechnik Bd 23:117–144

Shama MA (1974) A simplified procedure for calculating torsional stresses in container ships. Unpublished

Timoshenko S (1945) Theory of bending, torsion and buckling of thin-walled members of open cross section. J Franklin Inst 239(3):201–219

Vlasov VZ (1961) Thin-walled elastic beams. National Science Foundation, Washington DC

Budynas R (1977) Advanced strength and applied stress analysis. McGraw-Hill, New York

Boris B, Lin TY, Scalzi JB (1970) Design of steel structures. Wiley Eastern Private Limited

B.V. Rules and regulations for the Classification of ships, 2006

G.L. Rules and regulations for the Classification of ships, 2006

L.R. Rules and regulations for the Classification of ships, 2006

Feodosyev V (1968) Strength of materials. Mir Publishers, Moscow

Goodman RA (1970) Wave excited main hull vibration in large tankers and bulk carriers. RINA 113:167–184

Hughes OF (1987) Ship structural design: a rationally based, computer-aided optimization approach. SNAME

Pedersen B (1968) Wave loads on the fore ship of a tanker. Eur Shipbuild 6

Shama MA (2007) Lecture notes, M.Sc. Post-Graduate Course 08611: Advanced ship structure analysis, Department of Naval Architecture and Marine Engineering, Faculty of Engineering, Alexandria University

Shama MA (1969) Effect of variation of ship section parameters on shear flow distribution, maximum shear stresses and shear carrying capacity due to longitudinal vertical shear forces. Eur Shipbuild 18(5):74–80

Shama MA (1969) Effect of ship section scantlings and trans. position of long. bulkheads on shear stress distribution and shear carrying capacity of main hull girder. Int Shipbuild Prog 16(184):357–369

Shama MA (1971) On the optimization of shear carrying material of large tankers. March, SNAME, J.S.R, pp 74–96

Shama MA (1972) An investigation into ship hull girder deflection. Bull Faculty Eng Alexandria Univ XI:501–535

Shama MA (1974) Calculation of sectorial properties, shear centre and warping constant of open sections. Bull Faculty Eng Alexandria Univ XIII:305–335

Shama MA (1975) Structural capability of bulk carries under shear loading. Ship symposium, Rostock University, Sept 1975

Shama MA (1975) Shear stresses in bulk carriers due to shear loading. J. S. R., SNAME 19(3):155–163

Shama MA (1976) Analysis of shear stresses in bulk carriers. Comp Struct 6:75–79

Shama MA (1977) Flexural warping stresses in asymmetrical sections. PRADS77, international conference on practical design in shipbuilding SNAJ, Tokyo, (A-14), pp 103–108

Shama MA et al (1995) Shear strength of damaged coastal oil taker under vertical shear loading. AEJ 34(2):1–9

Shama MA et al (1996) Shear strength of damaged seagoing oil taker under vertical shear loading. AEJ 35(1):A61–A73

Shama MA (1997) Impact on marine environment of ship structural failures and casualties. AEJ 36(1):A43–A53

Taylor JL (1924–1925) The theory of longitudinal bending of Ships. TNEC, p 123

Index

A
Angle of twist, 4, 9, 12, 13, 18, 50, 143
Average shear stress, 111, 114, 115

B
Boundary conditions, 50, 55, 75, 77, 105

C
Calculation of shear flow
distribution, 149, 158, 163, 169
Calculation of shear stress
distribution, 161, 162
Calculation of shear stresses
in tankers, 167, 175
Combined open and closed thin-walled
sections, 34–36
Combined open section with
one closed cell, 34, 35
Concentrated torque
at end of beam, 4, 5
Constrained warping, 41, 76, 77
Conventional coastal oil tanker, 153, 177
Conventional oil tanker, 151, 153, 170
Correcting shear flow, 168, 171, 206, 208

D
Direct torsion loads, 3
Distribution of torsion shear
flow and stress, 7
Distribution of torsional loading, 70, 76
Double bottom structure, 189, 190, 193
Dynamic Component, 198, 199

E
Effective thickness, 203
Enforced center of rotation, 102
Equivalent stress, 161

F
Finite Element Method (FEM), 62
Flexural warping
stresses, 46, 49, 51, 52, 55, 56,
59, 61, 75, 89, 90
Free warping, 41, 42, 53, 75

G
General equation of torsion, 47, 53
General solution of the
torsion equation, 76, 105
Geometrical, torsional, flexural
and sectorial characteristics
of sections, 64, 79, 91, 98

H
Hull girder loading, 189
Hull girder stresses, 82, 87

I
Idealization of a ship section, 79, 84, 107, 149,
153, 169, 202

L
Linearly varying torque, 57
Local stresses, 87

M
Mechanism of load
transmission, 192

P
Parabolic torque distribution, 74, 77, 78
Principal sectorial properties, 93

R
Rate of twist, 10, 11, 13, 17, 30

S

Sectorial properties, 49, 51, 61, 62, 75, 78, 82, 84, 91–93, 95, 96, 101
Shear center, 3, 61, 69, 82, 94, 95, 100, 121–126
Shear deflection, 127
Shear deformation, 129
Shear flow, 8, 31, 115, 121, 133, 135, 138, 139, 142, 145, 146, 149, 157, 160, 163, 164, 169, 202, 204, 205, 207, 208
Shear flow in bulk carriers, 109, 207
Shear lag, 130
Shear loading and stresses, 107, 108, 187
Shear stress, 13, 107, 113, 117, 120, 130, 161, 177, 202, 205
Shear stress in tankers experiencing a local damage, 177
Shearing force, 108, 190, 200
Simple beam theory, 61, 130
Single cell box-girder, 133
Stools of transverse bulkheads, 188

T

Torsion constant, 6, 7, 12, 61, 80, 101
Torsion equation, 54, 75, 77, 105
Torsion loading on ships, 65
Torsion of a beam element, 3
Torsion of a thin-walled variable section, 41
Torsion of open thin-walled girders, 6, 7
Torsion of thin-walled closed sections, 9, 10
Torsion shear stress, 13–15, 17
Torsion stresses in thin-walled multi-cell box-girders, 21
Torsion warping deformations and stresses, 41
Torsional deformation, 54
Total stress over the deck plating, 84, 89, 90

U

Uniform torsion loading, 54

W

Warping constant, 7, 8, 51, 82, 91, 98, 101, 102
Warping deformations and stresses in the deck structure, 41, 44, 72
Warping of thin-walled sections, 43, 47
Wave-induced component, 195
Warping of thin-walled sections, 43

CV of the Author

Prof. Shama graduated from the Faculty of Engineering, Alexandria University with B.Sc. degree (Honors) in Naval Architecture in 1960. He obtained his Ph.D. degree in Naval Architecture from Glasgow University, Scotland, UK in 1965. He has a Patent of Design of Frame Bending Machine, Scotland, UK (1965).

Prof. Shama published 85 Papers in the International, Regional and Local Scientific Journals in the fields of Ship Design, Ship Production, Ship Structural Analysis and Ship safety. In 2006, he published a book on "Buckling of Ship Structure", Faculty of Engineering, Alexandria University.

Prof. Shama was awarded the Egyptian State Prize for Engineering as well as the Science & Arts Medal of First Grade in 1974. Alexandria University nominated Prof. Shama in 2005 for the State Prize of Recognition.

Prof. Shama was Head of the Department of Naval Architecture and Marine Engineering (1988–1992), Vice Dean for Engineering Education and Students Affairs (1992–1995), Dean of the Faculty of Engineering, Alexandria University (1995–1998).

Prof. Dr. M. A. Shama is currently Prof. Emeritus of Naval Architecture and Shipbuilding, Department of Naval Architecture and Marine Engineering.

Prof. Shama was granted two years Sabbatical leave at the Research Department of Lloyds Register of Shipping, London, UK (1969–1971).

Prof. Shama participated in the Honorary Editorial Advisory Board (HEAB) of the Encyclopedia of Life Support Systems, EOLSS, January (1995–1996).

He is a member of the Supreme Egyptian State Committee for the Promotion of Professors since 1981. He was the Chairman of this committee (1998–2002). Prof. Shama was an External Examiner for Ph.D. Degrees, Maryland University (1998–2000). Prof. M. Shama was awarded the Alexandria University 2010 Taha Hussain Prize in recognition for his intellectual, managerial and development pioneering achievements.

FSC
www.fsc.org
MIX
Papier | Fördert gute Waldnutzung
FSC® C083411

Kolibri 360 GmbH
Ferdinand-Jühlke-Straße 7
99095 Erfurt, Deutschland
produktsicherheit@kolibri360.de